Viscosity of Nematic Liquid Crystals

Viscosity of Nematic Liquid Crystals

V.V. Belyaev

Viva Books

New Delhi | Mumbai | Chennai | Kolkata | Bengaluru | Hyderabad | Kochi | Guwahati

First Indian Edition 2011

VIVA BOOKS PRIVATE LIMITED

- 4737/23 Ansari Road, Daryaganj, New Delhi 110 002
  E-mail: vivadelhi@vivagroupindia.net, Tel. 42242200
- 76, Service Industries, Shirvane, Sector 1, Nerul, Navi Mumbai 400 706
  E-mail: vivamumbai@vivagroupindia.net, Tel. 27721273, 27721274
- Jamals Fazal Chambers, 26 Greams Road, Chennai 600 006
  E-mail: vivachennai@vivagroupindia.net, Tel. 28290304, 28294241
- B-103, Jindal Towers, 21/1A/3 Darga Road, Kolkata 700 017
  E-mail: vivakolkata@vivagroupindia.net, Tel. 22836381, 22816713
- 7, Sovereign Park Apartments, 56-58, K. R. Road, Basavanagudi, Bengaluru 560 004
  E-mail: vivabangalore@vivagroupindia.net, Tel. 26607409, 26607410
- 101-102 Mughal Marc Apartments, 3-4-637 to 641, Narayanguda, Hyderabad 500 029
  E-mail: vivahyderabad@vivagroupindia.net, Tel. 27564481, 27564482
- First Floor, Beevi Towers, SRM Road, Kaloor, Kochi 682 018, Kerala
  E-mail: vivakochi@vivagroupindia.net, Tel: 0484-2403055, 2403056
- 232, GNB Road, Beside UCO Bank, Silpukhuri, Guwahati 781 003
  E-mail: vivaguwahati@vivagroupindia.net, Tel: 0361-2666386

www.vivagroupindia.com

Published by arrangement with
Cambridge International Science Publishing
7, Meadow Walk, Great Abington
Cambridge Cb1 6AZ, UK

ISBN 978-81-309-0944-8

Published by Vinod Vasishtha for Viva Books Private Limited, 4737/23, Ansari Road, Daryaganj, New Delhi - 110 002.

Printed and bound by Saku Printers, New Delhi.

Contents

Chapter 1

Concepts of the hydrodynamics of liquid crystals

1.1. Anisotropy of the viscosity coefficients of liquid crystals

Nematic liquid crystals (NLC) are liquids characterised by the presence of the long-range oriented order and the absence of long-range translational ordering. The unit vector **n**, parallel to the specific direction formed in the ordered distribution of the major axes of the molecules of nematic liquid crystals, is referred to as the director. The degree of ordering (or the order parameter) of liquid crystals is characterised by the deviation of the direction of the major axes of the molecules from the direction of the director:

$$S = (3\langle\cos^2\theta\rangle - 1)/2, \tag{1.1.1}$$

where θ is the angle between the major axis of the molecule and the director. In a solid crystal $S = 1$, in an isotropic liquid $S = 0$, and in the liquid-crystal state $0.3\text{–}0.4 < S < 1$.

The presence of the specific direction of orientation of the molecules of liquid crystals leads to the anisotropy of almost all properties of these crystals, including viscosity. In this case, the viscosity coefficients which have no analog in the isotropic liquid must be considered.

The hydrodynamic theory of nematic liquid crystals has been studied most extensively in [5–7]. When studying an incompressible thermally insulated medium at constant temperature, the equations of motion include four conservation laws: mass, amount of motion, energy and the momentum of the amount of motion. The liquid inside the volume V, restricted by the surface A, has density ρ.

The main difference between the equations presented below for nematic liquid crystals in comparison with the appropriate equations for isotropic liquids is that, in addition to the translational movement of the liquid with the speed v_i (i, j, k are spatial coordinates), they include the orientation movement of the director **n**. Therefore, the equations of nematodynamics include the following variables: the fields of speeds v_i and of the director n_i, and also their gradients $v_{i,j}$, $n_{i,j}$ and derivatives with respect to time $\dot{v}_i, \dot{n}_i$. The equations for the conservation of mass and the amount of motion (continuity) have the following form [2, 6]:

$$\frac{d}{dt}\int_V \rho\, dV = 0, \tag{1.1.2}$$

$$\frac{d}{dt}\int_V \rho \upsilon_i\, dV = \int_V f_i\, dV + \int_A \sigma_{ji}\, dA_j, \tag{1.1.3}$$

where f_i is the force acting on the unit volume, σ_{ij} is the stress tensor. In the law of energy conservation:

$$\frac{d}{dt}\int_V \left(\frac{1}{2}\rho \upsilon_i \upsilon_i + U + \frac{1}{2}\rho_I \dot{n}_i \dot{n}_i\right) dV =$$

$$= \int_V (f_i \upsilon_i + G_i \dot{n}_i)\, dV + \int_A (\sigma_i \upsilon_i + s_i \dot{n}_i)\, dA,$$

ρ_I is the moment of inertia per unit volume, U is the internal energy, G_i is the density of the external rotational moment acting on the director, $\sigma_i = \sigma_{ij} v_j$ is the surface force on the unit area (v_j is the normal to the plane), $s_i = \pi_{ij} v_j$ is the moment of rotation per unit area. Taking into account the equation of conservation of the amount of momentum:

$$\frac{d}{dt}\int_V (\rho e_{ijk} x_j \upsilon_k + \rho_I e_{ijk} \dot{n}_j \dot{n}_k)\, dV =$$

$$= \int_V (e_{ijk} x_j f_k + e_{ijk} \dot{n}_j G_k)\, dV + \int_A (e_{ijk} x_j \sigma_k + e_{ijk} n_j s_k)\, dA,$$

the Ozeen equation is derived:

$$\int_V \rho_I \ddot{n}_i\, dV = \int_V (G_i + g_i)\, dV + \int_A \pi_{ij}\, dA_j,$$

where the right-hand part is the result of action of all external (G_i) and internal (g_i) volume and surface (π_i) forces acting on the director. The conversion and simplification of these integrals give the following system of differential equations:

$$\dot{\rho} = 0, \quad \rho \dot{\upsilon}_i = f_i + \sigma_{ji,j},$$
$$\dot{U} = \sigma_{ji} A_{ij} + \pi_{ji} N_{ij} - g_i N_i, \quad \rho \ddot{n}_i = G_i + g_i + \pi_{ji,i},$$

where

$$\sigma_{ji} - \pi_{kj} n_{i,k} + g_j n_i = \sigma_{ij} - \pi_{ki} n_{j,k} + g_i n_j, \quad N_i = \dot{n}_i - \Omega_{ik} n_k,$$

(in the vector form $\mathbf{N} = \dot{\mathbf{n}} - \omega \times \mathbf{n} = (\Omega \ \ \omega) \times \mathbf{n}$, $\omega = (1/2 \text{ rot } \mathbf{v})$)

$$N_{ij} = \dot{n}_{i,j} - \Omega_{ik} n_{k,j}, \quad A_{ij} = \frac{1}{2}\left(\frac{\partial \upsilon_j}{\partial x_i} + \frac{\partial \upsilon_i}{\partial x_j}\right), \quad \Omega_{ij} = \frac{1}{2}\left(\frac{\partial \upsilon_j}{\partial x_i} - \frac{\partial \upsilon_i}{\partial x_j}\right). \qquad (1.1.4)$$

Vector **N** is the angular velocity of the director in relation to the liquid.

The order of magnitude of the terms included in this system of equations and describing the oriented movement of the director, are comparable or also higher than the appropriate terms for the translational movement of the liquid, i.e. there may also be interpreted as the 'flows' or 'forces' [2]. As shown later, their contribution depends strongly on the orientation of the director in relation to the actual flows.

Information on the viscosity of NLC is included in the antisymmetric potential of viscous stresses σ_{ij} and the vector g_i corresponding to the volume internal force acting on the director. To determine their form, the first step is represented by the inequality for the free energy, starting with the increase of entropy of the system:

$$\sigma_{ji} A_{ij} + \pi_{ji} N_{ij} - g_i N_i - \dot{F} > 0. \qquad (1.1.5)$$

Subsequently, the free energy is expanded into a series with respect to director n_i, its gradient $n_{i,j}$, the angular velocity of the director N_i and the gradient of the velocity or, more accurately, with respect to the symmetric tensor A_{ij}. In tensor σ_{ij} and vector g_i we separate the components corresponding to isothermal static strain ($\sigma^0_{ji} = -p\delta_{ij} - (\partial F/\partial n_{k,i}) n_{k,i} \delta_{ij}$ and $g^0_i = \mu n_i - \beta_j n_{i,j} - \partial F/\partial n_i$, p and μ are the arbitrary constants) and the dynamic part of strain (σ'_{ji} and g'_i). The hydrodynamic components σ'_{ji} and g'_i are expanded into a series with respect to n_i, N_i and A_{ij}:

$$\begin{aligned} \sigma'_{ji} &= K^0_{ji} + K^1_{jik} N_k + K^2_{jikm} A_{km}, \\ g'_i &= L^0_i + L^1_{ij} N_j + L^2_{ijk} A_{jk}, \end{aligned} \qquad (1.1.6)$$

where K and L are functions of the director. The dimension of the coefficient of expansion of these functions with respect to n_i is viscosity. Taking into account the series of transformations and physical considerations, from (1.1.6) we derive the final expression for the hydrodynamic part of the tensor of viscous stresses with the Leslie viscosity coefficients α_i (i = 1–6):

$$\sigma'_{ji} = \alpha_1 n_k n_m A_{km} n_i n_j + \alpha_2 n_j N_i + \alpha_3 n_i N_j + \\ + \alpha_4 A_{ji} + \alpha_5 n_j n_k A_{ki} + \alpha_6 n_i n_k A_{kj}, \tag{1.1.7}$$

and the appropriate vector of the internal volume force:

$$g'_i \equiv \gamma_1 N_i + \gamma_2 n_j A_{ji}, \quad \gamma_1 = \alpha_3 - \alpha_2, \quad \gamma_2 = \alpha_5 - \alpha_6. \tag{1.1.8}$$

These equations can also be written in the form proposed by Leslie in 1966 [6]:

$$\sigma' = \alpha_1 \left(\mathrm{n} \cdot \mathrm{n} : \bar{\bar{A}} \right) \mathrm{n} \cdot \mathrm{n} + \alpha_2 \mathrm{n} \cdot \mathrm{N} + \alpha_3 \mathrm{N} \cdot \mathrm{n} + \alpha_4 \bar{\bar{A}} + \alpha_5 \mathrm{nn}.\bar{\bar{A}} + \alpha_6 \bar{\bar{A}}.\mathrm{nn}, \\ g' = \gamma_1 \mathrm{N} + \gamma_2 \bar{\bar{A}}.\mathrm{n}. \tag{1.1.9}$$

The following notations are used in [6]: (ab) has the components $(ab)_{\alpha\beta} = a_\alpha b_\beta$, ab: $A = a.A.b = a_\beta b_\beta A_{\alpha\beta}$, $(A.ab)_{\alpha\beta} = A_{\alpha\gamma} a_\gamma b_\beta$.

Taking into account the Parodi relationship [7]:

$$\alpha_2 + \alpha_3 = \alpha_6 - \alpha_5 \tag{1.1.10}$$

which was derived from the expression for the rate of increase of entropy (dissipation) taking into account the Onsager reciprocity relationships for irreversible processes, five viscosity coefficients of NLC are independent.

We examine the expressions (1.1.7) and (1.1.8) in greater detail. As indicated by the expression for σ'_{ji}, the fourth term in the right-hand part is determined only by the field of velocities of the liquid and does not contain any information on the director. Coefficient α_4 is the analog of isotropic viscosity and is always positive. The terms with the angular velocity of the director N_j contain viscosity coefficients α_2 and α_3. For nematic liquid crystals consisting of elongated rod-shaped molecules, these coefficients are negative which corresponds to the increase (not decrease) of the deformation of the director in the liquid flow (for more details see below). Both α_2 and α_3 are included in terms which do not contain the velocities and their gradients, i.e. are associated only with the reorientation of the director. In particular, this relates to the combination $\gamma_1 = \alpha_3 - \alpha_2$, referred to also as the rotational viscosity of the NLC or the Tsvetkov viscosity coefficient. As shown later, rotational viscosity describes the case of rotation of the director in the absence of any flows. The signs of the coefficient α_5 and α_6 in NLCs differ, and $\alpha_5 > 0$, a $\alpha_6 < 0$. Coefficient α_1 corresponds to tensile deformation. The coefficient for NLCs consisting of elongated rod-shaped molecules is negative. In addition, the inequalities linked with the increase of entropy (type (1.1.5)) give also relationships between the Leslie coefficients:

$$
\begin{gathered}
2\alpha_1 + 3\alpha_4 + 2\alpha_5 + 2\alpha_6 \geq 0, \\
2\alpha_4 + \alpha_5 + \alpha_6 \geq 0, \\
4\gamma_1(2\alpha_4 + \alpha_5 + \alpha_6) \geq (\alpha_2 + \alpha_3 - \gamma_2)^2.
\end{gathered} \tag{1.1.11}
$$

The united picture of the hydrodynamics of the systems with the long-range or short-range spatial ordering (crystals, smectics, nematics, conventional liquids and also glasses), increasing the number of coefficients in the systems describing transfer phenomena – heat conductivity and viscosity, was published in [8]. Table 1.1, taken from [8], gives the number of parameters included in the hydrodynamic theories of different systems.

It should be mentioned that the alternative description of the hydrodynamics of NLCs, taking into account the propagation of acoustic waves, was published by Forster et al [9]. In the approach proposed in [8–10], a nematic liquid crystal is characterised by the field of the tensor of the parameter of the orientation order $Q_{ij} = Q\ (n_i n_j - (1/3)\delta_{ij})$. The major axis of tensor Q_{ij} is the axis of orientation of the molecule, and $Q = S$. However, in the spectrum of fluctuations of the quantities examined by hydrodynamic theory, there are only low frequencies which are present only in the fluctuation spectrum of the director and not the order parameter. In Forster's theory, the tensor of viscous stresses has the following form:

$$
\begin{aligned}
\sigma'_{ij} = 2\nu_2 A_{ij} + 2(\nu_3 - \nu_2)[A_{ik}n_k n_j + A_{jk}n_i n_k] + \\
+ (\nu_4 - \nu_2)\delta_{ij}A_{kk} + (\nu_1 + \nu_2 - 2\nu_3)n_i n_j n_k n_l A_{kl} + \\
+ (\nu_5 - \nu_4 + \nu_2)[\delta_{ij}n_k n_l A_{kl} + n_i n_j A_{kk}].
\end{aligned} \tag{1.1.12}
$$

In contrast to the Eriksen–Leslie–Parodi approach, in equation (1.1.12), the quantities $A_{\alpha\beta}$ and n_α are present in the role of forces, and quantity $\sigma_{\alpha\beta}$ and N_α as flows. In this equation ν_1, ν_2, ν_3 are the shear viscosity coefficients, $\nu_4 - \nu_2$, ν_5 are the volume viscosity coefficients. For an imcompressible medium $\nu_4 = \nu_2$, $\nu_5 = 0$.

The Eriksen–Leslie–Parodi theory is explained in greater detail in a book by Chandrasekar [2] and the Forster theory in the seventh volume of Lectures in Theoretical Physics by Landau and Lifshits [11]. The authors of [12] published a system of linearised equations which take into account both the volume, shear and orientation deformations of the liquid crystal the presence of external pressure. It should also be mentioned that the requirements on the orientation function of distribution for using the Eriksen–Leslie–Parodi approach are presented in [13]. This function should be uniaxial or have the form of the δ-function. The equations of orientational equilibrium for two-dimensional liquid crystals and also for liquid crystals consisting of biaxial molecules, are presented in [14, 15].

Table 1.1. The number of some thermodynamic parameters of different monatomic (momomolecular) systems

Phase	Number of independent hydrodynamic variables	Number of viscosity coefficients η_{ijkl}		Transfer coefficients		Dissipative coefficients (for NLC $\zeta \sim \gamma_1^{-1}$)	Number of parameters describing the relationship of the order parameter with the flow
		Total	In the incompressible system	Heat conductivity κ_{ij}	ξ^a_i		
Simple liquid	5	2	1	1	0	0	0
Nematic	7	5	3	2	0	1	1
Smectic A and cholesteric	6	5	3	2	1	1	0
Smectic B (uniaxial)	8	5	3	2	2	2	0
Uniaxial crystalline	8	5	3	2	2	2	0
Smectic C	7	13	9	4	2	2	2
General crystalline	8	21	15	6	9	6	0
Cubic crystalline	8	3	2	1	1	1	0
Glass	8	2	1	1	1	1	0

1.2. Relationships of the processes of the shear flow of liquid crystals and rotation of the liquid crystal director

The main aim of experiments was not the determination of the Leslie coefficients α_i but determination of the viscosity coefficient in the flow of NLCs with a constant velocity gradient (or a constant shear rate) and also the viscosity coefficient, describing the rotation of the director.

Let is be that a liquid crystal flows between two parallel sheets and the orientation of the director is fixed using the external (for example, magnetic) field. Then, as shown in [3,16], viscosity is described by the following expression:

$$\eta(\theta,\varphi) = (\eta_1 + \eta_{12}\cos^2\theta)\sin^2\theta\cos^2\varphi + \eta_2\cos^2\theta + \eta_3\sin^2\theta\sin^2\varphi, \quad (1.2.1)$$

where θ is the angle between the direction of the flow and the director, φ is the angle between the projection of the director on the plane formed by the unit vectors $\nabla\mathbf{v}$ and $\mathbf{n} \times \nabla\mathbf{v}$ (v is the flow speed) and the speed gradient. Four different viscosity coefficients of the NLC can be measured for four different flow geometries (Fig. 1.2.1): η_1, if the director is parallel to the direction of the velocity gradient; η_2, if the director is normal to the flow speed; η_3, if the director is normal to the shear plane, i.e. the plane in which the vectors of the speed gradient $\nabla\mathbf{v}$ and speed $\mathbf{v}$ are located; $\eta_{45°} = 0.5\,(\eta_1 + \eta_2) + 0.25\,\eta_{12}$, if the director is situated in the shear plane and forms the angle of 45° with the directions (in these conditions, the contribution of tensile deformation is maximum).

The anisotropy of the viscosity coefficients in the shear flow was reported for the first time by Mesovich [17] and, therefore, the coefficients η_i ($i = 1, 2, 3$) are often referred to as the Mesovich viscosity coefficients. The relationship of the Mesovich viscosity coefficients η_i with the Leslie viscosity coefficients is determined by substituting the components of the director and the speed gradients into (1.1.7) and described by the equations:

$$\eta_1 = 0.5\,(\alpha_4 + \alpha_5 - \alpha_2),\ \eta_2 = 0.5\,(\alpha_3 + \alpha_4 + \alpha_6),$$
$$\eta_3 = \alpha_4,\ \eta_{12} = \alpha_1. \qquad (1.2.2)$$

Equation (1.1.7) shows that the viscous tensile stresses includes the terms describing the rotational movement of the director. In the equation of motion of the director

$$I\left(\frac{d\Omega}{dt}\right) = \mathbf{\Gamma}_F + \mathbf{\Gamma}_V \qquad (1.2.3)$$

the left-hand part is the inertia term (I is the inertia momentum related to the unit volume of the liquid), and term $\mathbf{\Gamma}_F$ in the right-hand part is the rotational moment per unit volume acting on the director and determined by elastic forces:

$$\mathbf{\Gamma}_F = \mathbf{n} \times \mathbf{h}, \qquad (1.2.4)$$

where $\mathbf{h}$ is the molecular field (for more details see [6]). The rotational

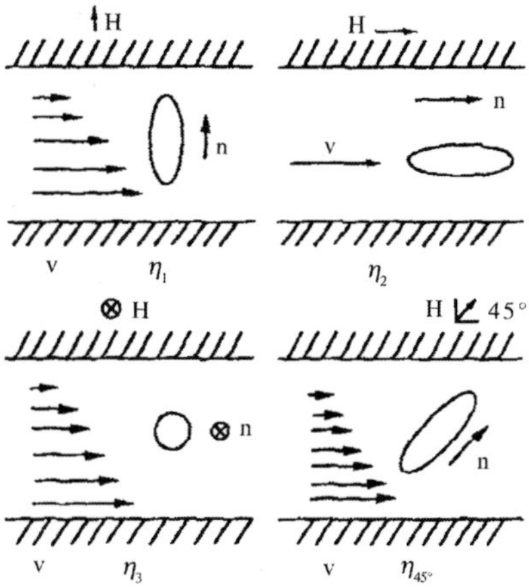

Fig. 1.2.1. Relative distribution of the speed vectors v, speed gradient $\nabla\mathbf{v}$, and director $\mathbf{n}$ in the determination of the Mesovich viscosity coefficients.

momentum Γ_V, acting on the director and determined by viscosity, is written in the following form:

$$\Gamma_V = \mathbf{n}\times(\gamma_1\mathbf{N} + \gamma_2\bar{A}\cdot\mathbf{n}), \quad (1.2.5)$$

where γ_1 is the rotational viscosity coefficient of the NLC (dissipative coefficient describing the speed of rotation of the director), γ_2 is the torque coefficient in the speed gradient. Writing the antisymmetric part of the tensor (1.1.7) and the rotational momentum Γ_V (1.2.5) for the case of flow of the liquid crystal at a constant speed gradient, we can determine the relationship of the coefficients γ_1 and γ_2 and the Leslie viscosity coefficients α_i:

$$\gamma_1 = \alpha_3 - \alpha_2,\ \ \gamma_2 = \alpha_2 + \alpha_3 = \alpha_6 - \alpha_5. \quad (1.2.6)$$

When solving the last problem, it may be shown that the flow of the liquid crystal is oriented by the director. If the director **n** is situated in the shear plane and has a stable orientation (no turbulence), the angle of orientation of the director in the flow θ is determined by the following expressions [6, 7]:

$$\cos 2\theta_0 = -\frac{\gamma_1}{\gamma_2} \text{ or } \mathrm{tg}^2\theta_0 = \frac{\alpha_3}{\alpha_2} = \frac{\gamma_2+\gamma_1}{\gamma_2-\gamma_1}. \quad (1.2.7)$$

The authors of [8] proposed an analogy between the coefficients γ_1 and γ_2 and the thermodynamic parameters of the conventional liquid.

Having the entropy increase for the slowly changing perturbation, equal to

$$\delta\dot{s} = \left(\frac{\partial s}{\partial T}\right)_\rho \delta\dot{T} + \left(\frac{\partial s}{\partial \rho}\right)_T \delta\dot{\rho}, \quad (1.2.8)$$

and using the energy conservation equation $Tds + (P/\rho^2)\,d\rho = d\,(\varepsilon/\rho)$, the continuity equations $\dot{\rho} = -\nabla_i v_i$, and also the heat conductivity equations, $\dot{T} = (k/\rho c_p)\,\nabla^2 T$, we obtain the equation

$$k^{-1}T\rho\delta\dot{s} - \left(\frac{\partial P}{\partial T}\right)_\rho k^{-1}\nabla_i\delta\upsilon_i = \left(\frac{c_\upsilon}{c_p}\right)\nabla^2\delta T, \quad (1.2.9)$$

where P is pressure, ε is energy, ρ is density, κ is the heat conductivity coefficient, and c_p, c_v are the heat capacities at constant volume and pressure, respectively. Denoting $k^{-1}T\rho = \bar{\gamma}_1$, $\left(\frac{\partial P}{\partial T}\right)_\rho k^{-1}T = \bar{\gamma}_2$ we obtain an equation identical with the equation for the director (1.2.5). This shows that $\bar{\gamma}_1^{-1}$ is the Onzager transfer coefficient, and $\bar{\gamma}_2$ is not. This gives the frequency of propagation of the temperature mode $\bar{\gamma}_1^{-1}$ $(c_v/T)^{-1}$, i.e. from the relationship of the parameters $\overline{\gamma}_1^{-1}$ and γ_1^{-1} we obtain the identical relationship between c_v/T and K_{ii}^{-1} (K_{ii} is the Frank elasticity coefficient), and also the relationship between $(dP/dT)_\rho$ and $(1-\gamma_2)/\gamma_1$.

The difference between $\overline{\gamma}_1, \overline{\gamma}_2$, and γ_1, γ_2 is manifested in different hydrodynamic modes. For example, for γ_1, γ_2 we cannot used the Kubo equation, although this equation can be written for γ_1^{-1}:

$$\gamma_1^{-1} = \lim_{\omega\to\infty} \lim_{q\to 0} \left(\frac{\omega^2}{2k_bT}\right) S_{n,n}(\mathbf{q},\omega), \tag{1.2.10}$$

where ω and $\mathbf{q}$ are respectively the frequency and spatial wave vector of the external perturbation (field). Without discussing in detail equation (1.2.10) it should be mentioned that in the book we discuss everywhere the viscosity coefficients measured at low time and spatial frequencies. This requirement is fully satisfied by the dimensions of the cells and control conditions studied in Chapter 2..

The relationship between the non-hydrodynamic processes (translational self-diffusion D) and viscosity η is described by the Stokes law:

$$D \sim \frac{k_bT}{6\pi\eta a}, \tag{1.2.11}$$

where a is the molecular dimension.

An identical relationship can be written for the orientation of the molecule in space, linking rotational diffusion D_{rot} and viscosity:

$$D_{\text{rot}} \approx \frac{k_bT}{\gamma_1 a^3}. \tag{1.2.12}$$

The relationship between these values was used in [18] for one of the conclusions on the temperature dependence of rotational viscosity.

Chapter 2 presents the equations of motion for different experimental geometries, obtained on the basis of equations (1.1.7), (1.2.1), (1.2.3)–(1.2.5). The rotational viscosity coefficient the liquid crystal describes only the speed of rotation of the director and, consequently, has no analogue in the isotropic liquid. It should be mentioned immediately that there is a phenomenon inverse to the orientation of the director by the flow of the NLC – the so-called inverse flow in which the reorientation of the director under the effect of some external field under specific conditions result in the hydrodynamic flow of the liquid crystal [19]. Both phenomena are linked with the anisotropic form of the molecules of liquid crystals and this results in different values of the coefficients α_3 and α_2 ($|\alpha_3| \ll |\alpha_2|$) and, consequently, different values of the rotational momentum at different orientations of the director [20]. The inverse flow reduces the value of the measured coefficient γ_1 [20, 21]. This will be discussed in greater detail in section 2.3. The value measured for transverse bending deformation is $\eta_s = \gamma_1 - \alpha_3^2/\eta_1$, for deformation by longitudinal bending $\eta_b = \gamma_1 - \alpha_2^2/\eta_2$.

The temperature dependences of the Leslie viscosity coefficient α_i, the Mesovich viscosity coefficients η_i (i = 1, 2, 3) and rotational viscosity γ_1 are presented in [22–25] (Fig. 1.2.2, 1.2.3). Although the viscosities η_i and γ_1 describe the dissipative processes of different nature (shear flow and the rotation of the director), the values are interlinked. In [26] it is shown that the relationship $\eta(\gamma_1/S^2)^{-1}$, where η is the dynamic viscosity measured in the flow of the liquid crystal through a capillary (Fig. 1.2.4), is almost independent of temperature and is determined by the molecular structure of matter. With the exception of a narrow range in the vicinity of the clearing point, this ratio is almost completely constant and its value is determined, generally speaking, by the molecular structure of the NLC [26] (for more details see sections 4.1 and 4.3 and chapter 5).

The relationship between the shear flow and the reorientation of the NLC is strictly justified in [27] where the equality:

$$\frac{\eta_1}{\eta_2} \equiv \frac{\eta_s}{\eta_b}. \qquad (1.2.13)$$

is proven. This identity enables evaluation of the value of the flow of the NLC on the basis of the results of measurements of viscosity coefficients in deformation of the director and, conversely, predict the rate of reorientation of the director in respect of capillary viscosity.

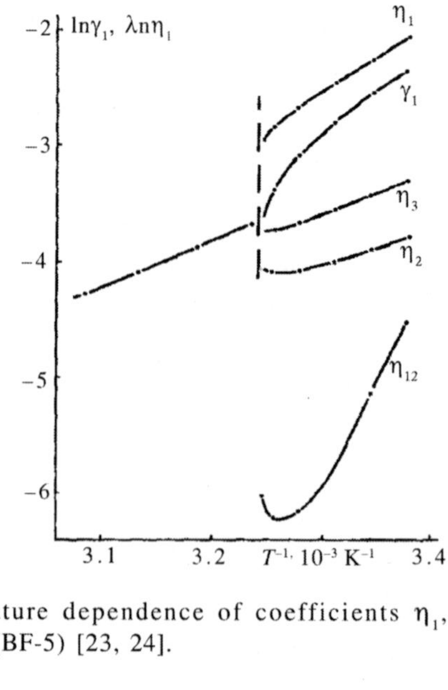

Fig. 1.2.2. Temperature dependence of coefficients η_1, η_2, η_3, η_{12} and γ_1 for pentylcyanobiphenyl (BF-5) [23, 24].

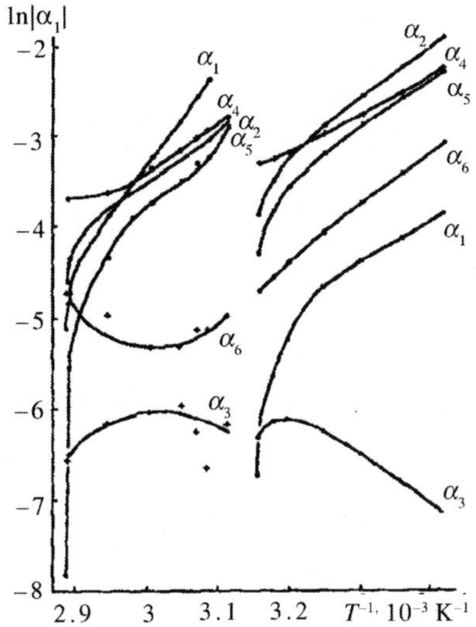

Fig. 1.2.3. Temperature dependence of the Leslie viscosity coefficient α_i for NBBA (right) and BBBA (left) [22].

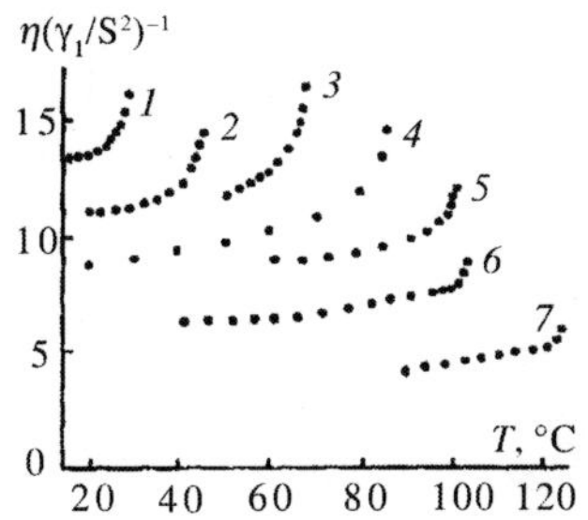

Fig. 1.2.4. Temperature dependence of the product $\eta_S\,(\gamma_1/S^2)^{-1}$ [26]: *1)* BF-6, *2)* MBBA, *3)* OBF-5, *4)* ZLI-1565, *5)* BTsO-5, *6)* C_4H_9O–B–N(O) = N–B–C_5H_{11}, *7)* C_5H_{11}–B–COOH.

Chapter 2

Methods for measuring the viscosity of nematic liquid crystals

All methods used for conventional fluids summarised in, for example, [28], can be used for the measurement of viscosity nematic liquid crystals.The methods based on measurement of the duration of passage of a nematic liquid crystal through a capillary at a given shear rate are used most widely because they are simple. It has been found that because of anisotropy, the measured value of viscosity is sensitive to a large number of parameters which are not always taken into account in conventional viscometry. It is the shear rate, the orientation the molecules on the walls of the capillary, the external magnetic or electrical field whose changes result in changes in the effective viscosity because of change in the orientation of molecules in the flow. The flow may become nonuniform even at a very low shear rate at a specific ratio of the Leslie coefficients. At the same time, because of the anisotropy of the properties of the nematic liquid crystals other methods of recording viscosity can also be used, for example, different optical ad capacitance methods. Viscosity is an integral part of the shear modulus and, consequently, ultrasound method can be used to measure it. Anisotropy of propagation and the absorption of ultrasound results in differences in the viscosity values measured by the ultrasound and capillary methods. The ultrasound method is used to determine the viscosity coefficient of the nematic liquid crystals in measurements of the spectrum of inelastic scattering of light on subsurface waves.

A specific feature of liquid crystals is that viscous stresses can form in the absence of flows because of the variation of the orientation director caused by the variation of the direction or amplitude of the external magnetic or electrical field applied to the specimen of the crystal. The value of the viscoelastic ratio is linked with the values of the relaxation time of the thermal fluctuations of the director. Measurements of the intensity of the spectrum of Rayleigh scattering yield a set of viscosity coefficients of the nematic liquid crystal.

In this chapter, the experimental conditions of measurements of the viscosity coefficients of the nematic liquid crystals are discussed, the effect of various parameters on the coefficient is dealt with, and the accuracy of measurements is evaluated.

2.1. The shear flow method

According to definition, viscosity is the ratio of the shear stress to the shear rate and, therefore, to determine viscosity, measurements are taken of the duration of passage of a given volume of the liquid crystal through a capillary at the given difference of the pressures determined by, for example, the natural weight of the fluid.

Because of viscosity anisotropy, the orientation of the director by the flow and interaction with the walls of the capillary, the value of the measured effective viscosity depends strongly on the experiment conditions.

The initial studies of the anisotropy of viscosity of the nematic liquid crystals were carried out by Mesovich [17, 29] and Tsvetkov and Mikhailov [30]. The studies by Mesovich are discussed on the basis of [31] using the notation η_i in accordance with Fig. 1.2.1. In this experiment, measurements were taken of the attenuation of oscillations of a thin glass sheet with the size of 32 × 22 mm, immersed in a liquid crystal and suspended using a glass filament with a diameter 0.1 mm to the arm of an analytical balance (Fig. 2.1.1*a*). The weight and shape of the sheet prevented irregular movement caused by heat flows. At a small amplitude of the oscillations (3 mm) and a relatively long period (5 s), the speed gradient was extremely low.

The sheet was placed in a rectangular vessel with a thickness of 6 mm with temperature control. The variation of angle φ between the direction of the magnetic field **H** and the normal to the sheet (Fig.2.1.1*b*), the viscosity

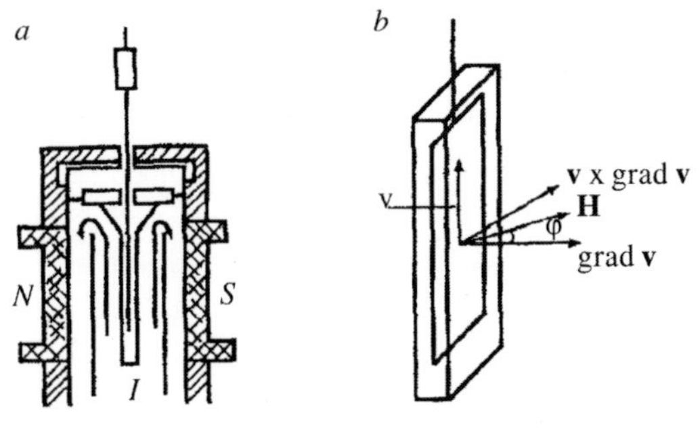

Fig.2.1.1. Measurement of the viscosity coefficient of NLC by the method of attenuation of vibrations of a glass sheet immersed in the vessel *1* with the NLC (a) and the relative position of the sheet in vessel *1* and of the speed vector, speed gradient and the magnetic field (*b*) [31].

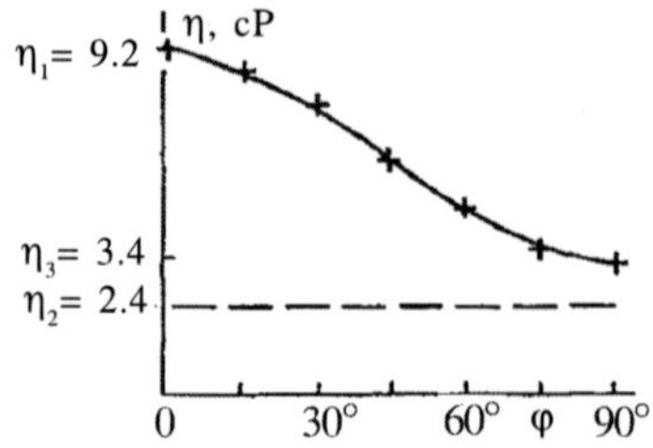

Fig.2.1.2. Dependence of the measured value of η_i on the angle φ between the direction of the magnetic field **H** and ∇ **v** [31].

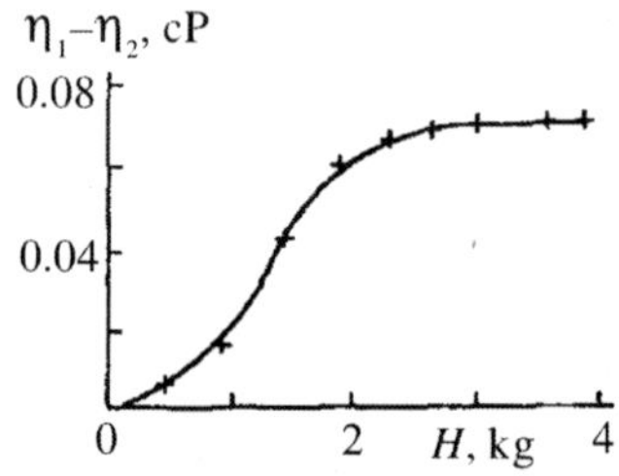

Fig.2.1.3. Diagram of dependence of the measured value of viscosity (at $H = 0$, $\eta = \eta_2$, for $H \to \infty$ $\eta = \eta_1$) on the strength of the magnetic field H [31].

of the nematic liquid crystal changed in accordance with the law $\eta(\varphi) = \eta_1 \cos^2 \varphi + \eta_3 \sin^2 \varphi$ (Fig. 2.1.2). The value of H was selected such as to ensure that with further increases of the strength of the magnetic field did not cause any changes in viscosity η_{ef} (Fig. 2.1.3). η_2 was measured in the absence of the magnetic field in the flow of the nematic liquid crystal through the capillary assuming that the director is completely reoriented by the flow. The latter, as shown later, is not completely accurate. This was improved in [32] with the main difficulty removed – recording of the amplitude of attenuating oscillations. The photodetector receives the laser beam light transmitted through two diffractive gratings with a period of 10 μm. One of them is stationary and the other one is connected with the oscillating sheet. The oscillations of the sheet are described by the equation

$$\frac{d^2 z}{dt^2} + 2\beta \frac{dz}{dt} + \omega_0^2 z = 0, \tag{2.1.1}$$

where ω_0 is the frequency of the free oscillations of the sheet, β is the parameter proportional to the viscosity of the NLC. The solution of (2.1.1) is the function:

$$z(t) = A\exp(-\beta t)\cos(\omega t + \varphi), \tag{2.1.2}$$

if $\omega_0^2 > \beta^2$, which holds for low-viscosity specimens (in [32] $\eta < 2$ P), or the function:

$$z(t) = A_1 \exp[-(\beta - \delta)t] + A_2 \exp[-(\beta + \delta)t], \qquad (2.1.3)$$

if $\omega_0^2 < \beta^2$. In (2.1.2) and (2.1.6) φ is the phase of oscillations, $\delta = (\beta^2 - \omega_0^2)^{1/2}$, $\omega = (\omega_0^2 - \beta^2)^{1/2}$. In [32], the amplitude of free oscillations was approximately 2 mm at a period of 14 s.

In contrast to the experiments carried out by Mesovich, Tsvetkov and Mikhailov [30] measured the variation of viscosity in the application of the magnetic field in the flow of a NLC through a capillary. This method was developed in [16, 24, 33, 34]. Special attention will be given to the results published in [24]. The diagram of experimental equipment is shown in Fig. 2.1.4. Each of the ends of the capillary *K*, placed between the holes of a magnet, is connected to a closed vessel filled with nitrogen. The increase of pressure in one of the vessels resulted in the flow of the liquid crystal through the capillary. Recording the values of the pressure difference between the vessels with nitrogen over the time period given by the pulse generator PG using the printer (P), the approximation equations are used to determine the effective value of viscosity η_{ef} related to the experimental conditions, with the value of the measured viscosity coefficient according to Mesovich η_i. The accuracy of measurements increases as a result of the efficient orientation of the liquid crystal due to the application of a strong magnetic field, low shear rate and a large thickness of the capillary, and also because of the large width of the longitudinal section of the capillary. The capillary consists of 20 sheets situated in the parallel direction (for measurement of η_2, η_3) or in the normal direction (for measurement of η_1) in relation to the direction of the magnetic field. The sheets are made of a non-magnetic metal (brass). The cross section of the capillary is

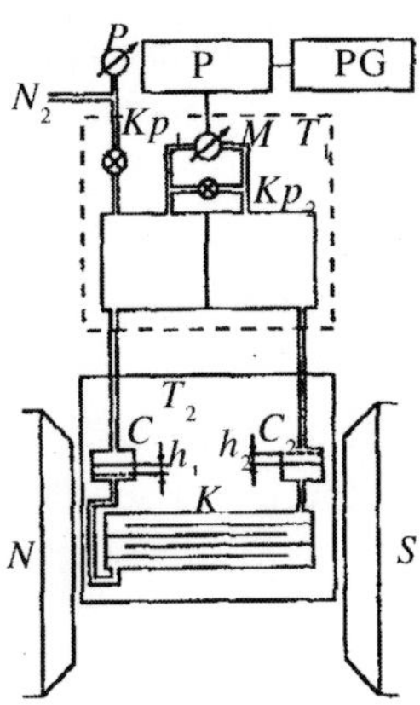

Fig.2.1.4. Diagram of experimental equipment for measuring coefficients η_1, η_2, η_3, η_{12}.

0.3×16 mm, total length 855 mm. These dimensions greatly differ from those given in [8]: 0.4 × 50 mm, and in [28] 0.3 × 4 × 50 mm. To reduce the effect of surface tension on the liquid crystals, glass cylinders C_1 and C_2 are placed in the ends of the capillary. The diameter of the cylinders was 15 mm and they were coated inside with a film of stannous oxide in order to ensure complete wetting. All the measurements were taken using a constant pressure of nitrogen as a result of the dependence of the compressibility of the gas on pressure. Each of the buffer volumes (85 cm^3 of N_2) consisted of 25 orifices with a diameter of 7 mm, drilled in a copper block in order to ensure isothermal expansion of the gas. The capillary block, the glass cylinders C_1 and C_2 and the brass pipes connecting them with the buffer volumes, were maintained at a measurement temperature T with the accuracy of 0.01 K in an oil bath, and the buffer volumes and the differential pressure gauge were kept in an aluminium box at a temperature of T_b = 22°C.

For the laminar flow in the long capillary, the flow speed through the volume of the fluid $\dot{V}$ with viscosity η, is linked with the pressure difference between the end of the capillary Δp_c by the relation:

$$\dot{V} \sim \frac{\Delta p_c}{\eta}. \tag{2.1.4}$$

The assumptions on the ideal nature of the buffer gas and proportionality of the displacement of the diaphragm of the pressure gauge to the pressure difference Δp gives the proportionality of the flow speed to the rate of variation Δp:

$$X\frac{d}{dt}\Delta p \sim -\frac{\Delta p_c}{\eta}, \tag{2.1.5}$$

where X is a multiplier which is a function of temperatures T and T_b, gas volumes and the displacement of the diaphragm of the pressure gauge. The quantity Δp differs from Δp_c by the difference of the hydrostatic pressure $\rho g \Delta h$ ($\Delta h = h_2 - h_1$, see Fig. 2.1.5):

$$\Delta p_c = \Delta p - \rho g \Delta h = \Delta p - (\Delta p_0 - \Delta p) A X \rho g, \tag{2.1.6}$$

where Δp_0 with the initial difference of pressures (at $t = 0$ and $\Delta h = 0$), A is a constant which depends on the cross section of the vessels C_1 and C_2. Consequently, equation (2.1.6) may be written in the form suitable for integration:

$$\frac{d\Delta p_c}{dt} = -\frac{K}{\eta X}(1 + AX\rho g)\Delta p_c, \tag{2.1.7}$$

where K is determined in calibration of the viscometer with a fluid with the known value of η. However, expression (2.1.7) can be used to determine only the effective value of viscosity η_{ef} which, because of the presence of the pressure gradient in the capillary, is lower than the appropriate Mesovich viscosity coefficient η_i. The fluidities (inverse viscosity) ϕ_{ef} and ϕ_i are linked by the relationship

$$\phi_{ef} = \phi_i + D\Delta p_c^2 \left(\eta_{ef}^{-1} = \eta_i^{-1} + D\Delta p_c^2\right), \tag{2.1.8}$$

which gives the following equation after substitution in (2.1.7) and subsequent integration:

$$\ln\left[\Delta pc\left(1 + D\eta_i \Delta p_c^2\right)^{-1/2}\right] = -\frac{K}{\eta_i X}(1 + AX\rho g)t + C. \tag{2.1.9}$$

Although coefficient D is a function of the experimental conditions and the viscosity coefficients of the nematic liquid crystals, its value is adjusted in order to obtain the maximum correlation coefficient for the experimentally measured dependence $\Delta p(t)$ (Fig. 2.1.5) which is linked with Δp_c by the relationship (2.1.6).

The relationship between η_{ef} and η_i is derived when examining the equations of motion and orientation included in the Leslie–Eriksen theory. The main results will be presented. The variation of the angle of inclination of the director in the thickness of the capillary for the parallel orientation of the director and the speed gradient (see the measurement of η_i, and geometry 1, Fig. 1.2.1) is described by the expression

$$\theta = \frac{\pi}{2} + \frac{\alpha_2 \mu_0}{\eta_1 \chi_a B^2}\frac{dP}{dz} = \frac{\pi}{2} + \beta_i x, \tag{2.1.10}$$

where $-a/2 < x < a/2$, a is the thickness of the capillary, χ_a is diamagnetic anisotropy, B is magnetic induction, dp/dz is the pressure gradient along the length of the capillary. The profile of the flow speed $\upsilon(x)$ and its mean value are described by the equations:

$$\upsilon = \frac{1}{2\eta_1}\frac{dp}{dz}\left[x^2 - \frac{a^2}{4} + \frac{\beta_1^2}{6}\left(x^4 - \frac{a^4}{16}\right)\right], \tag{2.1.11}$$

$$\langle\upsilon\rangle = -\frac{a^2}{12\eta_1}\frac{dp}{dz}\left(1 + \frac{\beta_1 a^2}{20}\right), \tag{2.1.12}$$

and effective viscosity

$$\eta_{ef}^{-1} = \eta_1^{-1}\left(1 + \frac{\beta_1 a^2}{20}\right). \tag{2.1.13}$$

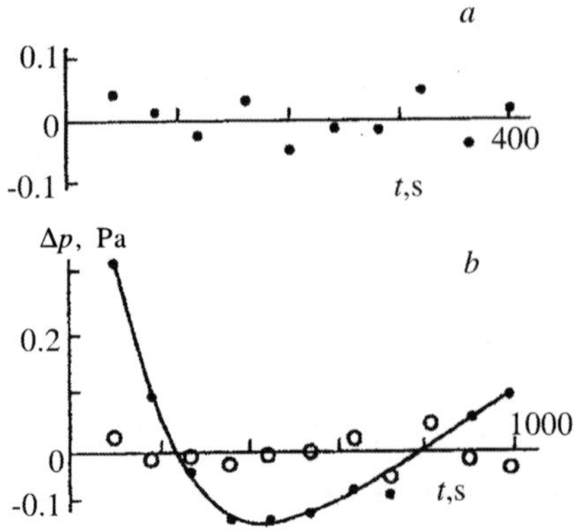

Fig.2.1.5. Time dependence of the measured (D = 0 in (2.1.67); •) and calculated ($D_{\eta 1}$= $4{\cdot}10^{-8}$ Pa^{-2}; o) the pressure difference Δp at the end of capillary tube [24].

Measuring η_2 (speed υ is parallel to the director, see the geometry 2, Fig. 1.2.1) in the presence of the magnetic field, the angle of orientation of the director is described by the equation

$$\theta = -\frac{\alpha_3\mu_0}{\eta_2\chi_a B^2}\frac{dP}{dz}. \qquad (2.1.14)$$

It should be mentioned that in the absence of the field, the value is determined by the expression (1.2.7). In this case, the mean flow speed is:

$$\langle\upsilon\rangle = -\frac{a^2}{12\eta_2}\frac{dp}{dz}\left[1-\frac{\beta_2 a^2}{20\eta_2}\left(\frac{\eta_1}{\eta_2}-1\right)\right], \qquad (2.1.15)$$

and effective viscosity.

$$\eta_{ef}^{-1} = \eta_2^{-1} - \frac{\beta_2 a^2}{20\eta_2}\left(\frac{\eta_1}{\eta_2}-1\right). \qquad (2.1.16)$$

When the director is perpendicular to both the speed and the gradient, speeds $\eta_{ef} = \eta_3$. Thus, the presence of the pressure gradient results in a decrease of the measured values of η_1 and η_2. The second source of errors is the presence of boundary layers with the width

$$\xi = \sqrt{K_{ii}/\chi_a\mu_0 H^2} \cong \sqrt{K_{ii}/\chi_a\mu_0 B^2}, \qquad (2.1.17)$$

which are not oriented by the magnetic field (K_{ii} is the elasticity coefficient). The profile of the Poiseuille flow in the boundary regions and in the centre of the cell differs (Fig. 2.1.6*a*) resulting in the dependence of η_i^{ef} on magnetic coherence length ξ:

$$\frac{1}{\eta_i^{ef}} = \frac{1}{\eta_i} + \left[6\frac{\xi}{a} - 12\left(\frac{\xi}{a}\right)^2 + 8\left(\frac{\xi}{a}\right)^3\right]\left(\frac{1}{\eta_{sat}} - \frac{1}{\eta_i}\right). \quad (2.1.18)$$

Ignoring the terms $(\xi/a)^2$ and $(\xi/a)^3$ and taking into account (2.1.17), we obtain

$$\frac{1}{\eta_i^{ef}} = \frac{1}{\eta_i} + \sqrt{\frac{K_{ii}\mu_0}{\chi_a}}\frac{6}{aB}\left(\frac{1}{\eta_{sat}} - \frac{1}{\eta_i}\right), \quad (2.1.19)$$

which shows that the value of Mesovich viscosity η_i is obtained by extrapolating the dependence of η_{ef}^{-1} on B^{-1} to infinitely high values of magnetic induction (Fig. 2.1.7). The data presented in this graph show that in a relatively strong field the measured values of η_2 and η_3 are close to true values. At the same time, even in the presence of fields $B = 1.1$ T, the measurement error of η_1 is 10%. It is possible that because of this reason and also due to high shear rates, the values of η_1, obtained by Gaehwiller [16] are too low. It should also be mentioned that because of the large thickness of the specimen (6 mm) and low shear rates, the viscosity value η_1 published by Mesovich in [29, 31] are likely to be accurate.

The effect of the orientation of the director on the surface of the wall of the capillary was investigated in [35]. If the orientation is given only by the magnetic field, the effective value of the coefficient η_1 is proportional to B^{-1} and its true value is obtained by the extrapolation $B^{-1} \to 0$. If 0.1 % of cetyl trimethylammonium bromide (CTAB) is added to a nematic crystal resulting in homeotropic orientation of the crystal in a flat capillary, the value $\eta_1^{ef} = \eta_1$ at the magnetic field of $B < 1.1$ T and the pressure difference at the end of the capillary of 100 Pa (Fig. 2.1.7). The flow profile is also influenced by the size of the pouring orifice. In the width of the orifice is considerably smaller than the width of the capillary, the liquid crystal at the edges of the capillary separates from the liquid crystal in the central section (Fig. 2.1.6*c*). Consequently, the tangent to the front edge of the flowing-in liquid crystal forms different angles ($+\alpha$ and $-\alpha$) with the long axis of the capillary; the liquid crystal at the front edge rotates in different directions in relation to the flow direction. In the twist cell this leads to the formation of regions with different twisting directions [36].

The effect of boundary conditions on the Kutta flow of a molecular liquid was investigated theoretically in [37]. A liquid crystal was placed between two sheets moving in the directions y with the speeds υ_s, $-\upsilon_s$ situated at the distances L and $-L$ from the centre of the layer. Assuming that the atoms do not slide on the surface, we can write the boundary conditions for the arbitrary orientation part of free energy k_bT with respect to the orientation tensor $a_{\mu\nu}$:

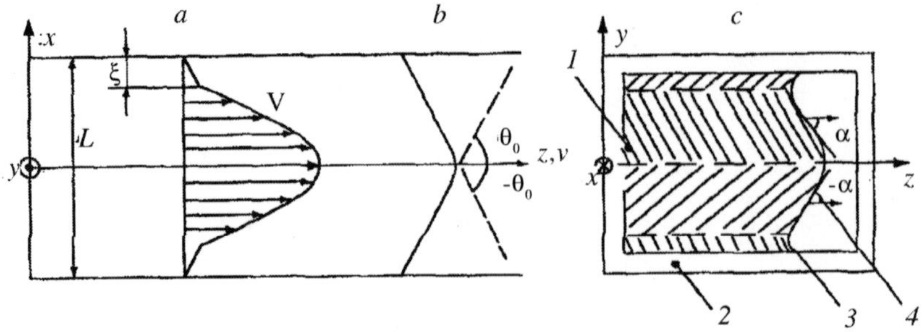

Fig. 2.1.6. Distribution of velocity v (a), the angle of inclination of the director θ (b) in the thickness of a flat capillary and the profile of the leading edge of the NCC flowing into the twist cell (c): ξ) the size of the wall region; L) the thickness of the capillary; θ_0, $-\theta_0$) the angle of inclination of the director in the upper and lower parts of the capillary, respectively; *1*) charging orifice; *2*) hermetic seal; *3*) movement front of the NLC; *4*) direction of orientation on one of the substrates, α, $-\alpha$, is the angle between the tangents of the movement front of the NLC and the orientation. Different directions of the primes indicate regions with different directions of twisting.

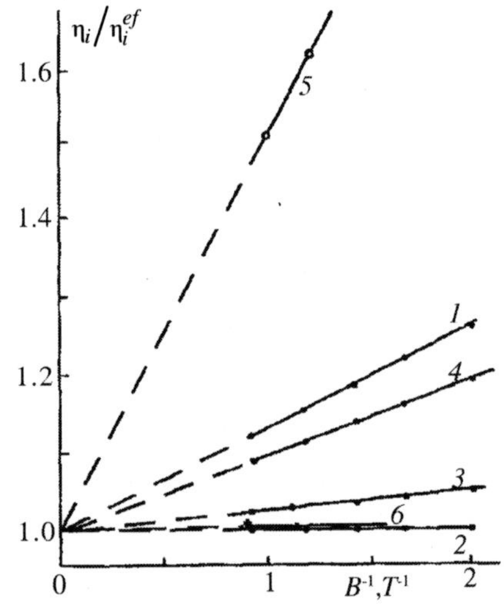

Fig. 2.1.7. Dependence of the measured value of η_i^{ef} (ratio η_i/η_i^{ef}) on the size of the magnetic field B in measurements of η_1 (*1*), η_2, (*2*), η_2, (*3*), η_{12}, (*4*) for NBBA [24] and η_1 (*5,6*) for dibutylazoxybenzol (DIBAB) [35]: 1–5) values of η_1/η_i^{ef} obtained without special processing of the surfaces of the capillaries; b) 0.1% of cetyl trimethylammonium bromide was added to DIBAB for homeotropic orientation.

$$\Sigma_{\mu\nu} = \frac{\partial \Sigma}{\partial a_{\mu\nu}} = Aa_{\mu\nu} = C_{\text{am}} \left(nk_b T\right)^{-1} k_\mu^{\tan} n_\nu + C_a C_0^{-1} n_\lambda b_{\lambda\mu\nu}. \tag{2.1.20}$$

Here $C_0 = (k_b T/m)^{1/2}$ is the thermal velocity of the molecule with the mass

m, n_ν is the component of the vector of the normal to the surface of the liquid, $k_{\mu\nu} = n_\nu \overline{P}_{\mu\nu}$, $P_{\mu\nu}$ is the component of the friction pressure tensor, $b_{\lambda\mu\nu} = D_a \nabla_\lambda a_{\mu\nu}$, D_a is rotational diffusion. The symbol with the line at the top relates to the symmetrical saddle-free traceless part of the tensor, for example $\overline{a_\mu b_\nu} = (a_\mu b_\nu + a_\nu b_\mu)/2 - a_\lambda b_\lambda \delta_{\mu\nu}/3$. The index 'tan' denotes the tangential component of the vector, for example $k_\mu^{\tan} = k_\mu - n_\mu n_\nu k_\nu$. The surface is characterised by the dimensionless coefficient $C_a \geqslant 0$, describing orientation. Effective viscosity η_{ef} is determined on the condition:

$$p_{xy}(L) = -\frac{\eta_{ef} \upsilon_s}{L}. \tag{2.1.21}$$

However, the real speed gradient γ_0 differs from the measured gradient $\gamma = \upsilon_s/L$ and is expressed using the functions linked with the state of orientation on the surface:

$$\gamma_0 = \frac{R\upsilon_s}{L}; \; R^{-1} = 1 + Q(l/L)\ \text{th}\ (L/l)[1 + \beta \text{th}\ (L/l)]^{-1}. \tag{2.1.22}$$

Q is linked with the ratio of the viscosity of the fluid at low and high shear rates (η and η_∞, respectively) and is the combination of the relaxation times, describing the translational ($\eta = nk_bT\,\tau_p$) and rotational ($\gamma_1 = 3nk_bT\,S^2\tau_a$) movement of the director and its twisting ($\gamma_2 = 2\cdot 3^{1/2} nk_bT \times S\tau_{ap}$, $\tau_{ap} = \tau_{pa}$ in accordance with the Onzager equation, for more details see section 3.4):

$$Q = \frac{\eta}{\eta_\infty} - 1 = \frac{\tau_{ap}\tau_{pa}}{\tau_a \tau_p}\left(1 - \frac{\tau_{ap}\tau_{pa}}{\tau_a\tau_p}\right)^{-1}. \tag{2.1.23}$$

The value l is linked with the characteristic length l_a which is the scale of the variation of orientation at spatial heterogeneity ($l_a^2 = D_a \tau_a$), by the relationship

$$l^2 = \left(1 - \frac{\tau_{ap}\tau_{pa}}{\tau_a\tau_p}\right) l_a^2. \tag{2.1.24}$$

It may easily be shown that $\tau_{ap}\tau_{pa} \times (\tau_a\tau_p)^{-1} = \gamma_1\eta/4\gamma_2^2$. Consequently, if $|\gamma_2| \approx \gamma_1 \approx 3\eta$, then $l \approx 0.96 l_a$ and $Q \sim 0.1$. The maximum possible value of Q for typical NLCs is determined from the ratios of the relaxation times (or the appropriate viscosities), is equal to 0.33 and $|\gamma_2| \approx \gamma_1 \approx \eta$, i.e. $(\eta/\eta_\infty)_{max}$ = 1.33. Generally speaking, this is slightly smaller than the actual variation η of the effective value of η for the flow equal to approximately $\eta_1/\eta_2 \approx 3$.

Only the parameter β is directly linked the surface orientation in (2.1.22)

$$\beta = C_a D_a (C_0 l)^{-1}. \tag{2.1.25}$$

The difference between the parameter R and unity is explained by the fact that the profile of the Kutta flow speed at a low value of β greatly differs from linear (Fig. 2.1.8) and is described by the expression:

$$\upsilon_x = R\upsilon_s\{y/L + Q(l/L)[\mathrm{ch}(L/l) + \beta\mathrm{sh}(L/l)\mathrm{sh}(y/L)]\}^{-1} \qquad (2.1.26)$$

Figure 2.1.9, taken from [37], gives the profile of the x–y component of the orientation tensor of the director. At low values of β or L/l and also in the vicinity of the sheets, the orientation on the director by the flow is not completed. Consequently, the measured value of the viscosity, described by the relationship (2.1.21) (Fig. 2.1.10), increases. For relatively large distances between the sheets ($L > 4l$), the relationship determined from (2.1.22):

$$\left(\frac{\eta}{\eta_{ef}} - 1\right) \approx Ql(1+\beta)^{-1}. \qquad (2.1.27)$$

is satisified. In the equations (2.1.18), (2.1.19), η_k is the viscosity measured in the flow of the nematic liquid crystal through a capillary at high shear rates. In early experiments [8, 9, 28], the value of η_k was assumed to be equal to η_2. If it is assumed that in the flow through the capillary the angle of orientation of the director in relation to the vector of the velocity tends to saturation (to zero) then for η_k we have [33]:

$$\eta_k = 0.25\alpha_2\left(1 - \frac{\gamma_1^2}{\gamma_2^2}\right) + 0.5(\alpha_2 + \alpha_4 + \alpha_5). \qquad (2.1.28)$$

The value of η_k differs from η_2 (see 1.2.1) by the addition equal to:

$$\Delta\eta_k = \eta_k - \eta_2 \left(1 + \frac{\alpha_1\alpha_2}{\gamma_2^2}\right), \qquad (2.1.29)$$

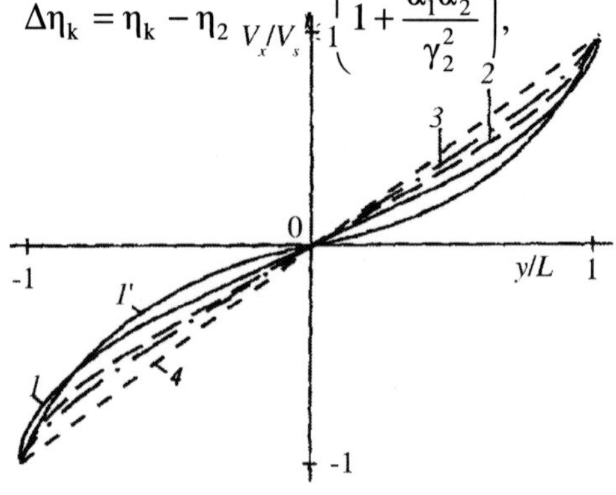

Fig. 2.1.8. Profile of the Kutta flow speed V_x of the NLC between two sheets moving at speed ± V_s at different values of coefficient β, describing the bonding of the molecules with the surface of the sheet: $B = 0$ (*1,1'*); 1(*2*); 2(*3*), ∞ (*4*), $Q = 8, 0$, L/l=16 (*1–4*); 4(*l'*) [37].

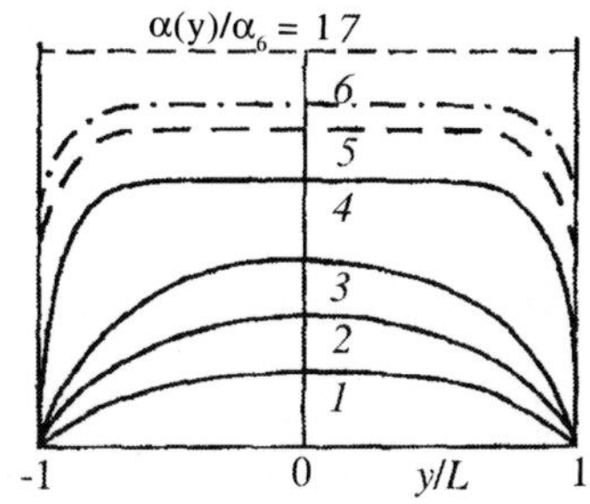

Fig.2.1.9. Profile of the orientation tensor of the NLC in Kutta flow at different values of β, *Q*, *L*/*l*: 1–4) β = 0,1) -*L*/*l* = 4, *Q* = 16, 2) *L*/*l* = 4, *Q* = 8; 3) *L*/*l* = 4, *Q* = 4; 4) *L*/*l* = 16, *Q* = 8; 4–7) *L*/*l* =16, *Q* = 8; 5) β = 1; 6) β = 2; 7) β = ∞.

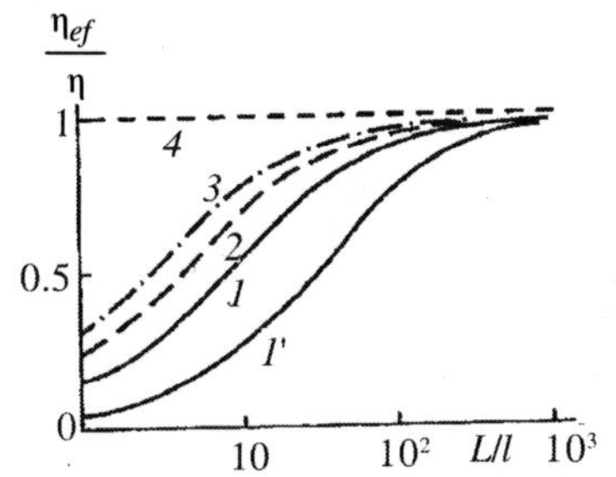

Fig.2.1.10. Dependence of the effective value of viscosity η_{ef}/η on the ratio of the microscopic *L* (2*L* is the distance between the Kutta sheets) to the diffusion length of different values of β: β=0 (*1,1'*); 1(*2*); 2(*3*); ∞ (*4*), *Q* = 8(*1-4*), 32(*1'*) [37].

i.e., by a very small value for the majority of nematic liquid crystals.

The measured value of η_k depends greatly on the radius of the capillary and the shear rate. Figure 2.1.11 [40] shows the temperature dependences of the apparent (i.e. effective measured in the experiments) viscosity η_{ef} for three values of the internal radius of the capillary: 175, 335 and 452 μm [40]. For the same shear rate, the apparent viscosity in a capillary with a small radius is 10% higher than in a capillary with the larger radius. The identical effect is detected if we measure the shear rate without changing the dimensionals of the capillary. Figure 2.1.12 shows the dependence of apparent viscosity η_{ef} on shear rate γ [35]. Measurements were taken using a method of measuring the viscosity of nematic liquid crystals situated between cylinders rotating with different angular velocities. The dependence $\eta_{ef}(\gamma)$ for a wide range of the values of γ may be described by the relationship

$$\eta_{ef} = \eta_2 + b|\gamma|^{-1} = \eta_2 + bG, \qquad (2.1.30)$$

where $G = |\gamma|^{-1}$ (Fig. 2.1.12). The examined independence becomes saturated at shear rates higher than 2000 s^{-1} (*para*-azoxyanisole, $T \approx 120°C$). For the homeotropic orientation the liquid crystal on the surface of the

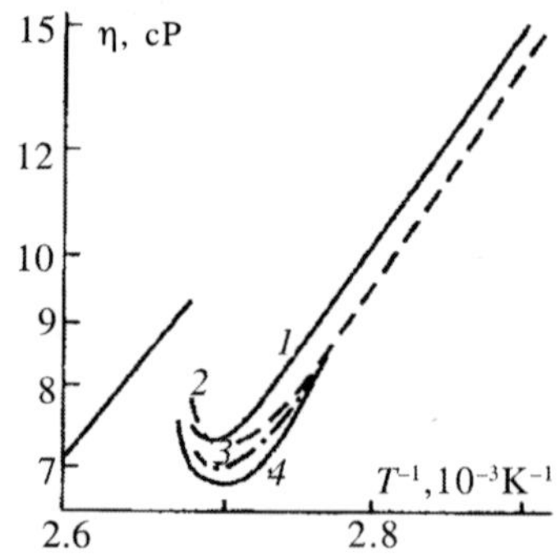

Fig.2.1.11. Viscosity of hexyloxybenzylidene aminobenzonitrile (HBAB) measured in capillaries with the internal radius of 175 μm (*1*), 355 μm, (*2*) 452 μm, (*3*) and measured by the method of the falling sphere (*4*) at the mean value of the gap between the sphere and the wall of the tube of 1400 μm.

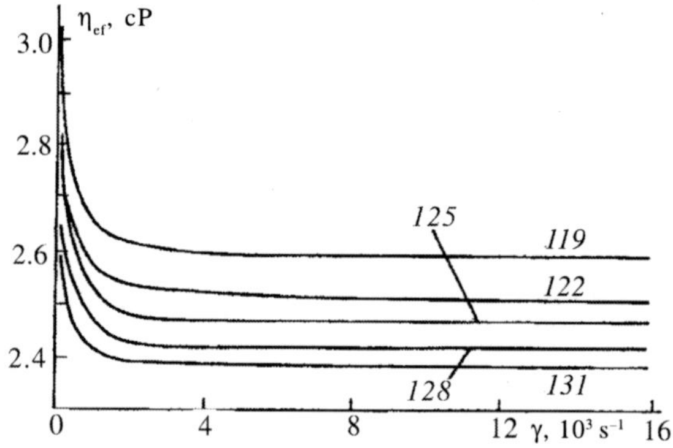

Fig.2.1.12. Effective value of the viscosity of para-azoxyanisole (PAA) in relation to the shear rate γ in measuring the viscosity between two rotating cylinders at different temperatures [41].

cylinders and a low shear rate $\eta_{ef} \rightarrow \eta_1$ [41]. In it is also shown that the heat flows formed during the flow of the nematic liquid crystal have almost no effect on the orientation of the director. It should be mentioned that at these high shear rates it is difficult to consider the homogeneity of the flow in the specimen.

The dependences on the radius of the capillary and the shear rate are a consequence of the dependence of the pressure gradient along the capillary $q = dp/dz$ on these parameters. In a general case, viscosity η_k is described by the equation

$$\eta_k = \frac{\pi q R^4}{8Q}, \tag{2.1.31}$$

where $Q = 2\pi \int_0^R \upsilon(r) r dr$ is the volume of the liquid crystal flowing per unit time.

Plotting the dependence of η_k on the Atkin number $4Q/\pi R$, which is the unambiguous function of the product qR^3 [42, 43], in the absence of the external electrical magnetic field, we determine the values of the capillary radius or shear rate at which the apparent viscosity ceases to depend on them. The data presented in Fig. 2.1.13 [40] show that the effective viscosity is almost completely independent of the capillary radius at $R > 300$ μm. this corresponds to the Atkin number $4Q/\pi R = 0.1 \div 0.2$ cm^2/s. This value of the Atkin number is also in agreement with the results obtained for *para*-azoxyanisole. Figure 2.1.14 shows the experimental results obtained by Fisher and Fredrikson [44] at different values of the radius of the capillary and the shear rate. They are in good agreement with the results of numerical solution of the continuity equations and the variation of the director in the flow of the NLC through a capillary with a circular cross-section and homeotropic boundary conditions on the capillary walls [39]. In the steady flow and in the absence of the magnetic field these equations have the form:

$$2f(\theta)\frac{d^2\theta}{dr^2} + \frac{df}{d\theta}\left(\frac{d\theta}{dr}\right)^2 + \frac{2f}{r}\frac{d\theta}{dr} - \frac{K_{11}\sin 2\theta}{r^2} + (\gamma_1 + \gamma_2 \cos 2\theta)\frac{d\upsilon}{dr},$$
$$d\upsilon / dr = (1/g(\theta)) \times (-qr/2 + b/r), \tag{2.1.32}$$

where

$$f(\theta) = K_{11}\cos^2\theta + K_{33}\sin^2\theta,\ g(\theta) = 2\alpha_1 \sin^2\theta\cos^2\theta + \\ + (\alpha_5 - \alpha_2)\sin^2\theta + (\alpha_6 + \alpha_3)\cos^2\theta + \alpha_4, \tag{2.1.33}$$

$$q = -dp/dz.$$

At $4Q/\pi R > 0.1$ cm^2/s both the theory and experiments confirm the conclusion on the steady orientation of the director in the flow of the nematic liquid crystal through the capillary. It should be noted that for highly viscous

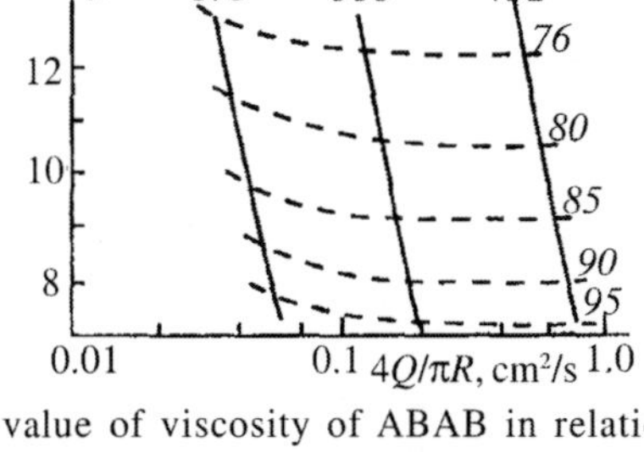

Fig.2.1.13. Measured value of viscosity of ABAB in relation to the Atkin number for different radii of the capillary. The broken line indicates the isotherms.

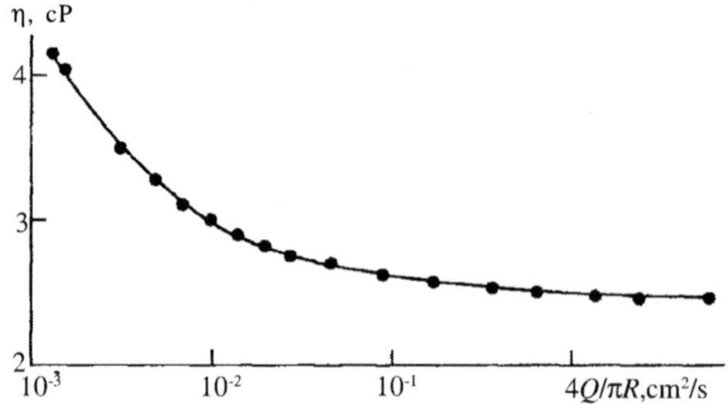

Fig.2.1.14. Dependence of effective viscosity of PAA η_k on the Atkin number. The points are the experimental data [44], solid line – calculated in [45].

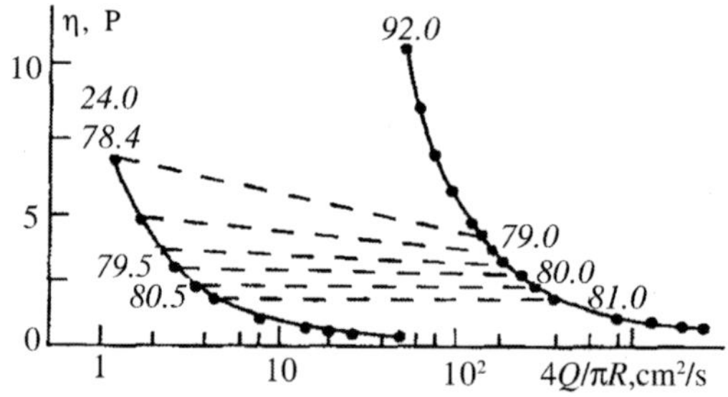

Fig.2.1.15. Dependence of effective viscosity η of cyanobenziledenoctyloxyaniline (CBOOA) of the Atkin number for different radii of the capillary in the vicinity of the NOC-Smectic A transition. The dotted line indicates the isotherms.

fluids, especially those containing smectic clusters, the steady orientation does not form even at higher Atkin numbers (Fig. 2.1.15, taken from [40]). In the application of the magnetic field, the values of the Atkin number at which effective viscosity reaches saturation decrease. Fig. 2.1.16, taken from [46], gives the results of numerical calculations of η_k ($4Q/\pi R$) for different values of $\chi_a H^2$ [46].

In the flow of the nematic liquid crystal through the capillary with the homeotropic boundary conditions on the walls, the effect of the boundaries extends to the region with thickness δ (Fig. 2.1.17, taken from [40]).

The value δ is determined by the equation

$$\frac{R-\delta}{R} = \frac{R_c}{R} = \frac{\eta_c}{\eta_l}\left(\frac{\eta_{ef}-\eta_l}{\eta_{ef}-\eta_c}\right)^{1/4}, \qquad (2.1.34)$$

where R_c is the radius of the central part of the capillary in which the nematic liquid crystal is reoriented, η_c is the viscosity of this part, η_l is the viscosity of the flow in the boundary layers [44]. Assuming that $\eta_c = \eta_2$, $\eta_l = \eta_1$ and using the experimental values of viscosity, for capillaries with

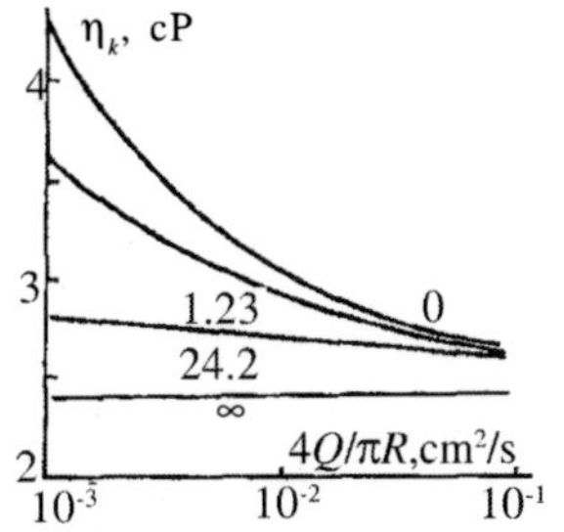

Fig.2.1.16. Dependence of the effect of viscosity of PAA η on the Atkin number for different values of the ratio $\chi_a H^2$.

different diameters $R_c/R = 0.93 \div 0.94$ which is in agreement with the estimates made when examining the flow of the nematic liquid crystal in a microscope.

It was shown in [47] that viscosity increases with an increase of the mechanical stress, indicating the ordering of the structure of the liquid crystal.

Different aspects of the measurement of viscosity in the flow of the nematic liquid crystals through a capillary have also been examined in [48]. If a nematic liquid crystal flows in a capillary only the under the effect of the gravitational force (Ostwald viscometer), the pressure difference at the end of the capillary is $\Delta p = \rho g h$, where h is the mean height of the fluid, ρ is density. The kinematic viscosity of a Newtonian fluid is determined from the Poiseuille equation describing the proportionality of the viscosity to the flow time taking into account corrections for the kinetic energy of the flow [49]:

$$\nu = Ct + \frac{B}{t}, C = \frac{\pi R^4 gh}{8LV}, \tag{2.1.35}$$

where L is the length of the capillary, V is the volume of the flowing fluid (Fig. 2.1.18*a*). Usually, the second term in (2.1.35) at $t > 200$ s is ignored and constant C is determined by measuring the flow time of the fluid with the known viscosity (the values of h and V are not available).

In the Ubellode viscometer (Fig. 2.1.18), the effect of gravitational force is compensated by the presence of the second arm of the capillary and by careful regulation of air pressure. Therefore, the duration of flow of the fluid through the capillary does not depend on the density of the liquid crystal and is proportional to dynamic viscosity η.

Figure 2.1.19 shows the design of the viscometer described in [50] for measuring the kinematic viscosity for small quantities of the NLC (0.4÷0.6 ml). The amount of the NLC is required for the micro-viscometer developed in [51] on the basis of a microcapillary viscometer is even smaller

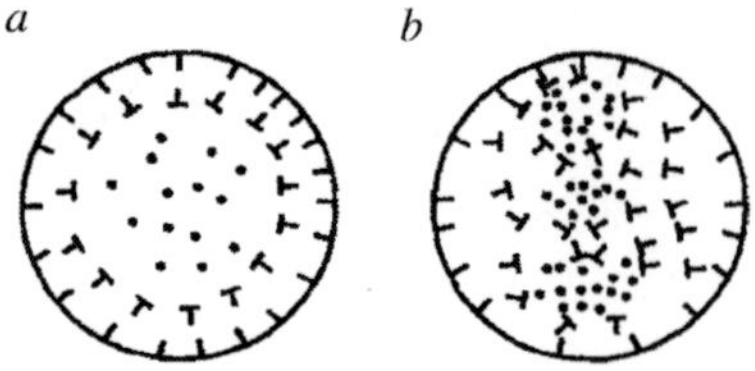

Fig.2.1.17. The structure of distribution of the director in flow of NLC through a capillary with $\alpha_3 > 0$ at low (a) and high (b) shear rates.

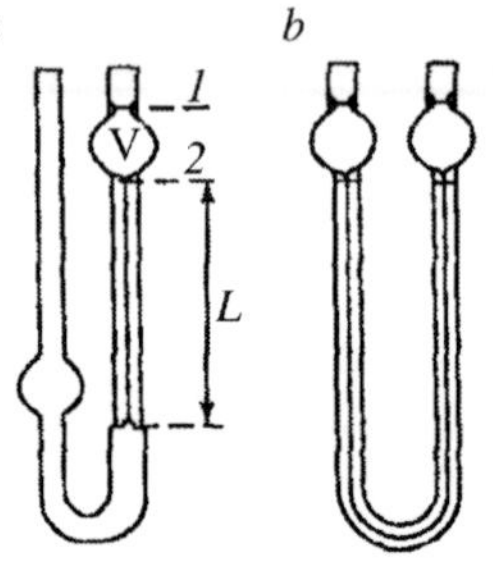

Fig.2.1.18. General view of Ostwald (a) and Ubellode (b) viscometers for measuring kinematic ν (a) and dynamic η (b) viscosity.

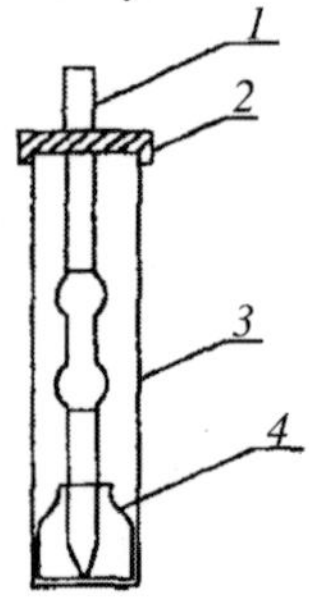

Fig.2.1.19. Design of a micro-viscometer for 0.4÷0.6 g of NLC.

[52]. It has the form of a precision pipe with an internal diameter of 0.2 to 0.4 mm and approximately 50 cm long, with temperature thermostatically controlled with the accuracy of 0.02 K The fluids travels under the effect of its own weight in the vertical capillary. The displacement of the level of the fluid is measured with a cathetometer. The absence of thickening in the capillary in filling with the fluid identical with that shown in Fig. 2.1.18 makes it possible to reduce the amount of the fluid for measurements to 50÷100 mg at the measurement accuracy of 1%.

As already mentioned, in the flow of the NLC through the capillary with homeotropic boundary conditions at low shear rates, the director deviates from the initial orientation by angle θ_0 determined by the ratio α_3/α_2 (or γ_1/γ_2, see (2.1.7)). If the sign of α_3 changes from negative to positive, in the application of optical methods of recording, the value of θ_0 decreases to zero and the stable laminar flow breaks up into irregular domains accompanied by the appearance of disclinations parallel to the flow direction [5]. However, the measurement of angle θ_0 by the capacitance method has shown no special features in the temperature dependence of θ_0 [53] [Fig. 2.1.20]. In both studies, investigations were carried out into the flow of hexyloxybenzyl idenamino benzonitrile ($C_6H_{13}O{-}C_6H_4{-}CH = N{-}C_6H_4{-}CN$, the nematic range 56÷101°C, $T(\alpha_3 = 0) = 91$°C). To explain this paradox, special attention will be given to the method of measurements used in [53] which is explained in Fig. 2.1.21. The brass cylinder *1* 19 mm long with a diameter of 18 mm rotates with the angular speed ω (1 rpm < ω/2π < 1 rpm). The cylinder is placed inside the glass cylinder *2* with a diameter of 22 mm. The measuring electrodes *3* (size 5 × 13 mm) are brazed into the diametrically opposite walls of the glass cylinder and are oriented parallel to the axes of the cylinders, together with the shielding electrodes *4*. The measuring bridge is used to determine the capacitance of the liquid crystal situated between the shortened electrodes *3* and the brass cylinder which is the second measuring electrode. The capacitance of the liquid crystal depends evidently on the angle of orientation of the director. The system is placed in a magnetic field whose direction (θ_1) is selected to ensure that

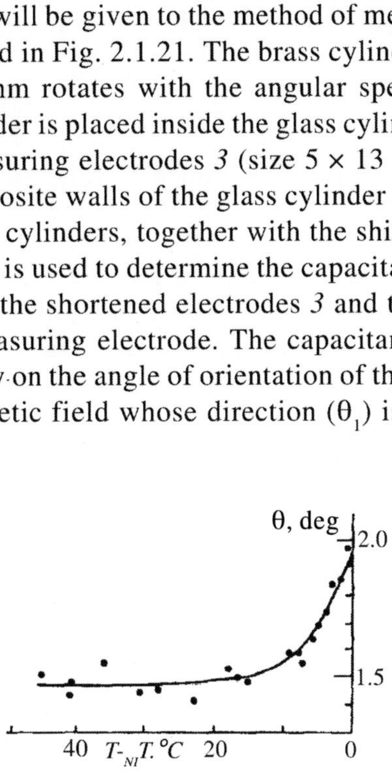

Fig.2.1.20. Angle of inclination of the director θ between the spinning cylinders, measured on the basis of the capacitance of GBAB (coefficient α_3, changing its sign) in relation to temperature.

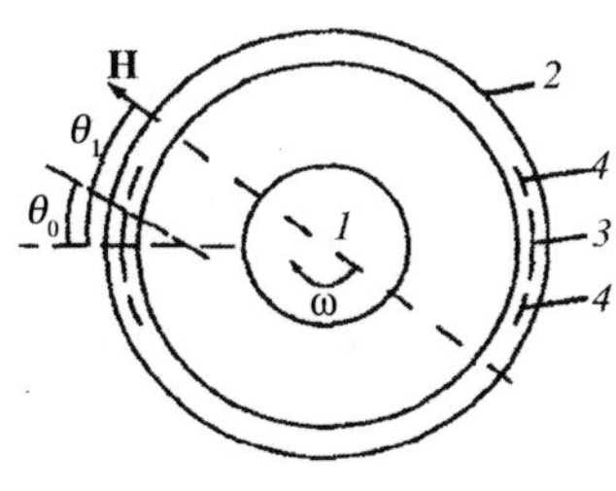

Fig.2.1.21. Diagram of measurement of the angle θ by the capacitance method.

the balance of the capacitances is not disrupted and the magnetic field is switched off. This is also carried when the cylinder *1* rotates in the opposite direction. The angle between the two directions determined in this manner **H** is equal to $2\theta_0$ taking into account the components of the corrections for the cylindrical configuration of the electrons. Although the variation of the sign of α_3 is accompanied by flow instability accompanied by the formation of inverse walls with the angle of orientation of the director in them of $180°-2\theta_0$, the measured capacitance does not change. The possible structure of the orientation of director at flow instability has been proposed in [40] (Figure 2.1.17*b*).

The heterogeneity of the distribution of the director in the layer of the liquid crystal during its flow may appear in the conditions described in [48]. If the angle of rigid orientation of the director on the substrate is θ_1, then the effect of constant shear stress τ_1 applied to one of the substrates is accompanied by deformation of the director in the volume. Depending on whether θ_1 its higher or smaller than θ_0 (see 1.2.7), where $\theta_0 = \text{arctg}\,(\alpha_3/\alpha_2)$, deformation of the type *1* or *2* takes place. (Fig. 2.1.22). The distribution of the vector in the layer of the liquid crystal at different shear stresses is the solution of the Euler–Lagrange equation:

$$K\frac{d^2\theta}{dz^2} = \tau\frac{\alpha_3\cos^2\theta - \alpha_2\sin^2\theta}{0.5(\alpha_3+\alpha_4+\alpha_6)-(\alpha_3+\alpha_2)\sin^2\theta}, \qquad (2.1.36)$$

where $K = (K_{11} + K_{22} + K_{33})/3$ is the mean elasticity coefficient. If $\theta_1 > -\theta_0$, the dependence $\theta_m(\tau)$ tends to θ_0 at $\tau \to \infty$, if $\theta_1 < -\theta_0$, $\theta_m(\tau) \to \theta_0 - \pi$ at $\tau \to \infty$ (Fig. 2.1.23). If $\theta_1 = -\theta_0$, deformation is of the threshold type and metastable conditions may occur. For an MBBA layer with a thickness of 10 μm the threshold is approximately 0.1 N/m^2. The nonuniform distribution of the director may distort the measured viscosity value. The effective viscosity of the layer of the liquid crystal shown in Fig. 2.1.22 is described by the relation:

$$\eta_{\text{ef}} = L\left\{\int_{-L/2}^{L/2}\left[\eta_2-(\alpha_3-\alpha_2)\sin\theta\right]^{-1}dz\right\}^{-1}. \qquad (2.1.37)$$

The nonmonotonic dependence $\theta(z)$ at $\theta_1 < -\theta_0$ (Fig. 2.1.23) may lead to

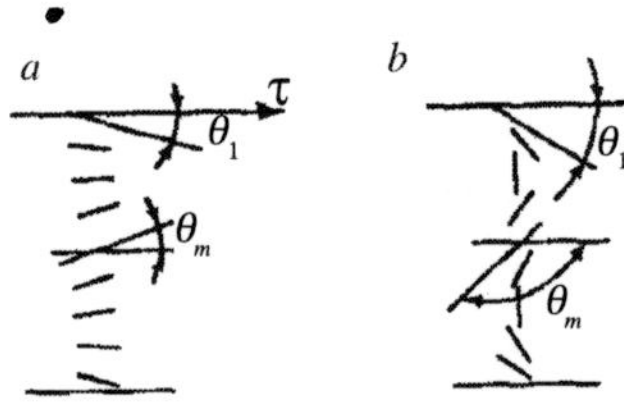

Fig.2.1.22. Possible deformation of the director in the sheer flow [48].

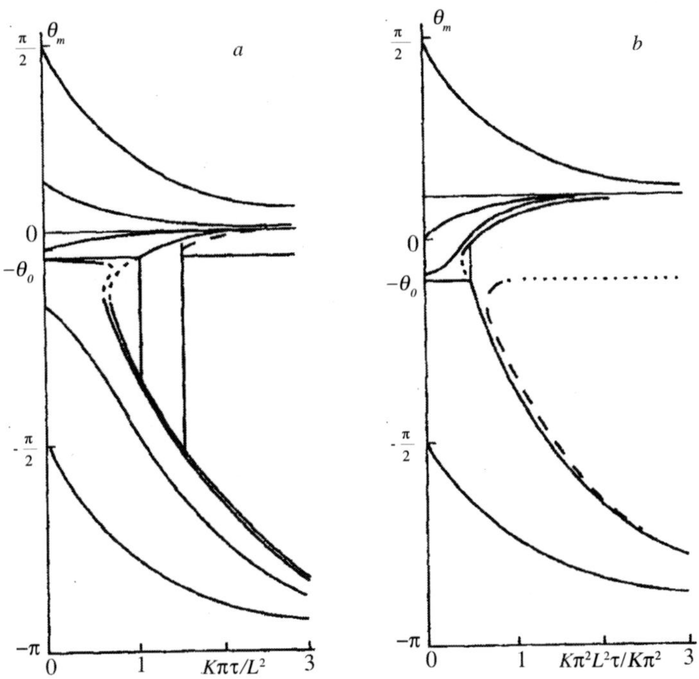

Fig.2.1.23. Dependence $\theta_m(\tau)$ for different values of the angles of surface orientation θ_1[48]; $\alpha_3/\alpha_2 = 0.01$, $\eta_2/\alpha_2 = -0.2$; $\alpha_3/\alpha_2 = 0.1$, $\eta_2/\alpha_2 = -0.2$.

the same dependence of η_{ef} on the shear stress and bistable dependence on temperature (Fig. 2.1.24).

For non-symmetric boundary conditions, for example, planar conditions on one substrate and homeotropic conditions on the other substrate, in calculation of the time dependence of the reorientation of the liquid crystal on time it is necessary to take into account the so-called surface viscosity of the liquid crystal [55]. Without discussing this problem in greater detail, we refer readers to studies [56–58]. The dimension of the surface viscosity corresponds to the product of viscosity by characteristic length (thickness of the cell or boundary layer). In [55] the estimates of surface viscosity gave $2.7 \cdot 10^{-7}$ N $\cdot$ cm^{-1}. If this viscosity is not taken into account, the values of the surface energy obtained from the static investigations and the dynamics of Fredericks transition differ by a factor of 20.

If a moving nematic crystal contains stationary disclinations, then depending on the intensity of the dislocations, the flow may be attracted or repulsed by the disclination [59] (Fig. 2.1.25). After all, this also results in a change of the measured viscosity values. The case in which the disclination itself moves will be examined later.

Previously, we examined the situation in which the viscosity of the

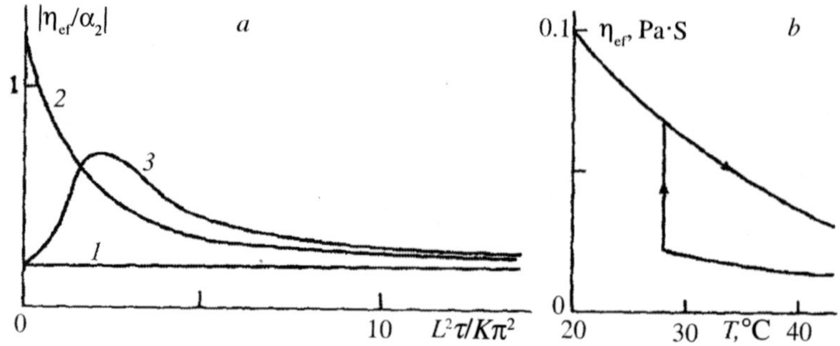

Fig.2.1.24. Dependence of effective viscosity η_{ef} on shear stress τ (*a*) and temperature (*b*) [48]: *a*) $\theta_1 = \theta_0$(*1*), $\pi/2$(*2*), -0.1 (*3*); $\alpha_3/\alpha_2 = 0.01$, $\eta_2/\alpha_2 = -0.2$; *b*) $\theta_1 = 0.18$, $\tau = 3\pi^2 K/L^2$.

nematic liquid crystal was determined from the characteristic time of flow of the fluid through the capillaries. At the same time, the orientation effect of the flow on the nematic crystal (relationship (1.2.7)) has been noted several times. Since deformation of the nematic crystal leads to large changes of its optical characteristics, the optical methods of recording are more suitable for measuring angle θ_0.

In [16, 35] the angle of inclination of the director θ_0 in the flow of the nematic liquid crystal in a capillary was measured by the optical method on the basis of the optical difference of the path $\Delta\Gamma$ of the unconventional and conventional beams. At some (not too high) shear rate, the value of $\theta(x)$ in the layer of the liquid crystal with the exception of the transition layer in the centre is a constant value equal to θ_0 at $x > 0$, $-\theta_0$ at $x < 0$. For low values of θ_0, which is almost always the case in the experiments, $\Delta\Gamma$ is equal to:

$$\Delta\Gamma = \int_{-L/2}^{L/2} \left[\frac{(n_e n_o)^{1/2}}{(n_e^2 \cos^2\theta + n_o^2 \sin^2\theta)^{1/2}} - n \right] dx = \\ = -\frac{1}{2} n_e L \frac{n_e^2 - n_o^2}{n_o^2} \sin^2\theta. \tag{2.1.38}$$

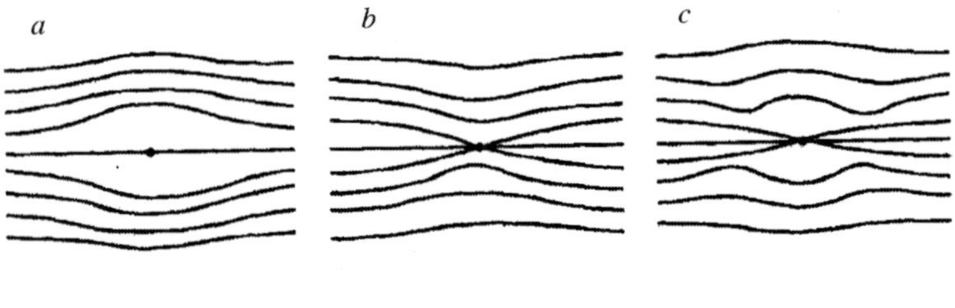

Fig.2.1.25. Distribution of the flow lines of the NLC in the vicinity of a linear disclination with force *a*) $S = +1$; *b*) $S = 21$; *c*) $S = -2$ [49].

The capillary proposed in [35] is flat with nozzles for inflow and outflow of the liquid crystal. The same form of the measuring cell is also used to obtain data on electrical conductivity and dielectric permittivity in [60].

An original method for determining the angle of inclination of the director in a wide range of shear rates in a single measurement and, consequently, the ratio γ_1/γ_2, was described in [61]. The measuring cell consists of two circular glass sheets with homeotropic orientation of the nematic liquid crystal on the surfaces. One of the sheets rotates with constant angular velocity ω. The point situated in at distance r from the axis of rotation has linear velocity $v = \omega r$. This makes it possible to study a wide range of shear rates (angular velocity changes in the range from $4.19{\cdot}10^{-5}$ to $4.19{\cdot}10^{-2}\ s^{-1}$). Since the increase of v is accompanied by the increase of the angle of inclination of the director in the centre of the layer, the birefringence of the layer the liquid crystal also changes and this lead to the formation of a system of concentric light and dark rings in illumination of the cell with monochromatic radiation (in [61] $\lambda = 546$ nm). The number of the ring is linked with the mean birefringence of the liquid crystal at distance r from the axis of rotation by the relationship:

$$\frac{m\lambda}{L} = \langle n_e(r) - n_o \rangle, \tag{2.1.39}$$

where L is the thickness of the specimen (40÷500 μm, the deviation from the mean thickness is less than 2 μm). The ratio of the viscosities is determined by determining radius r^* (or number m) at which the density of the rings is maximum (the distance between the rings is minimum). In this case, the ratio r^*/mL is the function of the physical parameters of the nematic liquid crystal and temperature. The dependence of the ratio m/L on the parameter $(vL)^{-1/2}$ has the form of a straight line and the angle of inclination of this straight line is determined by the ratio γ_1/γ_2:

$$\frac{\gamma_1}{\gamma_2} = 1 - \frac{2(1-\omega^{-2})}{1 - n_e^2/n_o^2}, \quad \omega = \frac{m}{L}\frac{\lambda}{n_e} + 1. \tag{2.1.40}$$

To determine the absolute values of the viscosity coefficient using this method, the liquid crystal is placed in a magnetic [62] or electrical [63, 64] field. It is shown in [64] that the method of the measurement of viscosity on the basis of the optical characteristics of the liquid crystal layer, situated between the rotating glass sheets, may be used for nematic liquid crystals characterised not only by positive but also negative dielectric anisotropy. In contrast to (2.1.40) the ratio of the number of the ring to the thickness of the field of the liquid crystal is a function of not only the parameters of the liquid crystal but also applied stress. Therefore, the viscosity values are determined from the coefficients by approximation

$$m\left[L(\upsilon L)^2\right]^{-1} = a_1 + a_2(\upsilon L)^2, \tag{2.1.41}$$

$$a_1 = \frac{n_e}{960\lambda}\left(1-\frac{n_e^2}{n_o^2}\right)\frac{(\gamma_1-\gamma_2)^2}{K_{33}^2}\left[1+\frac{\varepsilon_a U^2(\gamma_1-\gamma_2)}{8K_{33}}\right]^{-2}\times$$
$$\times\left[1-\frac{17}{14}\frac{\varepsilon_a U^2}{K_{33}\eta_2}\left(4-\frac{\gamma_1-\gamma_2}{\eta_2}\right)\right], \tag{2.1.42}$$

$$a_2 = \frac{n_e}{630}\left(1-\frac{n_e^2}{n_o^2}\right)\frac{(\gamma_1-\gamma_2)^4}{K_{33}}\left[1+\frac{\varepsilon_a U^2(\gamma_1-\gamma_2)}{8K_{33}}\right]^{-4}\times$$
$$\times\left[\frac{36}{11}\frac{\varepsilon_a U^2}{K_{33}\eta_2}\left(4-\frac{\gamma_1-\gamma_2}{\eta_2}\right)-\frac{17(K_{33}-K_{11})}{K_{11}}-\frac{4}{3}\right]. \tag{2.1.43}$$

The equations (2.1.41)–(2.1.43) were obtained under the condition of low deformation of the director in the flow of the nematic liquid crystal. The relationships, described by (2.1.41), are shown in Fig. 2.1.26 [64]. At the zero field $U = 0$ we can determine the ratios $(\gamma_1 - \gamma_2)/K_{33}$ and K_{33}/K_{11}. Processing the approximations we obtain the values of K_{33}, K_{11}, $(\gamma_1-\gamma_2)$, η_2. This means that the described method can be used to determine the absolute values of γ_1, γ_2, η_2. The restrictions for the effective stress in the layer of the liquid crystal are given in [64]. At $\varepsilon_a < 0$ the values of applied stressshould be lower than $\left[\frac{48K_{33}\eta_2}{|\varepsilon_a|(\gamma_1-\gamma_2)}\right]^{\frac{1}{2}}$. For MBBA the limiting value of U is 17 V. At a sufficiently high frequency this is lower than the threshold of formation of electrohydrodynamic instability. However, even larger restrictions on the nematic liquid crystal with any sign of ε_a follow from the small value of deformation angle θ. If $\theta = \theta_{max}\,[1 + (\varepsilon - 1)\,4z^2L^{-2}]$, where θ_{max} is the maximum deformation angle, ε is a low parameter, the value of U should not be greater than $U_1 = \left[\frac{8K_{33}\varepsilon}{|\varepsilon_a|\left(4-(\gamma_1-\gamma_2)\eta_2^{-1}\right)}\right]^{1/2}$. For pentylcyanobiphenyl $U_1 \sim \varepsilon^{1/2}$ which at $\varepsilon \sim 1$ is equal to 0.4 V. It is assumed in [64] that the experiments can be carried out at higher stresses. To substantiate this, it is necessary to obtain a numerical solution of nonlinear equations describing the dynamics of nematic liquid crystals. Although the method has not been used widely for the measurement of the viscosity of the nematic liquid crystal, it has been used efficiently for measuring coefficients α_2, α_3 of

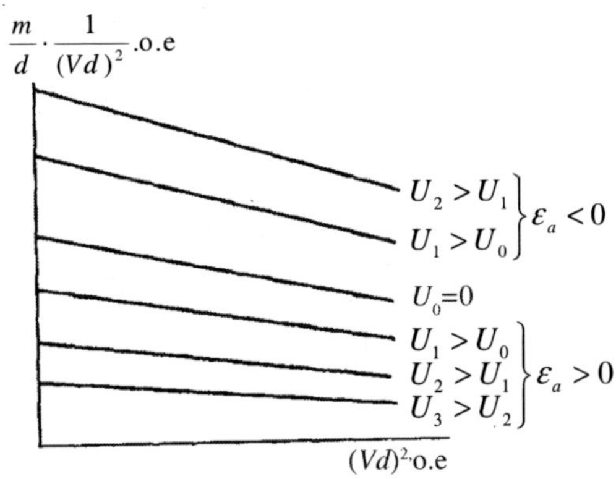

Fig.2.1.26. Determination of coefficients a_1 and a_2 at different stresses for NLC with positive and negative dielectric anisotropy.

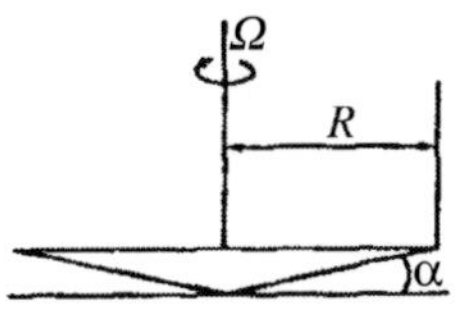

Fig.2.1.27. Rotational viscometer [49].

polymeric liquid crystals [65].

In the method described previously, we determine the set of the viscosity coefficients of the nematic liquid crystal but the commercially produced rotational viscometers, consisting of a rotating cone with the radius of the base R and the angle of the tip of 180°–2α and of a flat sheet (Fig. 2.1.27) [49, 66], are used far more frequently. In this case, the moment acting on the sheet is measured. The advantage of the method is that for low values of α both the stress and shear rate are independent of radius R.

It is important to mention the application of the falling sphere method in the measurement of the viscosity of liquid crystals. Viscosity η_{FB} measured by this method is lower than in the flow of the nematic liquid crystal in the capillary (Fig. 2.1.11) and very close to the Mesovich viscosity value [40]. The value of η_{FB} does not depend on the width of the gap between the sphere and the walls of the cylinder (Fig. 2.1.28). In the temperature range in which the coefficient α_3 changes the sign, the value of η_{FB} slightly increases. This is evidently associated with the formation of turbulence [40]. The falling sphere method is highly suitable for experiments in which the shear flow cannot be realised, for example, in samples subjected to high pressures [67, 68].

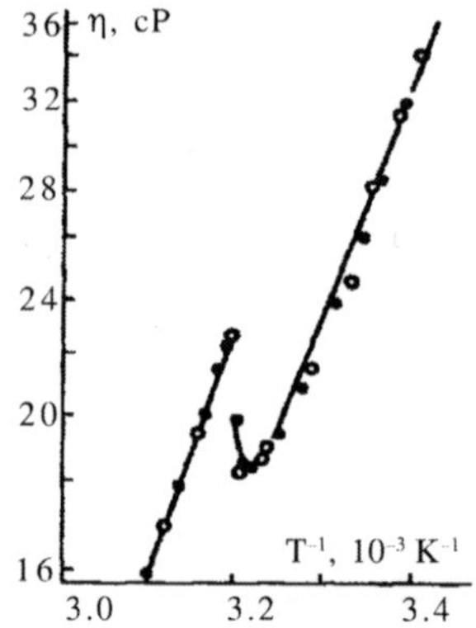

Fig.2.1.28. Temperature dependence of viscosity η MBBA measured by the method of the falling sphere at the mean width of the gap between the sphere and the pipe wall of 60μm (o) and 1400 μm (•).

2.2. Ultrasound methods of measuring the viscosity of nematic liquid crystal

The values of the viscosity coefficient similar to those measured in the shear flow of the nematic liquid crystal from the capillary can be obtained in measuring the passage of ultrasound shear waves through the liquid crystal layer [69, 70]. Depending on the mutual orientation of the director **n**, ultrasound displacement **u** and the wave vector of the ultrasound wave **k** (Fig. 2.2.1) we determine one of the three viscosity coefficients:

$$\eta_A = \frac{\alpha_4}{2}, \eta_B = \frac{\alpha_4 + \alpha_6 - \alpha_3 \gamma_2 / \gamma_1}{2}, \eta_C = \frac{\alpha_4 + \alpha_5 - \alpha_2 \gamma_2 / \gamma_1}{2}. \quad (2.2.1)$$

It may be easily shown that $\eta_B \equiv \eta_C$ (Rapini equality). The value η_B differs from η_2 by:

$$\Delta\eta_{B2} = \eta_B - \eta_2 = -0.5\alpha_3(1 + \gamma_2 / \gamma_1). \quad (2.2.2)$$

Taking into account the small value of $|\alpha_3|$ and the approximate equality $\gamma_2/\gamma_1 \approx -1$, the difference between the η_B and η_2 is very small.

The ultrasound methods used for measuring the values of the viscosity coefficients η_A, η_B, η_C have been described in detail in [70]. The pulsed-phase method of the determination of the viscosity of the nematic liquid crystal, proposed in [69], is based on the variation of the amplitude and phase of the radiation reflected from the surface of the measuring element with the oriented nematic crystal placed on the element in a specific direction. The measuring element is a trapezoidal sheet of fused quartz. The emitter of ultrasound oscillations which may also work in the reception regime is bonded to the end of the sheet. The flow diagram of the pulsed-phase

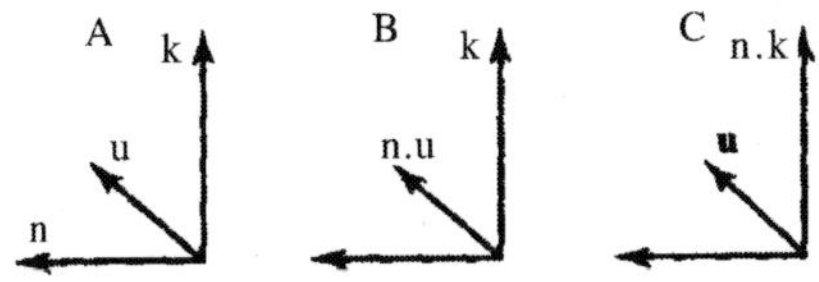

Fig.2.2.1. Relative position of the vectors of ultrasound displacement **U**, wave vector **k**, director k in the measurement of the viscosity by ultrasound methods.

method, described in [69, 70], is shown in Fig. 2.2.2 [70, 73]. The advantage of the method is that the viscosity of a relatively small amount of the liquid crystal (0.5÷1 g) may be measured in a very wide range ($10^{-3}\div10^{12}$ Pa·s). However, the signal/noise ratio is impaired: it is important to note that the mechanical parameters of the measuring element depend strongly on temperature and time, the quality of bonding of the emitting element and also the need for orders of reflection with a high number for low viscosity fluids ($\eta < 0.1$ Pa·s) in the measurement of the reflection coefficient and phase shift. The accuracy of the method is 5÷10% and for low-viscosity fluids it is 20%. It cannot be used for express analysis.

The values of viscosity in the range typical of the nematic crystals (0.005÷5 Pa·s) may be obtained using the low-frequency (24÷270 kHz) resonance method of torsional oscillations [70, 71]. The high accuracy of the method (1÷2%) is due to the fact that it is quite simple to measure the variation of the resonant frequency Δf and of the half-width of the resonance curve δf of the quartz element immersed in the medium. The measured characteristics of the resonantor are linked by a simple relationship with the impedance of the medium:

$$Z = R + iX = K - 1(\delta f + i\Delta f). \qquad (2.2.3)$$

The components of the complex shear modulus $G = G' + i\omega\eta$ are linked with the components of the impedance by the equation:

$$G' = \frac{R^2 - X^2}{\rho}, \eta = \frac{RX}{\pi f \rho}. \qquad (2.2.4)$$

The authors of [70, 71] describe the design of the measuring element for eliminating the damping effect of fixing on the efficiency of the resonator and shunting by the conducting fluid and its vapours of the electrodes of the resonator (Fig. 2.2.3). The four cross-connected electrodes, sprayed on the cylindrical surface of the quartz, excite torsional oscillations of the resonator. When the part of the resonator free from the electrodes is immersed in the investigated medium, the moment of inertia increases as a result of connection of the mass of the liquid to the side and top surfaces of the electrode and this also results in the variation of resonator frequency

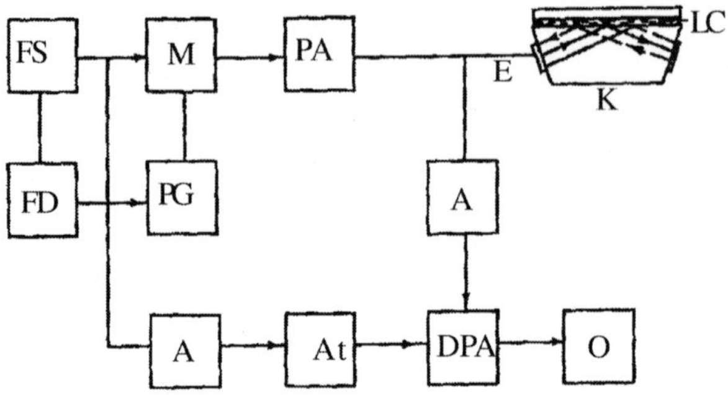

Fig.2.2.2. Flow diagram of the high-frequency pulsed-phase method: FS – frequency synthesizer, FD – frequency divider, PG – pulse generator, M – modulator, PA – power amplifier, A – amplifier, DPA – differential preamplifier, O – Oscilloscope, At – Attenuator, K – quartz sheet, E – emitter, LC – layer of liquid crystal.

Δf. The circuit diagram of recording the resonance parameters, based on the method of the amplitude–phase balance, is shown in Fig. 2.2.4. The range of the measured viscosity values may be widened using the regime of the freely attenuating oscillations of the resonator loaded with the investigated fluid (the value of η up to $2 \cdot 10^2$ Pa·s [70, 72]), or using a resonator with an internal cylindrical cavity. It is shown in [70] that the sensor with a height of 60 mm, the external and internal diameter of 20 and 10 mm respectively, the depth of the internal cylindrical cavity of 30 mm, may be used to measure the viscosity of the fluid prior to the glass transition.

The large (to 0.1÷0.5 cm^3) decrease of the investigated volume of the matter and also of the dimensions of the resonator element is ensured using a thin quartz sheet as the resonator of shear oscillations (Fig. 2.2.5) [70, 73]. The centre of the polished quartz sheet with the thickness $L = 0.34$ mm and the diameter $D_1 = 2.5$ mm contains gold electrodes with the diameter of $D_2 = 6$ mm. The structure is produced with the electrodes at a large distance from the edge of the sheet in order to localise the energy of oscillations in the region below the electrode and eliminate the effect of fixing on the efficiency of the resonator. The lower edge of the resonator is

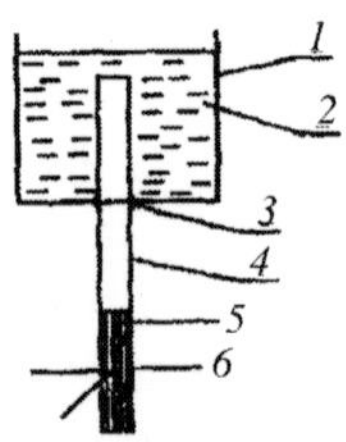

Fig.2.2.3. Design of the measuring element for the low frequency resonance method of torsional oscillations: 1) glass vessel, 2) investigated fluid, 3) quartz resonator, 4) fused cylindrical quartz sheet, 5) crystalline quartz, 6) electrodes.

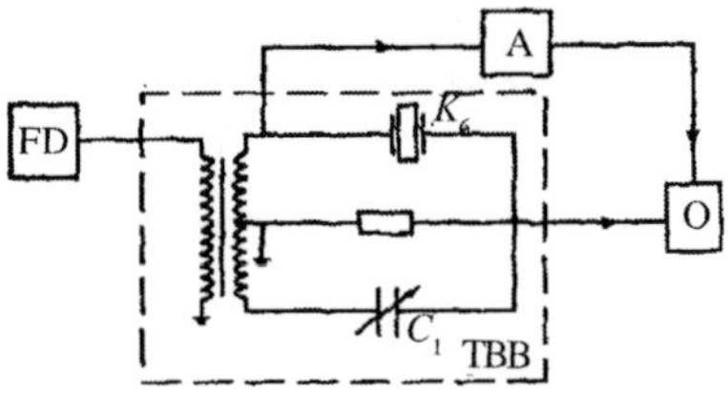

Fig.2.2.4. Diagram of the resonance method of torsional oscillations: TBB – transformator balance bridge [70, 71], FS – frequency synthesizer Ch-31 (decadic generator GZ-49), L – attenuator (type TT-4115, Hungary), O – oscilloscope with a differential amplifier.

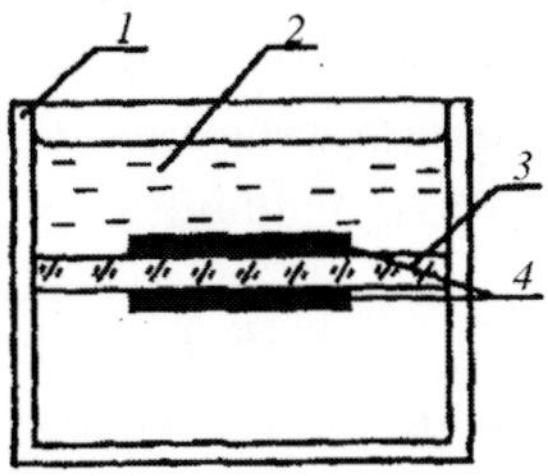

Fig.2.2.5. Design of the measuring element using the high-frequency resonance method: 1) restricting vessel, 2) investigated fluid, 3) quartz resonator, 4) electrode [70].

insulated from the external atmosphere. The operation of the resonator at the harmonics from the first to the seventh covers the frequency range 4.9÷33.9 MHz.

The circuit diagram of the measuring equipment is shown in Fig. 2.2.6. The planar orientation of the nematic liquid crystal on the surface of the resonator is obtained by spreading of a phenyltrichlorosilane film of across or along the direction of oscillations after immersing the sheet in a 5% solution of toluene followed by drying. The homeotropic orientation is produced by processing the surface with a 5% solution of dimethyldichlorosilane in toluene followed by drying of the resonator. Typical temperature dependences of the active *R* and reactive *X* components of the shear acoustic impedance for different orientations of the nematic liquid crystal in relation to the direction of oscillations of the resonator are presented in Fig. 2.2.7. Figure 2.2.8 shows the values of viscosity calculated from equation (2.1.24) for three different geometries [74]. It may be seen that $\eta_B \neq \eta_C$, i.e., the Rapini relationship is not fulfilled. It is possible that this is associated with the difference in the relaxation time of the orientation and translational movement of the molecules (see, for example (2.1.23)). The fact that the Rapini equality is not fulfilled at high frequencies does not indicate that the Parodi relationship is violated.

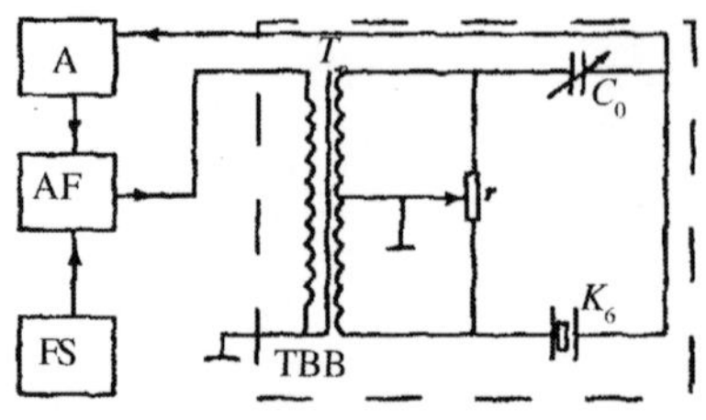

Fig.2.2.6. The flow diagram of the high-frequency resonance method of measuring viscosity: TBB – transformer balance bridge, AF – device for measuring amplitude–frequency characteristics (X1-27), FS – frequency synthesizer (SCh6-31), A – wide-band amplifier [74].

The temperature dependences η_A, $\eta_B = \eta_C$ and the differences $\Delta\eta = \eta_c - \eta_B$ (η_c is the viscosity in the flow through the capillary (see (2.1.28)) [39], obtained by the pulsed-phase method for pentylcyanobiphenyl, are shown in Fig. 2.2.9.

The viscosity coefficients η_A and η_B can be obtained by measuring the spectrum of inelastic scattering of the light on the capillary waves formed on the free surface of the nematic liquid crystals [75]. A thick specimen of the liquid crystal with a large diameter (5 mm and 25 mm respectively) is placed in a magnetic field with $H \sim 2$ kOe. The laser light beam is reflected from the free surface of the liquid crystal. Part of the beam is scattered under angles determined by the wavelength of the thermally excited capillary ways. The movement of the waves results in the Doppler shift of the frequency of incident light. The values of the frequency and the attenuation factor are determined by surface tension σ and the viscosity of the nematic liquid crystal η and can be measured by the method of spectroscopy of optical shift. The accurate equations for the density of the spectral energy of scattered light are presented in [76, 77]. For low viscosity values, the spectrum is Lorentzian with the centre at the frequency

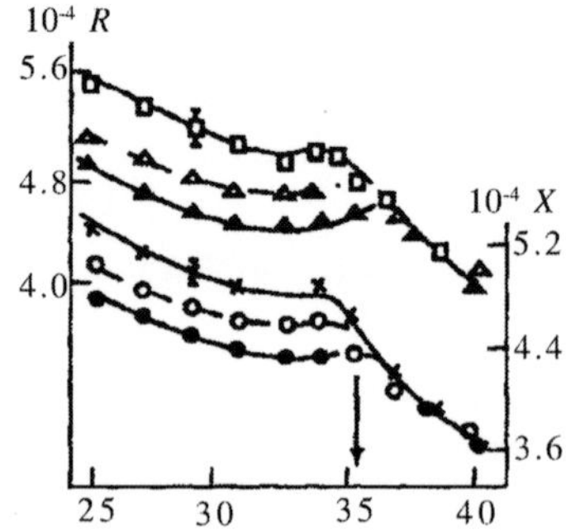

Fig.2.2.7. Dependence of active (R) and reactive (X) components of shear acoustic impedance on temperature: f = 33.9 MHz [74].

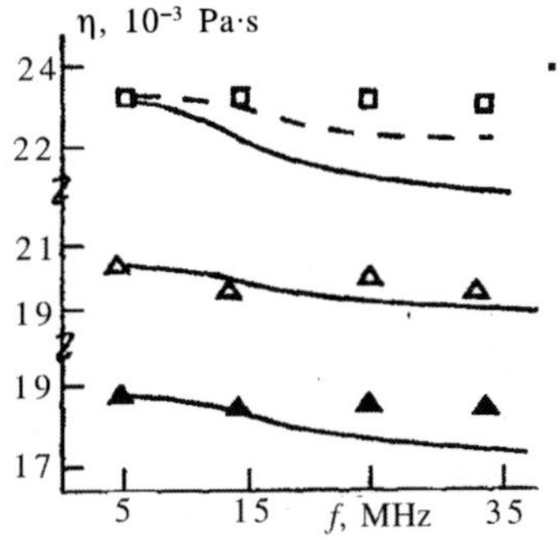

Fig.2.2.8. Dependence of the dynamic viscosity η of BF-5 measured by the high frequency resonance method, on the frequency of acoustic oscillations *f*. Solid lines – dependences calculated from the single relaxation model, dotted – using the two-relaxation model [74].

$$\nu_q = (2\pi)^{-1}(\sigma q^3)^{1/2} \tag{2.2.5}$$

and the half width

$$\Delta\nu_\perp = \eta A \frac{q^2}{\pi\rho}(q \perp H),\ \Delta\nu_\parallel = \left(\eta_B + \frac{\eta_D}{8}\right)\frac{q^2}{\pi\rho}(q \parallel H), \tag{2.2.6}$$

where σ is surface tension, ρ is the density of the nematic liquid crystal, q is the wave vector indicating the direction of propagation of the capillary waves normal or parallel to the magnetic field, $\eta_D = \alpha_1 + \gamma_2^2/\gamma_1$. Using the mathematical processing of the form of the spectrum for $q \parallel H$ it is possible to determine the values of η_B and η_D.

2.3. The method of the rotating magnetic field

The experiment with a liquid crystal specimen placed in a rotating magnetic field was formulated for the first time in 1939 by Tsvetkov [78, 79] and subsequently as a result of the increased interest in the liquid crystals at the end of the 60s and the beginning of the 70s it was used almost without any change for measuring the coefficient of rotational viscosity γ_1 [80–83]. In some cases, the magnet is not rotated but the specimen suspended between the poles of the magnet is rotated [84, 85]. The experiment geometry is shown schematically in Fig. 2.3.1. The equation of motion of the director (1.2.6) taking into account the presence of a constant magnetic field **H**, the absence of fluxes and the uniformity of the orientation is reduced to the following form:

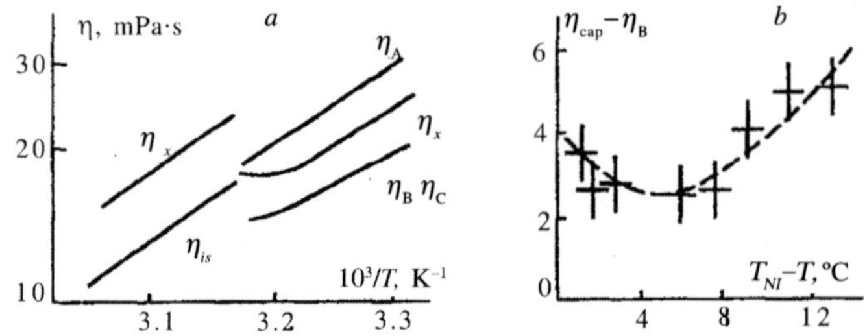

Fig.2.2.9. Temperature dependence of viscosity η_k, measured in flow of BF-7 through the capillary η_A, η_B, η_C, (in the nematic phase), η_{is} (in the isotropic phase), measured by the high-frequency pulsed-phase method (*a*) and the difference $\eta_{cap} - \eta_B$ (*b*) [39].

$$\ddot{\varphi} - \gamma_1\dot{\varphi} + 0.5\rho\chi_a H^2 \sin 2(\varphi - \varphi_H) = 0, \tag{2.3.1}$$

where φ and φ_H are the angles of rotation of the director and of the magnetic field, respectively; I and ρ is the moment of inertia and the mass of the unit volume. We examine the processes of establishment and the relaxation of the orientation of the director at the beginning and end of rotation of the magnetic field and also in the case of small oscillations of the direction of the magnetic field [85]. Assuming that the difference $\varphi - \varphi_H$ is small and reducing the equation (2.3.1) to the linear form, we obtain its solution:

$$\varphi = \varphi_0 + \omega_0 t + \int_0^t f(t')\psi_n(t - t')dt', \tag{2.3.2}$$

where $\varphi_0 = \varphi(t = 0)$ in $\omega_0 = \dot{\varphi}(t = 0)$ are the initial conditions, $n = -1.0$ and $+1$, and $f(t)$ and $\psi_n(t)$ are given by the expressions:

$$f(t) = -\frac{2I\chi_a H^2}{\gamma_1}[\varphi_H(t) - \varphi_0 - \omega t] - 2\omega_0,$$
$$\psi_n(t) = \left[\frac{sh(\alpha t/\tau)}{\alpha} + nch(\alpha t/\tau)\right]\exp(-t/\tau) \tag{2.3.3}$$

(α is a dimensionless constant linked with the value of the critical field $H_c = -\gamma_1/(4I\rho\chi_a)^{1/2}$ by the relationship $\alpha = (1 - H^2/H_c^2)^{1/2}$, $\tau = -2I/\gamma_1$ is the attenuation time). When measuring the rotational viscosity by the method of the rotating field, measurements are usually taken of the density of the momentum of forces Γ acting on a wire from the side of the unit volume of the specimen, and Γ is linked with the speed of rotation by a simple relationship:

$$\Gamma = -\gamma_1\dot{\varphi}, \tag{2.3.4}$$

which after substitution of (2.3.2) with (2.3.3) taken into account leads to:

$$\Gamma = -2\rho\chi_a H^2 \left[\varphi_0 \psi_0(t) + \frac{1}{\tau} \int_0^t \varphi_H(t')\psi_{-1}(t-t')dt \right] + \gamma_1 \omega_0 \psi_{+1}(t). \quad (2.3.5)$$

If the magnetic field with the constant speed of rotation ω, $\varphi_H = \omega_t$ is activated at the initial moment of time when $\varphi_0 = 0$, $\omega_0 = 0$, and at the moment acting on the wire is:

$$\Gamma = -\gamma_1 \omega [1 - \psi_1(t)]. \quad (2.3.6)$$

The uniform rotation of the director is established after a relatively long period of time.

The inertial members in (2.3.1) influence only the time to establishment of the uniform orientation of the director. Therefore, below the critical angular speed ω_c

$$\omega \leqslant \omega_c = \chi_a H^2 / 2\gamma_1 \quad (2.3.7)$$

equation (2.3.2) has a very simple form:

$$\varphi = \omega t - \varepsilon, \ \sin 2\varepsilon = \omega / \omega_c. \quad (2.3.8)$$

At $\omega < \omega_c$ the distance of the director from the field is expressed by the constant angle ε and the problem of measuring the values γ_1 is reduced to determining its value. At $\omega > \omega_c$ the solution of equation (2.3.1) is [84]:

$$tg\left(\varepsilon - \frac{\pi}{4}\right) = \left(\frac{\omega - \omega_c}{\omega + \omega_c}\right)^{1/2} tg\left[\left(\omega_2 - \omega_c^2\right)^{1/2} t - t_0\right], \quad (2.3.9)$$

where

$$t_0 = \text{arctg}\left(\frac{\omega - \omega_c}{\omega + \omega_c}\right)^{1/2}. \quad (2.3.10)$$

Consequently, the director rotates and the angular velocity of:

$$\Omega = \left(\omega^2 - \omega_c^2\right)^{1/2}. \quad (2.3.11)$$

If a specimen of the liquid crystal with the volume V is suspended on an elastic suspension (wire), it is possible to measure the elastic moment acting on the specimen equal to

$$M = \Gamma V = \gamma_1 \omega V \text{ or } \omega < \omega_c, \quad (2.3.12)$$

$$M = 0.5 V \chi_a H^2 \text{ and } \omega = \omega_c, \quad (2.3.13)$$

$$M = \frac{\gamma_1 V \omega_c^2}{\omega + (\omega^2 - \omega_c^2)^{1/2}} \text{ and } \omega > \omega_c, \quad (2.3.14)$$

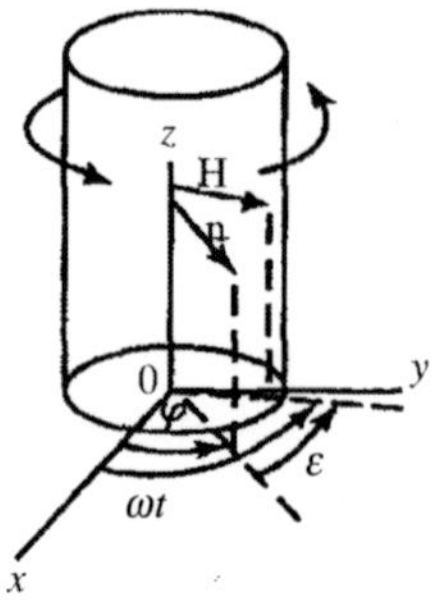

Fig.2.3.1. Relative position of the vectors **H** (phase $\omega\tau$) and **n** (phase φ) in measurement of γ_1 by the method of the spinning magnetic field.

The relationships (2.3.12)–(2.2.14) are confirmed in Fig. 2.3.2 which shows the dependence of the rotating momentum acting on a specimen of *para*-azoxyanisole at different temperatures, on the angular speed of rotation of the magnetic field determined by Tsvetkov [78]. The left parts of the curves ($\omega < \omega_c$) are characterised by a quantitative agreement with theoretical description (2.3.12), and the experimental values of M for the right-hand section slightly exceed the theoretical values. The best agreement between the experiments and theory for the right-hand part of the dependence $M(\omega)$ was obtained in [82] where the rotating momentum was measured using a solid Teflon cylinder immersed in the specimen (instead of using metallic wire).

It was shown in [86] that the increase of the speed of rotation of the magnetic field results in changes in the structure of the specimen. This is indicated by a decrease of the intensity of the satellite lines of the spectrum of nuclear magnetic resonance, and at $\omega \gg \omega_c$ – by their disappearance (Fig. 2.3.3).

The values of the inertial terms may be evaluated on the basis of the measurement of the rate of relaxation of the force momentum Γ when arresting the rotation of the magnetic field or in an oscillating magnetic field [85]. In the former case $\varphi_H = 0$, $\varphi_0 = \gamma_1\omega/\rho\chi_a H^2$ and equation (2.3.5) is transformed to

$$\Gamma = -\gamma_1 \omega \psi_1(t). \qquad (2.3.15)$$

The decrease of Γ is monotonic at $H \leqslant H_c$ and in the form of attenuating oscillations at $H > H_c$. In this case, the value of φ_1 is determined from the value of H_c or by mathematical processing of the attenuation curve.

Experimental equipment will now be described briefly and the measurement errors which may form when using the method of the rotating magnetic field will be examined. Figure 2.3.4 shows the diagram of equipment taken from [84]. The pipe 9 (diameter 5 or 30 mm) is filled with the liquid crystal 11 to the height of approximately 80 mm (this height is required to avoid the effect of the bottom of the pipe), suspended from the glass bar 8 and the twisted wire 2 (iron wire, length 1 m, diameter 30 or 100 μm). The pipe with the specimen is rotated using the step motor 1 (500 strokes per revolution). The specimen is placed between the poles of the electromagnet 10 (the diameter of the core 200 mm, the gap 50 mm,

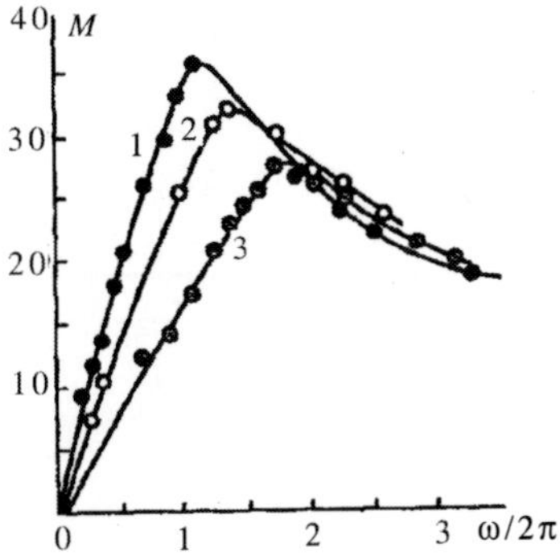

Fig.2.3.2. Dependence of the spinning moment *M*, acting on the specimen of PAA, on the angular velocity of rotation of the magnetic field w at temperature of 110°C (*1*), 115°C (*2*), 120°C (*3*) [78].

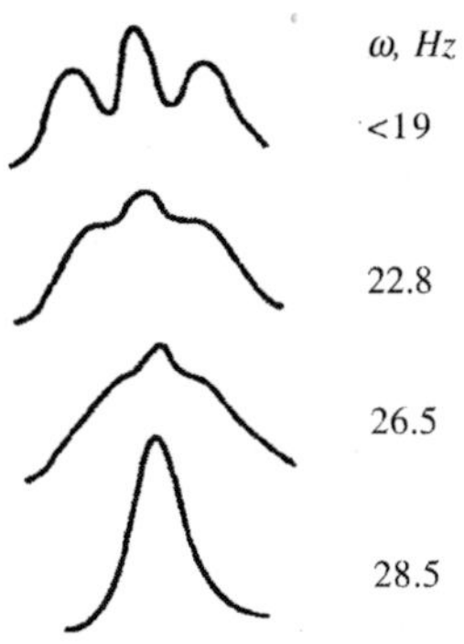

Fig.2.3.3. Form of the line of absorption of proton magnetic resonance of PAA at different angular velocities ω, characterising the transition from synchronised to asynchronised regime; T = 117°C[86].

$B = 0 \div 1.3$ T) and is surrounded by the two-wall glass pipe 7 maintained at a constant temperature. The coefficient of rotational viscosity is determined from the angle of twisting of the wire according to the equation:

$$\gamma_1 = \frac{D\alpha}{\dot{\theta} V}, \tag{2.3.16}$$

where *D* is the constant of twisting of the wire, $\dot{\theta}$ is the angular velocity of the step motor. The twist angle α is determined from the phase delay formed

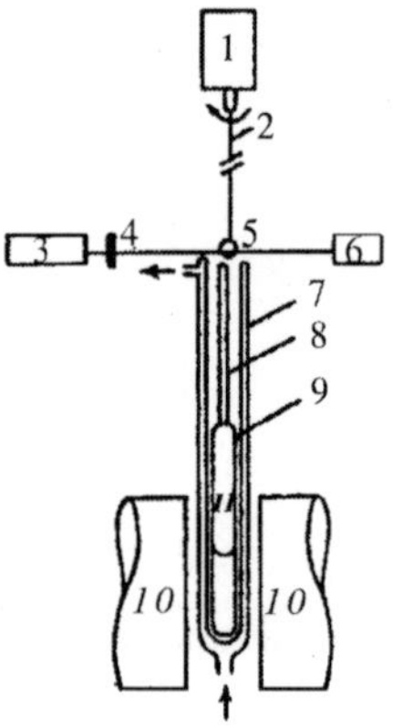

Fig.2.3.4. Experimental equipment for measuring the viscosity γ_1 by the method of the spinning magnetic field.

between the step motor and the specimen after activating the magnetic field. The system for detection of the phase delay is formed by the laser 2, the cylindrical lens 4, the mirror 5 and six photodiodes 6 distributed through every 60° in the horizontal plane. The detection system is connected with a computer averaging the differences in the phase delay in several rotations. The deviations from Hooke's law at large angles of twisting of the wire are taken into account, and the residual twisting with the magnetic field applied and formed as a result of the humidity of air inside the thermostatically controlled pipe, and the errors formed as a result of the non-coaxial position of the specimen, are also considered.

Other errors which considerably greater than the errors mentioned previously, may form as a result of the effect of walls, i.e. the reversed flow, inversed walls and elastic moments associated with incomplete orientation of the liquid crystal by the magnetic field. However, this effect can be attributed to the usual inverse flow [19, 20] because of different directions of the flow. In all likelihood, it is caused by the reversed flow between the wall of the cylinder and the free surface in the upper part of the specimen whose speed is higher than that of the conventional flow. The identical effect was theoretically discussed in [21]. Figure 2.3.5 shows the profile of the flow of a nematic liquid crystal in a cylindrical specimen under the effect of the rotating magnetic field. In the boundary condition $v = 0$ on the wall of the cylinder as a result of the presence of elastic forces, the magnetic field does not completely determine the orientation of the specimen at distances of the of the magnetic coherence length [87]:

$$\xi = \pi H^{-1}\left(\frac{K_{ii}}{\chi_a}\right). \qquad (2.3.17)$$

Therefore, the effective radius of the rotating specimen of the crystal and, consequently, the torque of the wire decrease. The effective value of the

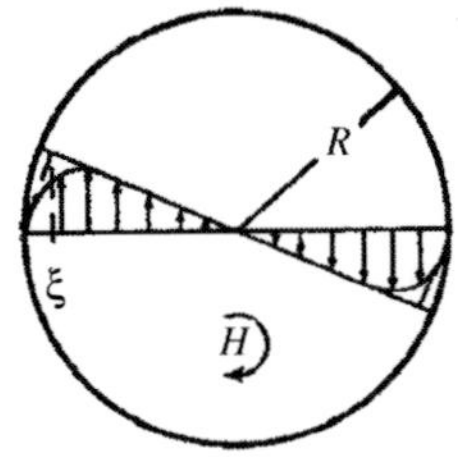

Fig.2.3.5. Profile of the flow speed of the ONLC in a cylindrical specimen formed under the effect of a spinning magnetic field [64].

measured and rotational viscosity is equal to [75]:

$$\gamma_1^{ef} = \gamma_1\left(1-\frac{2\xi}{R}\right). \qquad (2.3.18)$$

An even stronger effect on the measured values of γ_1 is exerted by the inversion walls of the type of Bloch walls in ferromagnetics, separating, inside the specimen of the liquid crystal, the regions with different orientation of the director [88, 99]. The problems of stability of these walls and their topology have been examined in [84, 89]. The relaxation time of the walls may be determined from the balance of the viscous, magnetic and elastic energies and is:

$$\tau \cong \frac{\pi\gamma_1}{\chi_a B^2}, \qquad (2.3.19)$$

which for 4-methoxybenzilidene-4′-butylaniline (MBBA) at room temperature and $B = 1$ T is equal to approximately 0.1 s. Since the relaxation time is comparable with the duration of rotation of the specimen in the magnetic field, this mechanism has a strong effect on γ_1^{ef}.

In the first approximation, the part of the specimen that is occupied and does not provide any contribution to the rotational moment M, is taken into account by replacing the coherence length ξ in the equations (2.3.18) by higher values $\xi^* = c\xi$. The numerical calculations show that for two inversion walls $\xi^* = 9\xi$. The error introduced by the presence of the inversion walls or, if these walls are not present, by elastic forces, can be reduced using stronger magnetic fields and by increasing the diameter of the pipe for the specimens (Fig. 2.3.6). In [84] in the measurement of γ_1 by the method of the rotating magnetic field, the accuracy was approximately 0.3%, but the amount of material required to reach such accuracy is large (~70 g) which is not always an option. Many authors have used small specimens for similar measurements [82, 83] but the value γ_1 determined by these authors are lower because of the previously mentioned reasons in comparison with the results of other measurements obtained in the methods described in section § 2.4, 2.5.

The phase delay between the direction of the magnetic field and the direction of the director (and, consequently, the value of γ_1) can be measured in the examination of the EPR spectrum of a paramagnetic marker dissolved in the nematic liquid crystal rotating in a magnetic field [86], and also by acoustic methods – in measurements of the anisotropy of the speed of ultrasound [90, 91] or the 'anisotropy' of the coefficient of absorption of ultrasound [92, 93]. In the latter two cases, measurements may be carried out using the anisotropy of diamagnetic susceptibility and the available values of rotational viscosity.

The ultrasound methods of recording are highly suitable in cases in which the direct measurement of the torque is not possible or the optical characteristics of the nematic liquid crystals cannot be determined, for example, in experiments with specimens subjected to high pressure [94].

Measurements of the viscosity of the nematic crystals with the variation of external pressure make it possible to determine the relationship of viscosity with molecular packing and the free volume because increasing the pressure reduces the molecular spacing. The authors of [67] measured dynamic viscosity by the method of the falling sphere, and in [94] measurements were taken of the rotational viscosity by the method of the rotating magnetic field. Figure 2.3.7 shows schematically the chamber for measurement of the rotational viscosity at high pressure [94]. The device with the precision pipe T is connected through the capillary K (diameter 0.8 mm) and the cylinder C with the wire (diameter 30 μm, length 70 cm) with the known constant of twisting D, identical with that described in section § 2.3. The two pistons P, bonded with the glass capillary C, separate the volume of the liquid crystal from the oil O, transferring the pressure from the high-pressure cylinder HPC. The temperature is maintained constant

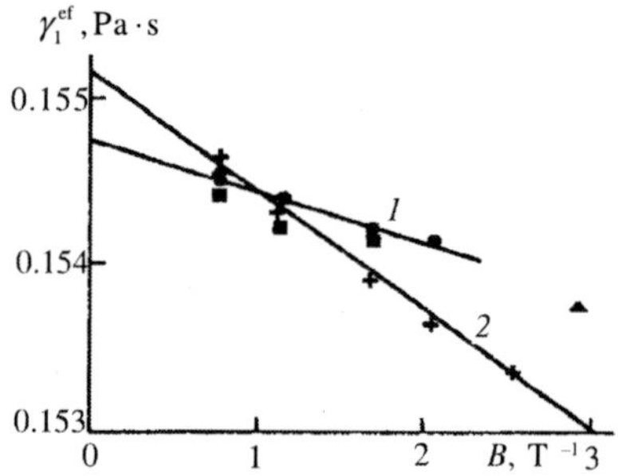

Fig.2.3.6. Dependence of the effective rotational viscosity η_1^{ef} of the MBBA on the induction of the magnetic field B in specimens with a diameter of 30mm (*1*), 5 mm (*2*) at the frequency of rotation of magnetic filed of 0.05 Hz (+, Δ), 0.1 Hz (■ ,•) [64].

by the oil circulating through the brass shell BS. The specimen of the liquid crystal is positioned between the poles of the permanent magnet NS ($B \sim 0.3$ T; gap 4 cm; the diameter of the pole terminals 11 cm) setting the orientation of the director and rotating the latter with velocity ω_H. The angle of rotation of the director φ is fixed on the basis of the reflection of the beam of laser light from mirror M_1, fixed to the end of the cylinder C. The beam is reflected from the semi-transparent mirror M_2. In the experiments, the wire is rotated in such a manner that at the initial moment of time, the plane of mirror M_1 is perpendicular to the light beam. The magnet NS is rotated to ensure that the angle of rotation of the director is 180°. Subsequently, experiments are repeated in the reversed direction. Taking into account (2.3.12) and (2.3.16) gives:

$$\gamma_1 = [(\varphi_r - \varphi_1)D][(\omega_r - \omega_1)V]^{-1} = 2\pi D[(\omega_r - \omega_1)V]^{-1}, \tag{2.3.20}$$

where ω_r and ω_1 are the speeds of rotation of the magnet in the right and left directions.

Taking into account the complicated nature of the determination of the rotational moment with a high accuracy, the value γ_1 may be determined in the measurement of the anisotropy of absorption of ultrasound $\varepsilon = \Delta\alpha/f^2$ in the pulsating or rotating magnetic field [95–99] and also in a static magnetic field ($\omega = 0$) [100–103]. In the first case, the form of the time response of the quantity $\Delta\alpha/f^2$ is described by the equation [95]:

$$\Delta\alpha(t) = \alpha_{\|}(t) - \alpha(0) = c(\langle\cos^2\theta(t)\rangle - 1/3) + b(\langle\cos^4\theta(t)\rangle - 1/5), \tag{2.3.21}$$

where the angle of the formation of the director $\theta(t)$ and the duration of the increase of the magnetic field in the electromagnet τ_H, which depends on the induction of the field ($H(t) = H_m \times [1-\exp(-t/\tau_H)]$), determine the following functional relationships:

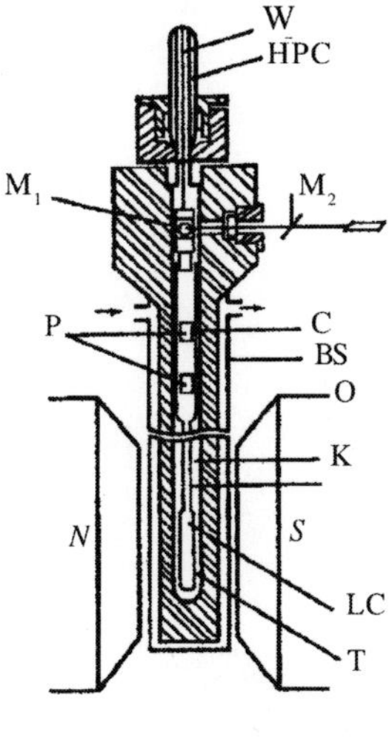

Fig.2.3.7. Diagram of equipment for measuring rotational viscosity at high pressure [94].

$$\langle\cos^2\theta(t)\rangle = \left\{1 - \left[\frac{A}{(1-A)}\right]^{1/2} \arccos A^{1/2}\right\}(1-A)^{-1},$$

$$\langle\cos^4\theta(t)\rangle = \left\{1 + \frac{A}{2} - \frac{3}{2}\left[\frac{A}{(1-A)}\right]^{1/2} \arccos A^{1/2}\right\}(1-A)^{-2},$$

$$A = \exp\left\{-\frac{\chi_a H^2}{\gamma_1}\left[2t - 3\tau_H + 4\tau_H \exp\left(-\frac{t}{\tau_H}\right) - \tau_H \exp\left(-2\frac{t}{\tau_H}\right)\right]\right\}.$$

Here c and b are coefficients describing the dependence of the anisotropy of absorption of ultrasound on the angle β between the wave vector and the director:

$$\alpha(\beta) = \alpha_0 + c\cos^2\beta + b\cos^4\beta \tag{2.3.22}$$

or, not taking into account the small contributions of heat conductivity and heat capacity [10]

$$\alpha(\beta) = \frac{\omega^2}{2\rho C_0}[(\nu_2 + \nu_4) + 2(-\nu_2 + 2\nu_3 - \nu_4 + \nu_5)\cos^2\beta + \\ + 2(\nu_1 + \nu_2 - 2\nu_3)\cos^4\beta], \tag{2.3.23}$$

where ν_i are the Forster viscosity coefficients, ρ is the density of the nematic liquid crystal, C_0 is the speed of propagation of ultrasound. The condition of incompressibility of the medium gives $\nu_2 = \nu_4$, $\nu_5 = 0$. In conventional acoustic measurements ($f = 620$ MHz, see [104, 105]), the nematic liquid crystal may be regarded as a quasi-incompressible medium. Consequently, the coefficients ν_1, ν_2 and ν_3 may be determined directly from the high-frequency measurements of the coefficient of absorption of ultrasound [105]:

$$\nu_1 = \left(\frac{\alpha^{\parallel}}{f^2}\right)_{620} \frac{\rho C_0^3}{4\pi^2};$$

$$\nu_2 = \left(\frac{\alpha^{\perp}}{f^2}\right)_{620} \frac{\rho C_0^3}{4\pi^2}; \tag{2.3.24}$$

$$2\nu_3 = \nu_1 + \nu_2 = \left(\frac{\Delta\alpha}{f^2}\right)\frac{\rho C_0^3}{4\pi^2}.$$

When the field is deactivated, the characteristic duration of disorientation is estimated from the strength of the saturation field H_{sat} using the following equation:

$$\tau_c = \frac{\gamma_1}{\chi_a} H^2_{Hac}. \tag{2.3.25}$$

When using the rotating magnetic field, the magnitude of the phase shift of rotation of the director recorded by the ultrasound methods in the synchronous regime ($\omega_H < \omega_c$) on the condition $\varphi(t = 0) = 0$ is described by the following equation [96]:

$$\begin{aligned}\varphi = \operatorname{arctg}\{(1-\exp[-2\omega_H t(\varepsilon^2-1)^{1/2}])\times \\ \times([\varepsilon+(\varepsilon^2-1)^{1/2}]-[\varepsilon-(\varepsilon^2-1)^{1/2}]\times \\ \times\exp[-2\omega_H t(\varepsilon^2-1)^{1/2}])^{-1}\},\end{aligned} \tag{2.3.26}$$

where $\varepsilon = \chi_a H^2/2\gamma_1\omega_H$. The expression for the anisotropy of absorption of ultrasound in the asynchronous regime ($\omega > \omega_c$) was published in [96]. As indicated by the equations (2.3.21) and (2.3.26) in the recording of the rotation of the vector by the ultrasound methods, the ratio γ_1/χ_a changes. Therefore, to determine the value of γ_1, it is necessary to know the value of diamagnetic anisotropy χ_a which is more difficult to measure than the density of the nematic liquid crystal. The volume of the working chamber is 5.5 cm^3 which at the given frequency of ultrasound of 2.8 MHz and the duration of the pulse makes it possible to reduce the errors associated with the length of the spectrum of the pulses, expansion and interference of the direct and reflected beams. Figure 2.3.8 shows that the pressure dependence of the measured time.

In nuclear magnetic resonance, measurements are often taken of the characteristics at slight oscillations of the direction of the magnetic field around some mean direction. In this case, $\varphi_H = \sigma \exp ikt$, where $\sigma \ll 1$. Substituting these and the initial conditions $\varphi_0 = 0$ and $\omega_0 = 0$ into (2.3.5) for the rotational moment of the forces, gives [85]:

$$\Gamma = \rho\chi_a H^2 \sigma\left[-\frac{2}{\rho}\int_0^t \exp(ikx)\psi_{-1}(x)\,dx\right]. \tag{2.3.27}$$

The oscillations of the momentum of the forces Γ, lagging in the phase behind the direction of the magnetic field by the angle $\theta \pm \pi$ are established after a relatively long period of time:

$$\Gamma = \rho\chi_a H^2 \sigma\cos\theta\exp[i(kt-\theta)]. \tag{2.3.28}$$

The delay angle is θ determined from the equation:

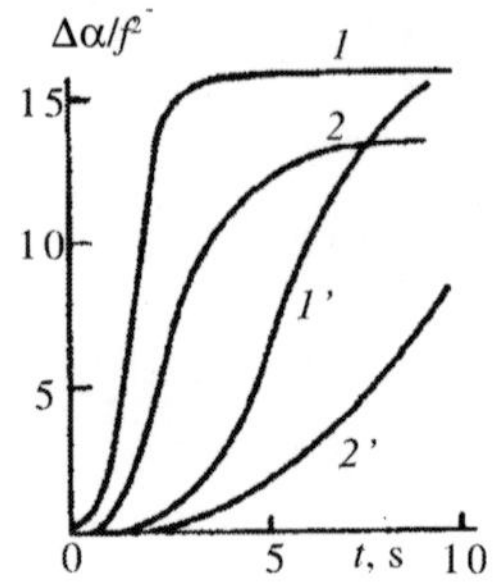

Fig.2.3.8. Dependence of the anisotropy of the coefficient of absorption of ultrasound on time in MBBA at $T = 30.5°C$ (*1*,*2*) and $T = 24.4°C$ (*1'*, *2'*) at $\tau_H = 1$ s (*1*, *1'*) and $\tau_H = 10$ s (*2*,*2'*) [95].

$$tg\theta = \frac{\rho\chi_a H^2}{\gamma_1 k}\left[1-(k/k_c)^2\right]. \tag{2.3.29}$$

Evidently, at $k = k_c = (\rho\chi_a H^2/I)^{1/2} = H/H_c\tau$, the momentum of the forces Γ changes synchronously with the direction the magnetic field **H** and, consequently, k_c is referred to as the angular velocity of the synchronisation regime. The density of the momentum of inertia I may be determined from the amplitude of phase delay of the momentum of forces.

In [106], Gerber determined the capacitance method of recording the re-orientation of the director at a small variation of the direction the magnetic field. The measurement diagram is shown in Fig. 2.3.9. The capacitor C and two solenoids S, generating a small re-orientation field $H_1 \sim 0.001$ T, are placed in the gap of the electromagnet which generates the magnetic field of up to 0.8 T. The directions **H** are $\mathbf{H}_1$ at perpendicular, the normal to the plane of the capacitor forms the angle of 45°C with the direction of both magnetic fields. The value of H and the thickness of the gap in the capacitor are selected to ensure that the magnetic coherence length is minimised (2.3.17). With the large variation of the direction of the field $\mathbf{H}_1$ of the solenoids, controlled by the generator of rectangular pulses G, the direction of the resultant field $\mathbf{H} + \mathbf{H}_1$ changes by several milliradians. The generator is positioned in the same circuit with the power source ε, the voltage converter CO, and the amplifier A. This changes the orientation of the director by the angle θ and, consequently, changes the capacitance C:

$$\frac{1}{C}\frac{dC}{d\theta} = \frac{2(\varepsilon-\varepsilon_\perp)}{\varepsilon+\varepsilon_\perp}. \tag{2.3.30}$$

The relaxation time of the recorded variation of capacitance is:

$$\tau = \frac{\gamma_1}{\chi_a H^2}. \tag{2.3.31}$$

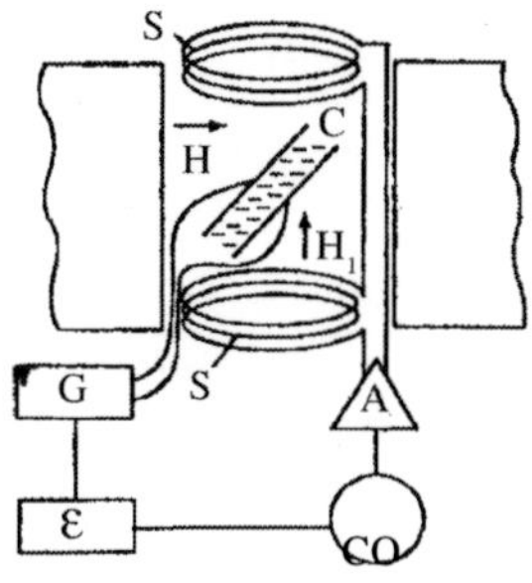

Fig.2.3.9. Diagram of the experimental equipment for measuring the rotation of viscosity with recording of the rotation of the director with small changes in the direction of the magnetic field [106].

The error of measurement of γ_1 using this method is approximately 6% and is determined not only by the inaccuracy of recording but also by the determination of the value of χ_a on the basis of the variation of the threshold voltage U_F of the Fredericks transition in the magnetic field [107], and also by the approximation of the temperature dependence χ_a by the temperature dependence of the birefringence Δn.

2.4. The methods associated with the examination of the dynamics of reorientation of the liquid crystal in the magnetic or electrical field

If a magnetic or electrical field is applied from the outside to the layer of a liquid crystal, placed between two substrates then, depending on the boundary conditions on the substrates and the magnitude and the sign of the diamagnetic or dielectric permittivity one observers deformation characterised by the director orientation. This effect is termed as the Fredericks effect who observed the effect for the first time [108]. A similar description of the Fredericks transition is presented in [1, 109]. Three partial cases are important for practice and suitable for experimental examination; these cases differ in the orientation of the director and the external field in relation to the substrates (Fig. 2.4.1). The strains, formed after applying magnetic or electrical fields to the layer, with the strength of the field higher than some critical value-threshold strength H_F or $E_F = U_F/L$ (L is the thickness of the layer) are referred to as the strains of transverse twisting, bending and longitudinal bending or, according to the terminology proposed in [1], as *S*-, *T*-, and *B*-effects. The threshold strength H_F (or threshold voltage U_F) is given by the relationship:

$$H_F = \frac{\pi}{L}\left(\frac{K_{ii}}{\chi_a}\right)^{1/2}, U_F = 2\pi\left(\frac{\pi K_{ii}}{\varepsilon_a}\right)^{1/2}. \tag{2.4.1}$$

Here χ_a or ε_a is the anisotropy of diamagnetic or dielectric permittivity, K_{ii} are the Frank elasticity moduli: $K_{ii} = K_{11}$ for the *S*-effect, $K_{ii} = K_{22}$ for the *T*-effect, and $K_{ii} = K_{33}$ for the *B*-effect. In the examination of the dynamics of the Fredericks transition we can measure the rotational viscosity γ_1. In this case, we usually measure the viscoelastic ratio of the nematic liquid crystal and, for this determination, it is necessary to know the value of the appropriate elasticity coefficient which, however, can be measured on the same specimen. The main equations of the of the orientation of the director inside the layer of the nematic liquid crystal for the three strains were published in [110].

The translational movement of the centres of gravity of the molecules are not observed only in the case of torsional deformation. Therefore, the equation of motion has the simplest form [35]:

$$K_{22}\frac{\partial^2\varphi}{\partial z^2} + \chi_a H^2 \sin^2\varphi\cos\varphi = \gamma_1\frac{\partial\varphi}{\partial t}, \tag{2.4.2}$$

where φ is the angle of twisting of the director. For the field whose strength is slightly higher than the strength of the threshold field H_F, the value of φ is small and the solution satisfying the boundary conditions $\varphi(z = \pm L/2)$, has the following form:

$$\varphi = \varphi_m \cos\frac{\pi z}{L}. \tag{2.4.3}$$

The equation (2.4.3) is valid up to the values $\varphi_m \approx 40°$ [111]. If the field H is deactivated, the strains relax in accordance with the exponential law:

$$\varphi(z,t) = \varphi(z,0)\exp(-t/\tau_o), \tag{2.4.4}$$

where the relaxation time τ_0 is given by the equation:

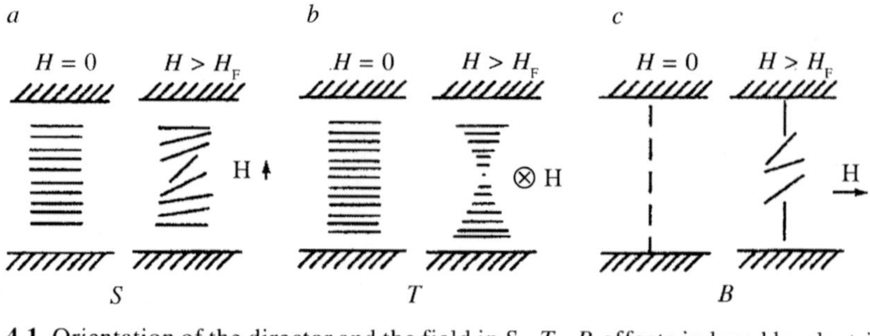

Fig.2.4.1. Orientation of the director and the field in *S*-, *T*-, *B*-effects induced by electrical or magnetic field.

$$\tau_0 = \frac{\gamma_1 L^2}{\pi^2 K_{22}}, \tag{2.4.5}$$

and $\varphi(z, 0)$ by the expression (2.4.3).

The variation of $\varphi(z, t)$ is determined on the basis of the rotation of the conoscopic figure formed in examination in a polarisation microscope of a planar-oriented layer of the liquid crystal to which the magnetic field **H** is applied in the direction parallel to the plane of the layer and normal to the director **n** (Fig. 2.4.1 *b*) [111, 112]. In twisting deformation, the conoscopic figure rotates through the angle δ, given by the following expression [112]:

$$tg2\delta = \frac{\langle \sin 2\varphi(z) \rangle}{\langle \cos 2\varphi(z) \rangle} = \frac{2\sin\varphi}{2E(1/2\pi, \sin\varphi_m) - F(1/2\pi, \sin\varphi_m)}, \tag{2.4.6}$$

where F and E are the total elliptical integrals of the first and second kind, respectively. However, if $\mathbf{H} \perp \mathbf{n}$, the conoscopic figure may rotate in both the clockwise and anticlockwise direction and this is also observed in the experiments and makes the accurate measurement impossible [113]. The problem is solved if **H** is oriented in relation to **n** by the angle $(90-\varphi_0)$. In this case, the equation (2.4.6) is complicated [111]:

$$\mathrm{tg}2\delta = \frac{2(\sin^2\varphi_m - \sin^2\varphi_0)^{1/2}\cos 2\varphi_0 - H(\psi_0,\varphi_m)\sin 2\varphi_0}{2(\sin^2\varphi_m - \sin^2\varphi_0)^{1/2}\sin 2\varphi_0 - H(\psi_0,\varphi_m)\cos 2\varphi_0}, \quad \sin\psi_0 = \sin\varphi_0 \sin\varphi_m, \tag{2.4.7}$$

$$H(\psi_0,\varphi_m) = 2\int_{\psi_0}^{\pi/2} d\psi(1-\sin^2\varphi_m\sin^2\psi)^{1/2} \int_0^{\pi/2} d\psi(1-\sin^2\varphi_m\sin^2\psi).$$

The angle δ depends in a linear manner on φ_m up to the values $\varphi_m \approx 40°$ [111]. The typical dependence of the value of δ on time t which can be used to calculate (if the value of the elasticity coefficient K_{22} is known) the value of γ_1, is shown in Fig. 2.4.2.

For the *S*- and *B*-effects, the situation is likely to be slightly complicated by the fact that the reorientation of the director is accompanied by the formation of the reversed flow [19], and measurements give not the rotational viscosity γ_1 but the effective value of the rotational viscosity γ_1^{ef}. The corrections for γ_1^{ef} formed in this case differ for the initial planar and homeotropic orientation is because of the differences in the value of the rotational moments, acting on the elementary volume of the liquid crystal [20]. For the *S*-effect

$$\gamma_1^{ef} = \eta_S = \gamma_1 - \frac{\alpha_3^2}{\eta_1} \approx \gamma_1, \tag{2.4.8}$$

and for the *B*-effect

$$\gamma_1^{\text{ef}} = \eta_B = \gamma_1 - \frac{\alpha_2^2}{\eta_2}. \tag{2.4.9}$$

Comparison of (2.4.8) and (2.4.9) shows that the measurement of the rotational viscosity in deformation by the transverse bending (Fig. 2.4.1*a*) gives a small error not exceeding 1%, since $\alpha_2 \ll \gamma_1, \eta_S$ [16, 24]. This error is usually smaller than the error caused by experimental equipment and processing of the measurement results. In deformation by longitudinal bending (Fig. 2.4.1 *c*) the correction not only greatly decreases the measured value of γ_1 but also significantly changes the form of its temperature dependence. Therefore, in the measurement of γ_1 on the basis of relaxation of *SB*-deformation we usually consider the initial planar orientation, and the variation of the angle of inclination of the director is the recorded on the basis of the variation of the phase delay $\Delta\Phi$ [114] or the capacitance [115] of the layer of the nematic liquid crystal. In both these cases, the deformation of the layer of the liquid crystal is caused by the external electrical field. Of the equation of relaxation of the director when the field is switched off, according to [114], has the following form:

$$\gamma_1^{\text{ef}} \frac{\partial\theta}{\partial t} = (K_{11}\cos^2\theta + K_{33}\sin^2\theta)\frac{\partial^2\theta}{\partial z^2} + (K_{33} - K_{11})\sin\theta\cos\theta\left(\frac{\partial\theta}{\partial z}\right)^2. \tag{2.4.10}$$

For low strains $\theta \ll 1$ and taking into account (2.4.8), the equation is simplified:

$$\eta_S \frac{\partial\theta}{\partial t} = K_{11}\frac{\partial^2\theta}{\partial z^2}. \tag{2.4.11}$$

For θ, the equation has one exponential solution, identical to the solution of the equation (2.4.4), with the time constant

$$\tau_0 = \frac{L^2}{\pi^2}\frac{\eta_S}{K_{11}}. \tag{2.4.12}$$

In the optical examination of deformation of transverse (or longitudinal) and bending the variation of the angle of inclination θ of the director results in the change of the effective value of the birefringence of the layer and, consequently, the phase delay $\Delta\Phi$, expressed in the oscillating nature of the transmission of the cell between the crossed polaroids in relation to the applied voltage or on time, as the voltage is switched off. The variation of transmission from minimum or maximum (or vice versa) corresponds to the variation of $\Delta\Phi$ by π rad. If the application of the voltage $U > U_F$ results in the change of the value of the phase delay by δ_0 times, then, recording the moments of time corresponding to the extreme values of transmission after disconnecting the voltage, it is possible to record the time dependence of the variation of the phase delay $\delta = (\Delta\Phi_{max} - \Delta\Phi)$ in comparison with the phase delay of the non-deformed layer (Fig. 2.4.2*c*). The curve of relaxation of the phase delay is described by the equation:

$$\ln\left(\frac{\delta_0}{\delta}\right) = \frac{2\pi^2}{L^2}\frac{K_{11}}{\gamma_1}t, \qquad (2.4.13)$$

i.e., the phase delay δ relaxes twice as fast as the angle of inclination of the director. The data in Fig. 2.4.3 show that when the deformation of the nematic liquid crystal ($U \gtrsim 3U_r$) increases, the approximations described by the equations (2.4.11), (2.4.13) do not hold.

This method of measurement of the relaxation time makes it possible to estimate, using a special cell, the variation of the viscoelastic ratio η_B/K_{33} in relation to pressure p (Fig. 2.4.4) [116, 117]. With the increase of the pressure from 10^5 Pa to $5 \cdot 10^7$ Pa, the viscoelastic ratio increases 1.7 times (from 2.3 $m^2\ s^{-1}$ to 3.9 $m^2\ s^{-1}$) because of the decrease of the free volume and the increase of the order parameter. This is shown in Fig. 2.4.5 where the ordinate gives the relaxation time proportional to the described viscoelastic ratio. The value of the ratio η_B/K_{33} for MMBA at 35°C is in agreement with the literature data [117].

The equation, identical with (2.4.13), is also obtained for the relative variation of the capacitance of the plane-oriented layer of the nematic liquid

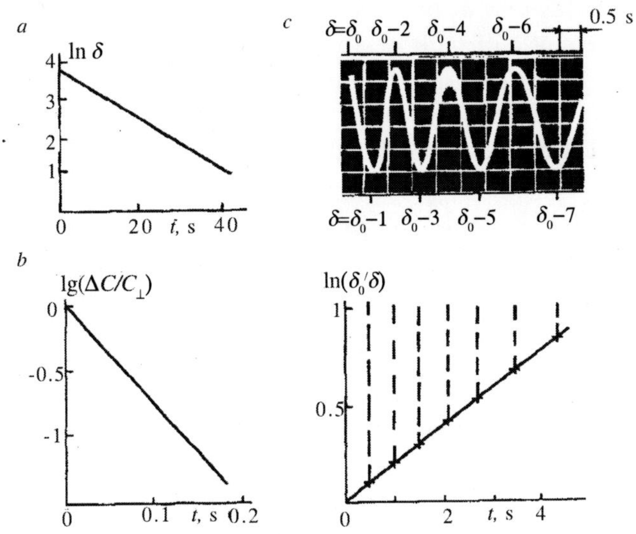

Fig.2.4.2. Relaxation of the angle of rotation of the conoscopic figure at *T*-deformation [111] (*a*) reduced capacitance $\Delta C/C_\perp$ [115] (*b*), the phase delay at *S*-deformation and transmission of light in crossed polaroids (*c*). The dotted lines indicate the positions of extrema of transmission of the cell.

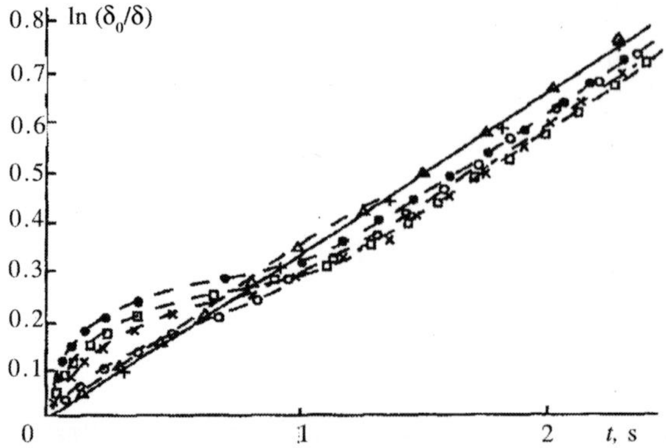

Fig. 2.4.3. Relaxation of phase delay δ, of planar-orientated layer of BF-5 (L=66 μm, T=23.5°C) at different deformation of the director: U = 0.98 V, δ_0 = 11.3π (+); U = 1.28 V, δ_0 = 20.4π (Δ); U = 2.93 V, δ_0 = 32.5π (o); U = 6.86 V, δ_0 = 35.7π (×); U = 16.2 V, δ_0 = 36.05π (□); U = 50 V, δ_0 = 36.4π (•).

crystal [115]:

$$\frac{\Delta C(t)}{C_\perp} \sim \left(\frac{\varepsilon_a}{2\varepsilon_\perp}\right)\theta_m \exp\left(\frac{-2t}{\tau_0}\right), \tag{2.4.14}$$

where ε_a is dielectric anisotropy; $\varepsilon_\perp$ and $C_\perp$ is the dielectric constant and the capacitance measured in the direction normal to the direction of the director; θ_m is the angle of inclination of the director in the centre of the layer; $\Delta C = C(t) - C_\perp$ is the variation of the capacitance of the layer after switching off the voltage.

The dependence of the time dependences of capacitance is shown in Fig. 2.4.2*b*.

Equation (2.4.14) also describes the variation of the capacitance of the twist-cell (the orientation of the director on the opposite substratesof the cell is orthogonal) at $U > U_F$, when the distortions of the orientation in respect of the azimuthal are small [118].

The value of the viscoelastic ratio $\eta_S \times K_{11}^{-1} \approx \gamma_1/K_{11}$ may be determined directly from the relaxation time of the electro-optical response of the twist effect T_{off}. In this case, we measure of the duration of variation of the intensity of the light passed through the twist-cell, plays between the parallel polaroids, from the maximum of the moment of this connection of voltage to reaching the level of 0.1. Figure 2.4.6 shows the values of $\eta_S/K_{11} \simeq \gamma_1/K_{11}$ for a number of Russian and foreign materials on the abscissa obtained in the measurement of the relaxation time of the phase delay in the planar-orientation layer, and the ordinate gives the time of this connection of the electro-optical response of the twist effect [119, 120]. All the points fit,

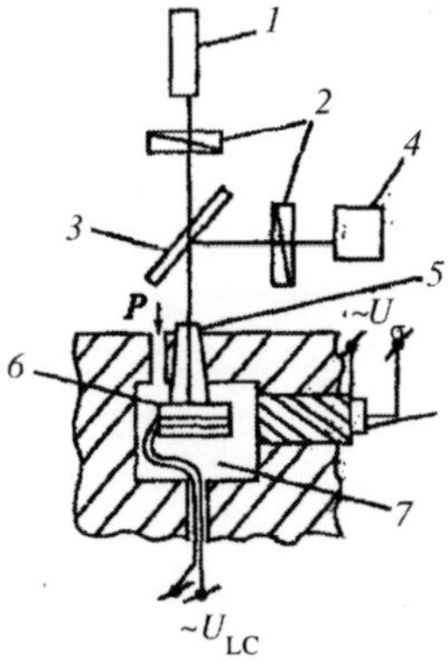

Fig.2.4.4. Chamber for measuring electro-optical characteristics of the liquid crystal under the effect of high pressure [117]: *1*) laser, *2*) polariser and analyser, *3*) light dividing sheet *4*) photodetector, *5*) conical light, *6*) LC, *7*) LC, *8*) piezotransducer.

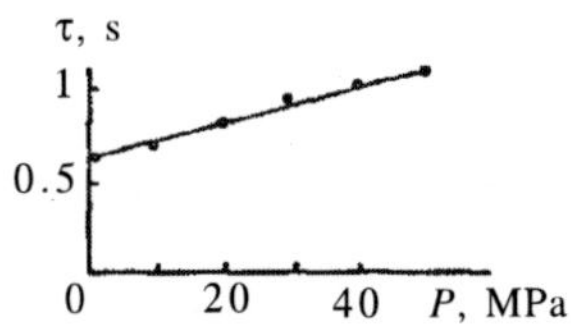

Fig.2.4.5. Dependence of the duration of orientational relaxation on MBBA on pressure in the absence of electrical voltage [117].

with sufficient accuracy, a straight line, defined by the equation:

$$T_s = \frac{0.05L^2\eta_S}{K_{11}}, \qquad (2.4.15)$$

which is in agreement with the equation describing the relationship of the relaxation time of the phase delay with the viscoelastic ratio since $(2\pi^2)^{-1} \approx 0.05$ (see (2.4.12)).

The relaxation time of the orientation of the director of the liquid crystal may be determined by measurement of not only the capacitance but also the resistance of the layer of the liquid crystal [121]. To eliminate the effect of the near-electrode layers, thick (1 mm) specimens of *para-azoxyanisol* were used. Planar orientation was produced on the surface of the electrodes (gold sheets). With reorientation of the liquid crystals, the resistance changed from R_t (the director is perpendicular to the field) to R_a (the director is oriented along the field). The resistance R at an arbitrary voltage is linked with the angle of deformation φ, counted from the normal to the surface of the substrates (along the axis z), by the relationship:

$$R = R_t \int_0^1 (1 + \sigma' \cos^2 \varphi)^{-1} \, d\xi, \qquad (2.4.16)$$

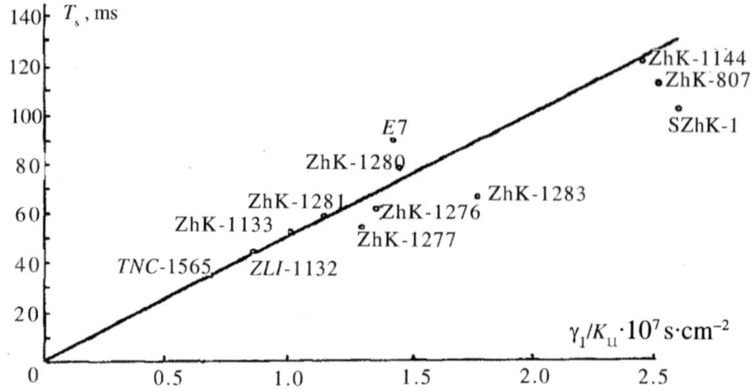

Fig.2.4.6. The switch off time of the optical response of the twist cell T_s as a function the viscoelasticity ratio γ_1/K_{11} for different liquid crystal materials.

where σ' is a constant determined by the values of the elasticity coefficients of the nematic liquid crystal, $\xi = 2z/L$ is the coordinate normalised for the thickness L (the boundary conditions $\varphi = 90°$ at $\xi = \pm 1$; $d\varphi/d\xi = 0$ at $\xi = 0$, i.e. in the middle of the cell). At the initial moment of time $t = \tau = 0$ ($t = (L^2\gamma_1/2K_{33})\tau$) the resistance R_0 is determined by the mean value of the angle of deformation φ_0:

$$\frac{R_t - R_0}{R_a - R_0} \times \frac{R_a}{R_0} = \cos^2 \varphi_0. \qquad (2.4.17)$$

Consequently, the relaxation of the resistance value after disconnecting the electrical field is expressed by the equation:

$$\beta_0 = \frac{R - R_0}{R_t - R_0} = \frac{1 + \sigma' \cos^2 \varphi_0}{\sigma' \cos^2 \varphi_0} \int_0^1 (1 + \sigma' \cos^2 \varphi)^{-1} d\xi \text{-} 1. \qquad (2.4.18)$$

The dependence of the relaxation of the angle of deformation φ (ξ, t) is determined by the numerical solution of the equation (2.4.10). Substituting $\varphi(\xi, t)$ into (2.4.18), specifying the value of φ_0 and using the tabulated values of K_{11}/K_{33}, it is possible to calculate the values of β_0 which are subsequently compared with the experimental values. Comparison of these relationships and also of the values of γ_1, for *para*-azoxyanisol (γ_1 (121.9°C) = 0.072 P) with the results obtained by other authors indicate the suitability of the method for measuring the coefficients determining the rate of reorientation of the director of the nematic liquid crystal.

It was shown in [122] that in the measurement of the current flowing through the cell it is possible to measure not only the rotational viscosity of the nematic liquid crystal but also the angle of inclination on the substrates.

The authors of [19] presented the solution of the equation of type (2.4.12) for the increase and decrease of the deformation of the layer of the liquid crystal at the value of the magnetic field H greater than or smaller than the threshold value H_c (it is assumed that $H/H_c - 1 \ll 1$). Assuming that the angle of deformation $\theta(z)$ (z is the coordinate on the axis, normal to the plane of the cell) is small, it may be assumed with sufficient accuracy that:

$$\theta(z,t)=\theta_m(t)\cos\left(\frac{\pi z}{L}\right), \tag{2.4.19}$$

where θ_m is the value of the deformation angle of the layer of the liquid crystal. Consequently, equation (2.4.10), describing the reorientation of the layer of the liquid crystal is reduced to the form:

$$\frac{\gamma_1^{\mathrm{ef}}}{\chi_a H^2}\frac{d\theta_m}{dt}=\left[1-\left(\frac{H_c}{H}\right)^2\right]\theta_m. \tag{2.4.20}$$

The solution of this equation is:

$$\theta_m^2(t)=\theta^2(\infty)\left\{1+\left(\frac{\theta^2(\infty)}{\varepsilon^2}-1\right)\exp\left[-\frac{2\chi_a H_c^2}{\gamma_1}\left(\frac{H^2}{H_c^2}-1\right)t\right]\right\}^{-1}, \tag{2.4.21}$$

Here $\theta(\infty)=\theta_m$ $(t\to\infty)$, $\varepsilon^2=\langle\theta_m^2(t=0)\rangle$ is the level of fluctuation of the orientation in the non-perturbed state, associated with the strength of the magnetic fields by the relationship:

$$\varepsilon^2=\frac{k_b T}{2\pi LK}\ln\left(1+\frac{H^2/H_c-1}{1-H^2/H_l^2}\right). \tag{2.4.22}$$

In equation (2.4.22) H_l is some critical field which takes into account the finite dimensions of the specimen. The fluctuations determine the increase of θ at the start of the reorientation process:

$$\theta_m^2(t)=\varepsilon^2\exp(2t/\tau), \tag{2.4.23}$$

where τ is the time constant, equal to:

$$\tau^{-1}(H)=\frac{\chi_a H_c^2}{\gamma_1}\left(\frac{H^2}{H_c^2}-1\right) \tag{2.4.24}$$

and having the positive or negative sign at different values of H/H_c (evidently at $H<H_c$ $\theta(\infty)=0$).

If monochromatic polarised light propagates along the axis z in the liquid crystal layer, then because of the variation of the phase difference between the non-conventional and conventional beams with the variation of the angle of the formation $\theta_m(t)$ the time dependence of the intensity of light behind the analyser is oscillating. The number of oscillations $N(t)$ is proportional to $\theta_m^2(t) = \theta_m^2(0)\exp(t/\tau)$ (Fig. 2.4.2*c*). Evidently, the time period between the two adjacent extrema with the numbers $N+1$ and N (for the planar orientation of the layer of the liquid crystal, the small number corresponds to smaller deformatiom, i.e. the high value of the phase delay) is equal to:

$$\tau N = \tau \ln \frac{N+1}{N} \text{and } H > H_c \text{ and } \tau_N = \tau \ln \frac{N}{N+1} \text{and } H < H_c. \quad (2.4.25)$$

Taking into account (2.4.24), equation (2.4.25) shows that the quantity γ_1^{ef}/χ_a or γ_1^{ef}/K_{ii} can be determined from the time period τ_N. However, the accuracy of measurements increases if we examine the total process of relaxation of the director.

For a number of problems it is important to know the effective value of viscosity for an arbitrary angle of deformation of the layer of the liquid crystal. This is especially the case when studying the regime of partial reorientation of the nematic liquid crystals in the operation of liquid crystal matrix devices or spatial-time light modulators. In these cases, the reorientation of the nematic liquid crystals does not start from the zero angle $\theta = 0$ and is not completed at $\theta = 90°$ and is determined by the angles, given by the applied initial and final displacement stresses U, and is described by the equation obtained assuming that $K_{11} = K_{33} = K$ and $\varepsilon_a \ll \varepsilon_\perp$ [21]:

$$\gamma_1^{ef} \frac{\partial\theta}{\partial t} = K\frac{\partial^2\theta}{\partial z^2} + \frac{\varepsilon_a E^2}{4\pi}\sin\theta\cos\theta - R(t)\frac{\varphi}{\psi}, \quad (2.4.26)$$

where $R(t) = \int_0^L \frac{\varphi}{\psi}\frac{\partial\theta}{\partial t}dz \left(\int_0^L \frac{dz}{\psi}\right)^{-1}$. The effective rotational viscosity γ_1^{ef} is associated with the functions describing the rotation of the vector (φ) and the flow of the nematic liquid crystal (ψ), i.e., the inverse flow, by the relationships

$$\eta^{ef} = \gamma_1^{ef} = \gamma_1 - \frac{\varphi^2}{\psi},\ \varphi = \alpha_3\cos^2\theta - \alpha_2\sin^2\theta,$$
$$\psi = \alpha_1\cos^2\theta\sin^2\theta + 0.5(\alpha_3+\alpha_6)\cos^2\theta + 0.5(\alpha_5-\alpha_2)\sin^2\theta + 0.5\alpha_4. \quad (2.4.27)$$

The equation for ψ coincides with (1.2.1) when the azimuthal angle is equal to zero. In [21] when studying the conditions of partial reorientation of the layers of the nematic liquid crystal the with the planar and homeotropic orientation corresponding to changes of the phase the light of the passing light by π, it is shown that equation (2.4.26) described quite efficiently of the experimental results. This means that the angular dependence of the effective viscosity can be described by the expressions (2.4.27). It may easily be shown that at $\theta = 0°$ $\gamma_1^{ef} = \eta_S$, and at $\theta \pm 90°$ $\gamma_1^{ef} = \eta_B$. Figure 2.4.7 shows the dependence $\eta^{ef}(\theta)$, constructed for MBBA on the basis of the data on the Leslie coefficients α_i determined in [16].

Previously, we discussed the case of the spatial homogeneously orientation of the director which depends only on the distance to the substrate (coordinate z). The authors of [123] investigated the relaxation of nematic liquid crystals in which stationary heterogeneous orientation with the characteristic size L_a was produced by some mechanism. The correlation length of the order parameters ξ_0, the characteristic length of the variation of the spatial-heterogeneous orientation $l_a = (D_a\tau_a)^{1/2}$, where D_a is the diffusion coefficient, and τ_a is the relaxation time of the orientation. For typical nematic crystals $D_e \sim 4 \cdot 10^{-2}$ cm$^2 \times$ s^{-1}, $\tau_a \sim 10^{-7}$ s and, consequently, $l_a \sim 0.6$ μm. The following equation was derived from the relaxation of fluctuations of the vector:

$$\tau_{ef} \cong \tau_a \frac{L_a^2}{\xi_0^2}\left[1 + l_a^2 L_a^{-1}\right]^{-1}. \qquad (2.4.28)$$

Taking into account the relationship of length ξ_0 with the elasticity coefficient K and time τ_a with rotational viscosity γ_1 [123, 124]:

$$K = \frac{3\rho k_b T}{m} S^2 \xi_0^2, \qquad (2.4.29)$$

$$\gamma_1 = \frac{3\rho k_b T}{\mathrm{m}} S^2 \tau_a, \qquad (2.4.30)$$

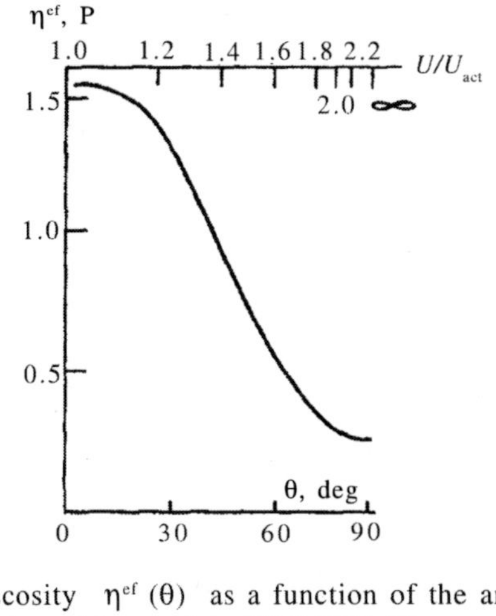

Fig.2.4.7. Effective viscosity $\eta^{ef}(\theta)$ as a function of the angle of orientation of the detector in the planar-oriented layer of the NLC [21]. The upper axis gives the values of bias voltage U/U_{act} at which the angle of reorientation in the centre of the layer of the NLCC is equal to θ.

equation (2.4.28) is transformed to the following form:

$$\tau_{ef} \cong \frac{\gamma_1 L_a^2}{K}\left[1+l_a^2 L_a^{-2}\right]^{-1}, \qquad (2.4.31)$$

which, without taking the multiplier π^{-2} into account, differs from the equation for the relaxation time of the homogeneous orientation (2.4.12) by the multiplier $\frac{L_a^2}{L^2}\left[1+\frac{l_a^2}{L_a^2}\right]$. In the condition $L = L_a = l_a = 10$ μm, the relaxation time of the heterogeneous orientation is half the relaxation time of the homogeneous orientation, and at $l_a \approx 0.6$ μm the effect of the heterogeneity is of the order of 0.5%.

The effect of the boundary conditions on the measured value of viscosity was also examined in [125]. For the finite value of binding energy W, characterised by the dimensionless parameter $g = WL/2K_{22}$, the equation of the re-orientation of the nematic liquid crystal (2.4.2) should be added up with the equation for the angle of inclination $\theta(z)$ which takes into account the orientation of the liquid crystal on the substrate (the angle of orientation and the surface energy for the coordinates $z = \pm L/2$):

$$K_{22}\left.\frac{\partial^2\theta}{\partial z^2}\right|_{\pm L/2} \pm W\sin\theta(\pm L/2) = \gamma_1\frac{\partial\theta(\pm L/2)}{\partial t}. \qquad (2.4.32)$$

It appears that with the decrease of the value of the parameter g^{-1} the time of both connection and disconnection increases. With the decrease of g from 10 000 to 5, the effective value of both times increases 1.4 times, and when the value of g is reduced to 2, the effective increase of the values is by a factor of 2 [125].

An interesting method which makes it possible to obtain the values of five viscosity coefficient $\alpha_1 \div \alpha_5$, has been proposed in [126, 127]. The interference of two powerful light beams in the volume of the liquid crystal results in the formation of a spatial-periodic structure because of the excitation of acoustic waves and heating. Since this is accompanied by the modification of the refractive index of the nematic liquid crystal, the diffraction efficiency of the resultant phase lattice may be changed using a low-intensity light beam. The relaxation time of diffraction efficiency is determined by the viscosity and elasticity coefficients of the nematic liquid crystal. Depending on the orientation of the director in relation to the polarisation of exciting radiation and the probing beam, there are five combinations of the viscosity coefficient. The restrictions, imposed on the duration of the physical processes, taking place in the liquid crystal, and also the equations of hydrodynamics of the nematic crystals for the investigated case, have been published in [127]. To ensure the accuracy of determination of the viscosity coefficient, the duration of the light pulse τ_{im}

should be short in comparison with the duration of propagation of heat, determined by the speed of propagation of sound $\tau_s \sim v_s^{-1}$ and the heat conductivity coefficient $\tau_T \sim K^{-1}$. In this case, it is necessary to ignore the effect of heat conductivity on the process of sound generation ($\tau_s \ll \tau_T$). The attenuation of sound as a result of viscosity should be comparable with the time to the establishment of constant temperature ($\tau_v \sim \tau_T$). All these processes should have no effect on the orientational relaxation of the nematic liquid crystal to the equilibrium value caused by the effect of Frank elasticity forces ($\tau_F = \gamma_1 q^2/K_{11} \gg \tau_T$). Thus, after determining the duration of the laser pulse $\tau_{im} \sim 50$ ns, it is possible to determine the time of relaxation of the orientation of the nematic liquid crystal. If $\tau_{im} \sim \tau_v \sim 1$ ms, acoustic waves are also excited in the volume of the nematic crystal and the attenuation of these waves is determined by another set of the viscosity coefficients. The direct effect of the electrical field of the light waves may be eliminated by selecting the direction of polarisation of these waves in relation to the director. If the polarisation of the excitation waves is parallel to the director ($e - e$), the unconventional probing wave is scattered into the unconventional wave ($e \rightarrow e$) and the direction of the director is parallel to the wave vector of the resultant thermal lattice **q** (angle $\angle(\mathbf{n}, \mathbf{q}) = 0$), and the values of η_1 and α_2^2/γ_1 are measured. The coefficients γ_1 and $\alpha_2/2 + \gamma_2 - \eta_1$ are measured at the same directions of polarisation of the excitation and probing waves, but at $\angle(\mathbf{n}, \mathbf{q}) = 45°$. At $\angle(\mathbf{n}, \mathbf{q}) = 45°$, normal to the direction of polarisation of the excitation waves and the director ($e - o$) and scattering of the conventional probing wave into the unconventional wave ($o \rightarrow e$) gives the values of $\eta_1 + \eta_3$. Thus, the method may be used to determine the values of all Leslie viscosity coefficients.

The liquid crystal can also be reoriented only by the lightwave field. If the polarised light beam propagates in the liquid crystal cell in the direction normal to the direction of the director, then at the value of the light intensity I exceeding the threshold value for the Fredericks optical transition I_F, the non-linear phase difference appears at the output of the cell. The dynamics of variation of the phase shift $\Phi(t)$ at $I > I_F$ was investigated in [128], and the authors of [129] describe the method of measurement of the viscoelastic ratio K_{33}/η_B using the Fredericks optical transition. The time dependence $\Phi(t)$ is described by the relationship

$$\Phi(t) = \frac{\Phi_E}{1 + (\Phi_E - \Phi_0 \exp(-2\Gamma_? t))/\Phi(t)}, \tag{2.4.33}$$

where $\Gamma_T = (K_{33}/\eta_B)(\pi/L)^2 (I/I_F - 1)$, and Φ_E and Φ_0 relate respectively to the values of the phase shift at the intensity slightly higher than the threshold value and to the so-called linear phase shift. The following equations were obtained in [129] for these parameters:

$$\Phi_E = \frac{\omega}{2c\sqrt{\varepsilon_\perp}} \frac{\varepsilon_a L}{1 - 9\varepsilon_a / 4\varepsilon_\parallel - (K_{33} - K_{11}) / K_{33}} \left(\frac{I}{I_F} - 1 \right) \qquad (2.4.34)$$

$$\Phi_0 = \frac{\omega L \varepsilon_a}{4c\sqrt{\varepsilon_\perp}} \theta_0^2, \qquad (2.4.35)$$

where ω and c is the frequency and the speed of light, $\varepsilon_\parallel$,$\varepsilon_\perp$ are the components of dielectric permeability at optical frequency, L is the thickness of the liquid crystal cell, K_{11}, K_{33} are the elasticity coefficients, θ_0 is the functional angle of deviation of the optical axis at which the phase transition of the second kind starts. Fixing the extrema of transmission at excitation and the relaxation of the homeotropically oriented cell and using the angle of inclination of the dependence ln $\Phi(t)$ it is possible to determine the viscoelastic ratio K_{33}/η_B [129].

Up to now, we discussed the reorientation of the homogeneous specimen of the liquid crystal in the field that is strictly perpendicular or parallel to the substrate plane. However, since the directions of the vector **n** and **–n** are equivalent, there are two equivalent directions of reorientation of the director with the angles of inclination θ and π–θ. The regions with the different angles of inclination are divided by the so-called inversion walls. Producing at least a small angle of inclination of the magnetic field in relation to the normal the plane of the substrate (angle φ), we obtain different values of the reorientation angle on both sides of the cell resulting in its migration (displacement). In [130], Lager discovered that the duration of migration of the wall is proportional to rotational viscosity γ_1 and inversely proportional to the magnetic field. The values of γ_1 for MBBA (0.86±0.2 P) and PAA (0.0 8±0.02 P), obtained by the author of [130] from the speed of displacement of the walls are in agreement with the available data. A more detailed examination of the movement of the inversion walls was carried out in [131]. The speed of migration of these walls v is described by the relationship:

$$\nu = \frac{3\sqrt{3}\pi\varphi K_{11}^* q}{8\gamma_1^* L(H/(H_c - 1))}, \qquad (2.4.36)$$

where γ_1^*, K_{11}^* are the effective coefficients of viscosity and elasticity, φ is the angle of inclination of the magnetic field (in the vicinity of the threshold of the Fredericks transition at $H/H_c = 1.1$ $\varphi < 3.9°$, $q = 3\varphi \sin\theta_M / 4\theta_M^4$, θ_M is the angle of inclination of the director in relation to the centre of the layer of the liquid crystal. For the specimen with MBBA ($\gamma_1^* = 0.85\gamma_1$, γ_1 = 0.76 P3, $K_{11} = 6 \cdot 10^{-7}$ dyne), with the length L = 100 μm at φ ~5°, the speed

v, calculated by Wang for different values of H/H_c using the equation (2.4.36) is in good agreement with the experimental results obtained by Wang (Fig. 2.4.8, [130, 131]). The movement of the inversion walls in the nematic liquid crystals is an efficient model of the propagation of the signal through a nerve, whose mielin has a liquid crystal structure. In both cases, the speed of movement is described by similar equations:

$$\frac{\partial \upsilon}{\partial t} = \frac{\partial^2 \upsilon}{\partial x^2} + f(\upsilon) \qquad (2.4.37)$$

(see, for example, (2.4.10)).

An analogue of the inversion walls are disclinations,separating the regions with the right and left twist in the twist cells. The speed of movement of the inclinations depends on the local radius of curvature and is described by the equation [132, 133] identical to (2.4.36):

$$\upsilon = c\frac{K_{11}}{\gamma_1}\frac{1}{R} = \frac{\Gamma}{R}, \qquad (2.4.38)$$

where Γ is the kinetic coefficient. The value of Γ can be determined from the dependence of the radii of curvature of the disclination loops on time (Fig. 2.4.9) [134]. When examining loops in a microscope, experiments were carried out to determine either directly the value of R or the correlation function describing the relaxation of the regions with the right and left twist in relation to the initial pattern, formed in the twist cell in transition between the isotropic liquid and the nematic liquid crystal. In both cases, the value of the measured parameter is proportional to $(\Gamma t)^{1/2}$ which is determined by the dependence of the length of the disclination on time. This dependence may be obtained on the basis of the u-field theory [135] which describes, in the order–disorder transition, a system consisting of randomly distributed interfacial surfaces of two stable phases.

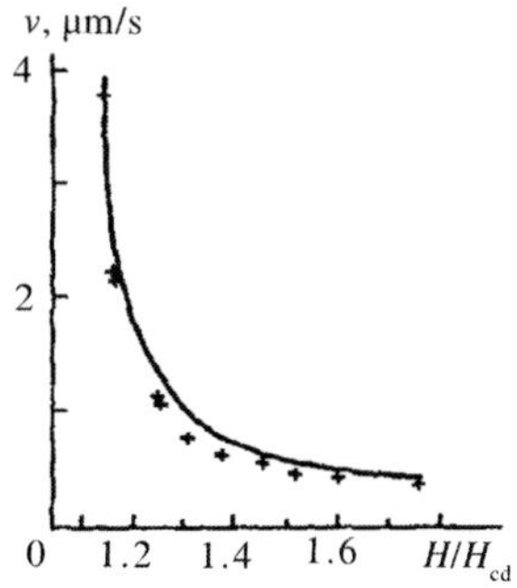

Fig.2.4.8. Dependence of the rate of migration of inversion walls on the structure of the magnetic field H/H_{cd} use the threshold strength. The solid line – calculated using equation (2.4.32).

The authors of [136] describe experiments whose geometry corresponds to the T-the effect [111]. A cell with a thickness of 25–250 μm with a planar orientation is placed between the poles of a magnet in such a direction that the direction of the field is parallel to the director. Subsequently, the cell is rapidly rotated through the angle of 90° during the time shorter than the duration of orientational relaxation. A transverse magnetic field can also be applied. In this case, periodic structures, the so-called twist-bend walls, form in the volume of the liquid crystal. It was shown in [137] that the period of the walls is linked with the strength of the magnetic field B when the strength is slightly higher than the threshold value of the Fredericks transition B_F and with the viscosity and elasticity coefficient of the nematic liquid crystal by the following relationship:

$$\left(\frac{B}{B_F}\right)^2 = \frac{1-\alpha}{\eta\alpha}kQ^4 + \frac{2k}{\alpha}Q^2 + \frac{1+k\eta}{\alpha}, \qquad (2.4.39)$$

Here Q is the ratio of the wave vectors of the cells and the walls, $Q = q_x/(\pi/L) = 2L/\lambda$, λ is the mean period of the wall, $\alpha = \alpha_2^2/\gamma_1\eta_1$, $\eta = \eta_1/\eta_3$, $k = K_{33}/K_{22}$. In [136] the values of the viscosity coefficient obtained by adjustment for an Mi-5 mixture (E. Merck GmbH) were equal to $\alpha = 0.86$ and $\eta = 0.23$.

Analysing the methods described in the section, it should be seen that all these methods require, in comparison with the method of the rotating magnetic field, a considerably smaller amount of the liquid crystal (this is very important in producing new substances), with a small decrease of accuracy, and simpler equipment can be used. It is preferred to use optical methods because considerable errors may form as a result of the

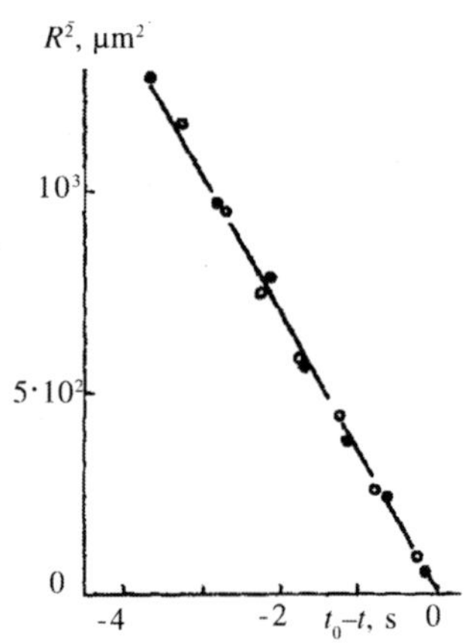

Fig.2.4.9. Time dependence of the radius of curvature R of the disclination loops in the twist-cell with the MBBA.

heterogeneities of the field and the thickness of the layer in the measurement of the capacitance of the layer of the liquid crystal. In addition to this, since the direct measurements of the anisotropy of the doubt magnetic susceptibility χ_a are difficult to carry out [3], it is better to use the electrical field, wherever possible.

2.5. The method of light scattering on thermal fluctuations of the director of liquid crystals

The local fluctuations $\delta\mathbf{n}$ of the director $\mathbf{n}$ of the liquid crystal result in extensive light scattering and, consequently, large specimens of the liquid crystal become turbid. The theory of statics and dynamics of the fluctuations of the director of the nematic liquid crystal was described in [138, 139]. A sufficiently detailed description is also presented in [2, 3, 87]. Therefore, here we confine ourselves to describing several practically important experiments [140–146].

In the analysis of the dynamics of the fluctuations $\delta\mathbf{n}$ of the director $\mathbf{n}$ it is convenient to expand $\delta\mathbf{n}$ into two mutually perpendicular components (modes). Using the results obtained in [139], we derive equations for the intensity and half width of the spectrum of light in every mode. The intensity of scattering is determined by the expression for the scattering section:

$$\sigma = \sum_{\alpha=1,2} \frac{(i_\alpha f_0 + i_0 f_\alpha)^2}{K_{33} q_\parallel^2 + K_{22} q_\perp^2}, \tag{2.5.1}$$

where i_α, f_α is the projection of the polarisation vector of the incident (**i**) and scattered (**f**) light on the unit vector e_α, selected as follows: $\mathbf{e} \perp \mathbf{n}$ and is situated in the plane of the director $\mathbf{n}$ and of the scattering wave vector $\mathbf{q} = \mathbf{k}_i - \mathbf{k}_s$ ($\mathbf{k}_i$, $\mathbf{k}_s$ the wave vectors of incident and scattered light); $\mathbf{e}_1 \perp \mathbf{e}_2$ and is perpendicular to the plane of the vectors $\mathbf{n}$, $\mathbf{q}$; $i_0 = (\mathbf{in})$, $f_0 = (\mathbf{fn})$; K_{ii} are the constants of Frank elasticity; $q_\parallel$ and $q_\perp$ are the components of the scattering wave vector in relation to the direction (Fig. 2.5.1). The expressions for the half width u and the spectrum of light scattering have the following form [146]:

$$S_{\Delta I}(\omega) = \beta^2 \left\{ \sum_{\alpha=1,2} \frac{G_\alpha^4(\theta)(2u\alpha)}{K_\alpha^2(q)[\omega^2 + 4u_\alpha^2} + \frac{G^2 G_\alpha^2(\theta)(u_1 + u_2)}{K_1 K_2(q)[\omega^2 + (u_1 + u_2)^2]} \right\}, \tag{2.5.2}$$

where β is the function of dielectric anisotropy at the optical frequency (the refractive index) of the scattering volume, the distance to the examination point, absolute temperature T. The transfer equations for different modes of the fluctuations of the director (δn_α ($\mathbf{q}$) $\ll$ 1) in the

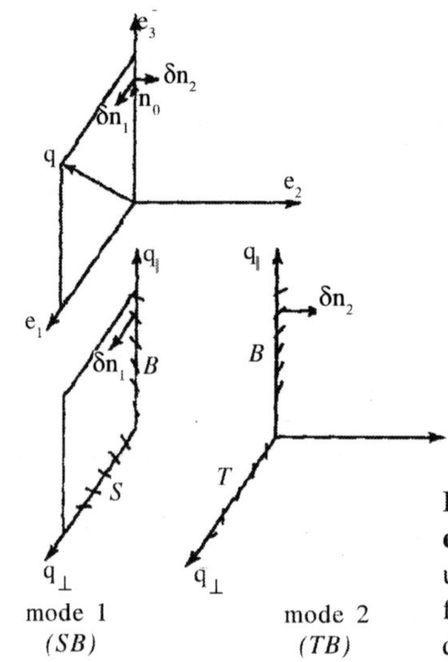

Fig.2.5.1. Construction of base unit vectors $\mathbf{e}_1$, $\mathbf{e}_2$, $\mathbf{e}_3$, $\mathbf{n}$ the basis of the wave vector and the director $\mathbf{n}$ used subsequently for expanding two non-correlated function modes $\delta\mathbf{n}_1$ and $\delta\mathbf{n}_2$, and disruptions of the director in these modes.

previously introduced system of coordinates have the form:

$$\frac{\partial}{\partial t}\delta n\alpha(q)+\frac{1}{\tau\alpha}\delta n\alpha(q)=0 \qquad (\alpha=1,2), \tag{2.5.3}$$

Here τ_α is the relaxation time:

$$\tau\alpha(q)=\left[\frac{\eta^{?\Phi}(q)}{K\alpha(q)}\right]. \tag{2.5.4}$$

The angular dependences of the effective viscosity are described by the relationships

$$\eta_1^{ef}(q)=\gamma_1-(\alpha_3 q_\perp^2-\alpha_2 q_\parallel^2)^2(\eta_2 q_\perp^4+ +(\eta_1+\eta_2+\eta_{12})q_\perp^2 q_\parallel^2+\eta_1 q_\parallel^4)^{-1}, \tag{2.5.5}$$

$$\eta_1^{ef}(q)=\gamma_1-\frac{\alpha_2 q_\parallel^2}{\eta_3 q_\perp^2+\eta_1 q_\parallel^2}, \tag{2.5.6}$$

and for elasticity:

$$K_\alpha(q)=K_{\alpha\alpha}q_\perp^2+K_{33}q_\parallel^2. \tag{2.5.7}$$

The equations (2.5.5) and (2.5.6) are greatly simplified if it is assumed that $q_\parallel \gg q_\perp$ or $q_\perp \gg q_\parallel$.

For both attenuating modes in the first case we have:

$$\frac{K_{33}q_{\|}^2}{u} = \eta_B = \gamma_1 - \frac{\alpha_2^2}{\eta_1}. \tag{2.5.8}$$

in the case in which $q_\perp \gg q_\|$ for the mode I:

$$\frac{K_{11}q_\perp^2}{u_1} = \eta_S = \gamma_1 - \frac{\alpha_3^2}{\eta_2}, \tag{2.5.9}$$

and for the mode 2

$$\frac{K_{22}q_\perp^2}{u_2} = \gamma_1. \tag{2.5.10}$$

here, u_i is the half width of the scattering spectra in rad/s, linked with the characteristic time of the function of correlation of the scattered light by the relationship $2u_i = \tau_i^{-1}$, η_S and η_B have the same meaning and form as in equations (2.4.8), (2.4.9). In most cases, the values of u_i equal 10–1000 rad/s, and the method of spectroscopy of optical adsorption should be used for the accurate determination of these values [147].

The intensity of excess noise determined by scattering of the thermal fluctuation of the director above the background noise for the heterodyne spectrum, measured in the displacement of the scattered light with the local oscillator (the initially split laser beam), is twice the value in the case of the homodyne spectrum (Fig. 2.5.2). Because of the Gaussian nature of the fluctuations, the half width of the heterodyne spectrum is half the width of the homodyne spectrum [142].

Equation (2.5.2) shows that the scattering spectrum consists of three lines of the Lorenz form with slightly different values of the half width. To increase the accuracy of the determination of the half width of the spectrum, the authors of [146] proposed to determine the frequency at which the value of $\omega S_{\Delta I}(\omega)$ is maximum. For the purely Lorenz form of the spectrum, the value ω_M coincides with the half width (Fig. 2.5.3, [146]). Figure 2.5.4 [146) shows the dependence $\omega_M(q_\perp^2)$. The relative error of determination of the viscoelastic ratio by this method is equal to 2%, and the adjustment procedure is only slightly influenced by the ratio K_{11}/K_{22}.

The electric field applied in the cell does not change the form of the scattering spectrum for the mode 2 (twist-bend), and the width of the line increases in proposed on to the square of the applied voltage (figure 2.5.5). In the case of mode 1, the value of $S_{\Delta I}(\omega)$ is situated slightly below the level given by the Lorenz form. Possibly, this is associated with the effect of electrical conductivity [148].

The diagram of experimental equipment for the measurement of the spectrum of light scattering shown in Fig. 2.5.6. The polarised laser light,

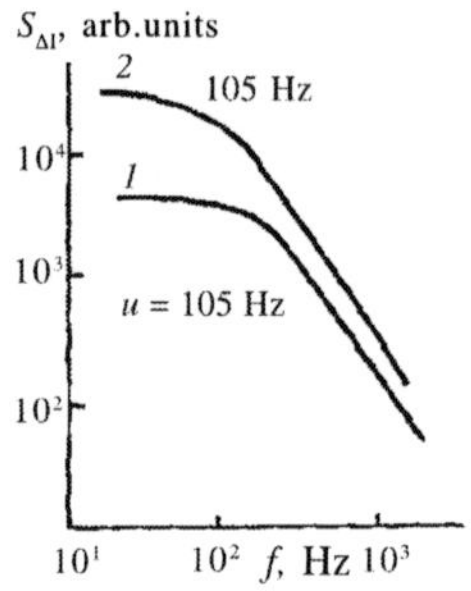

Fig.2.5.2. Determination of the half-width of the spectrum of scattering at fluctuations of the director at homodyne (*1*) and heterodyne (*2*) spectroscopy. Dibutyl azoxy benxol (DIBAB). The scattering angle $\theta = 9.5°$, $T = 22°C$.

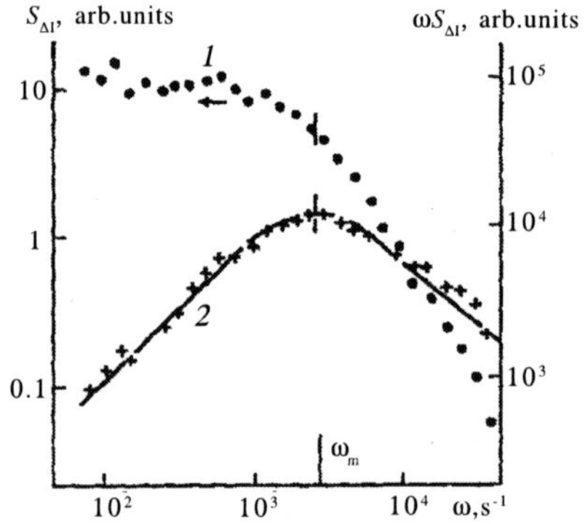

Fig.2.5.3. Spectrum of intensity of scattered light $S_{\Delta\tau}(\omega)$ (*1*) and $\omega S_{\Delta\tau}(\omega)$ (*2*) (see 2.5.2.). Anizilidene-*n*-aminophenyl acetate (AP-PA or MB1OA); $\tau = T/T_{NI} = 0.9988$; $\theta = 9.5°$.

scattered by the specimens of the liquid crystal under the angle θ, falls on the photodetector and from these it transferred supplied to the input of an analogue-pulsed correlator which determines the correlation function of fluctuation of the photos. In scattering on a single fluctuation mode, the correlation function has the form $\rho(t) \sim \exp(-t/\tau)$. Figure 2.5.7 shows the experimental geometry of examination of scattering on the attenuating *S*- and *B*-modes. The laboratory scattering angles θ_α were calculated using the values of the refractive index available for the investigated nematicliquid crystal in such a manner that one of the components of the vector **q** was converted zero. For example, for the geometry shown in Fig. 2.5.7*a*, $k_i = 2\pi n_o/\lambda$, $k_s = 2\pi n_e/\lambda$ (n_o and n_c by the values of the refractive index for the conventional and unconventional beam), $q_{\|} = 0$ and the laboratory angle $\theta_1 = 56°$ for the MBBA [143]. Comparison of Fig. 2.5.7*b* and 2.5.7*c* shows that the transition from the geometry in which the scattering takes place of

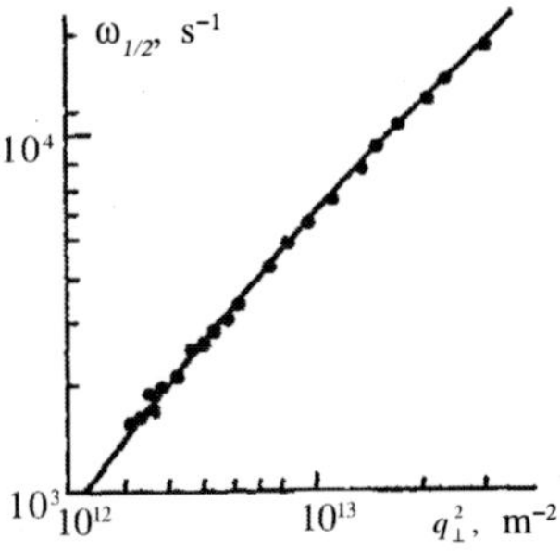

Fig.2.5.4. Dependence of the half-width of the scattering spectrum $\omega_{1/2}$ on the square of the wave length $q_{\perp}^{2}$; MB1OA; τ = 0.9988. Solid line are results processed using the condition $(dS/d\omega)_{\omega 1/2} = 0$.

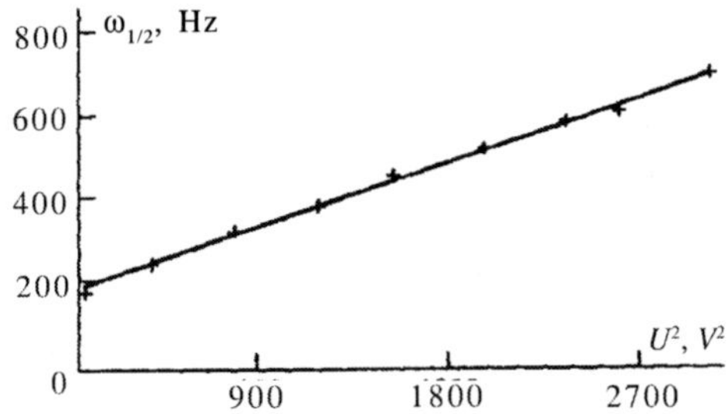

Fig.2.5.5. Dependence of the half width of the scattering spectrum $\omega_{1/2}$ for mode 2 (twist-bend) on the square of applied voltage U^2; BF-5; T = 20°C; θ = 10° [148].

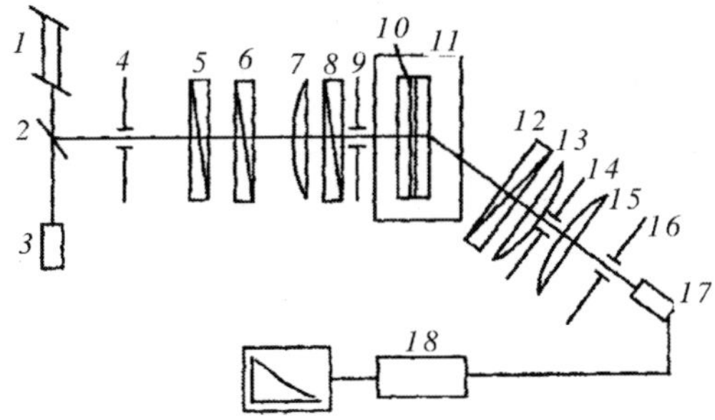

Fig.2.5.6. Experimental equipment for measuring viscoelastic characteristics of the NLC by the method of photon correlation spectroscopy: 1) laser, 2) light dividing element, 3, 17) photodetectors, 4,9,14,16) diaphragms, 5,8,12) polarizers, 6) quarter wave sheet, 7) lens for improving spatial coherence of the light beam, 10) LC cell, 11) thermal camera, 13,15) lenses of the collimator, 18) correlator with the oscillographic screen.

the *B*-mode, to the geometry in which scattering takes place of the *S*-mode takes place by rotation of the specimen of the nematic liquid crystal through the angle of 90°.

We examine in great detail the following geometry of the experiment

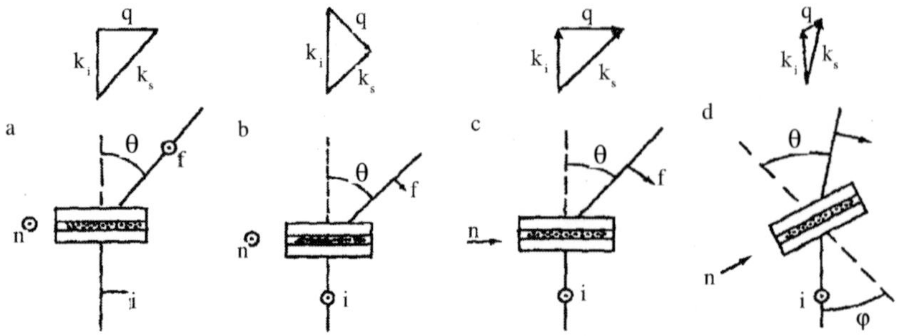

Fig.2.5.7. Geometry for examining light scattering in the attenuating *S*-mode (*a*,*b*) and *b*-mode (*c*,*d*) [143].

shown in Fig. 2.5.7*b*: vector **i** is parallel to the director **n** and normal to the scattering plane determined by the wave vectors $\mathbf{k}_i$ and $\mathbf{k}_s$. In this case, $k_i = 2\pi n_c/\lambda$, $k_s = 2\pi n_o/\lambda$, $q_{\|} = k_s \sin\theta$, $q_{\perp} = k_i - k_s \cos\theta$ and for the small scattering angles $\theta \to 0$, when $q_{\|} \to 0$. Thus, we have the scattering of the mode 2 whose dynamics in this case is determined by equation (2.5.10). However, the accuracy of determination of γ_1 for this observation geometry is not high because of the strong parasitic scattering (especially in the 'forward' direction) which forms on defects of ordering of the specimen. However, using the method of the optical displacement of the frequencies and carrying out measurements at different scattering angles θ, it is possible to measure quite accurately the viscoelastic ratios γ_1/K_{22} and η_S/K_{11} [145]. For the arbitrary angle θ, the spectrum of scattered light is a superposition of two Lorenz lines, and the correlation function of the scattered light is the superposition of two exponents. The normalised correlation function may be presented in the following form:

$$\rho(\tau) = \frac{[a_1 \exp(-t/\tau_1) + a_2 \exp(-t/\tau_2)]}{(a_1 + a_2)}, \tag{2.5.11}$$

$$a_\alpha = \frac{G_\alpha^2}{K_\alpha(q)} \quad (\alpha = 1, 2),$$

Here $G_\alpha = i_0 f_\alpha + i_\alpha f_0$ is the geometrical factor of scattering of the first and second mode; $K_\alpha(q) = K_{\alpha\alpha} q_\perp^2 + K_{33} q_\|^2$. It is efficient to determine the angle of inclination of the asymptote of the function ln $\rho(t)$:

$$\tau_{ef}^{-1} = \frac{d}{dt}(-\ln\rho(t))\bigg|_{t=0} \frac{(a_1/\tau_1 + a_2/\tau_2)}{(a_1 + a_2)}. \tag{2.5.12}$$

For the examined geometry of observation, the equation (2.5.12) is reduced to the form:

$$\tau_{ef}^{-1} = q_{\perp}^2 \frac{(G_1^2 / \eta_S + G_2^2 / \gamma_1)}{(G_1^2 / K_{11} + G_2^2 / K_{22})}. \quad (2.5.13)$$

Taking into account $\eta_S \cong \gamma_1$ and $G_1^2 + G_2^2 = 1$, the equation (2.5.13) is converted to the following form:

$$\tau_{ef} q_{\perp}^2 = \frac{\gamma_1}{K_{22}} \left[\left(\frac{K_{22}}{K_{11}} - 1 \right) G_1^2 + 1 \right]. \quad (2.5.14)$$

Figure 2.5.8 [145] shows the dependence of the relaxation time τ_{ef} of the value of G_1, which is a function of the scattering angle for two nematic liquid crystals. It may be seen that expression (2.5.14) describes efficiently the experimental results and that the ratio $\eta_S \cong \gamma_1$ is also efficiently fulfilled. At $G_1 = 0$, we obtain the values γ_1/K_{22}, at $G_1 = 1$ the values of γ_1/K_{11} for these substances.

Figure 2.5.9 shows the ranges of variation of the scattering angle of a specific mode [142]. Substituting the values of the elasticity constants, determined on the basis of the Fredericks transition threshold, into the resultant values of the viscoelastic constants it is possible to determine the values of rotational viscosity.

For the experimental geometry corresponding to Fig. 2.5.7*c* (the polarisation of incident light and the director are parallel to the scattering plane), we can determine η_B/K_{33} and γ_1/K_{33}, and also the ratio of the Mesovich coefficients η_1/η_3 [146]. The relationship of the measured half width of the spectrum ($\Delta\omega_{1/2}$) with the measurement angles is given, taking (2.5.4) and (2.5.8), by the relationship

$$B(q) = \frac{2\left(q_{\parallel}^2 + K_{22} q_{\perp}^2 / K_{33}\right)}{\Delta\omega_{1/2}} = \frac{1}{K_{33}} \left\{ \frac{1}{2}(\gamma_1 + \eta_B) + \frac{1}{2}(\gamma_1 - \eta_B) Z(q) \right\}, \quad (2.5.15)$$

where

$$Z(q) \equiv \frac{\eta_3 q_{\perp}^2 / \eta_1 - q_{\parallel}^2}{\eta_3 q_{\perp}^2 / \eta_1 + q_{\parallel}^2}. \quad (2.5.16)$$

At $Z = 1$ ($q_{\parallel} = 0$) $B = \gamma_1/K_{33}$, at $Z = -1$ ($q_{\perp} = 0$) $B = \eta_B/K_{33}$ (Fig. 2.5.10 [146]). The ratio η_1/η_3 can be determined either in adjustment of the dependence B (Z, u_S, K_{22}/K_{33}) or from the angle of inclination of the dependence $C(q) \equiv [B(q) - B_{Z=-1}] \times [B(q) - B_{Z=1}]^{-1} = -\eta_3 q_{\perp}^2 / \eta_1 q_{\parallel}^2$. It should be

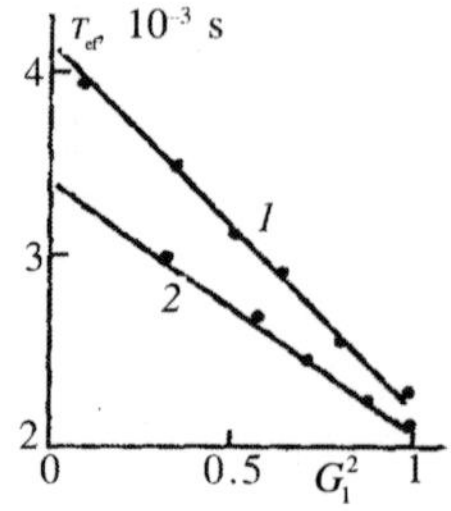

Fig.2.5.8. Dependence of effective relaxation time on $G_1^2 = (i_0 f_1)^2 = \cos^2(\theta)$: 1) HK-440, 2) MBBA; T = 21°C.

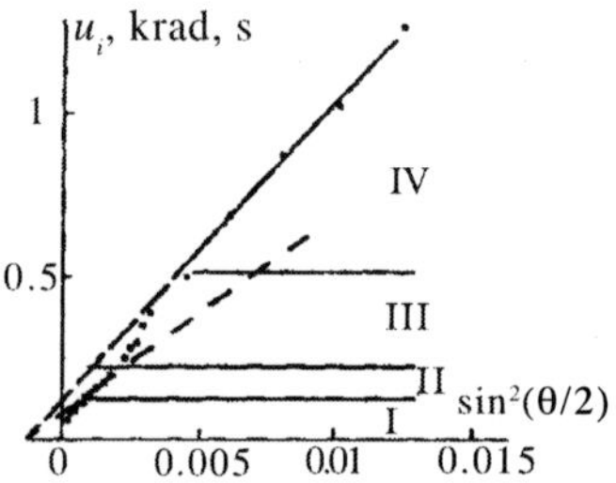

Fig.2.5.9. Determination of the range of scattering angles θ in which scattering takes place on mode *1* (I) on mixtures of modes *1* and *2* (II) and on mode *2*(III), with the displacement of the homodyne and heterodyne signals (IV); AB-4; T = 30.0°C; $q_{\|} = 0$.

mentioned that this ratio is equal to $\eta_1/\eta_3 \equiv \eta_S/\eta_B$ (see 1.2.13).

In the presence of the spatial heterogeneity of the director with the characteristic length l_a, the width of the line of the scattering spectrum increases by the multiplier $(1+l_a^2q^2)$ [123]. For the deformation induced by the external field, equation (2.4.31) is identical.

In this paragraph, we have discussed only the normal incidence of the light on the cell. The authors of [149, 150] calculated and measured the characteristic relaxation times for different scattering and incidence angles so that it is possible to widen greatly the range of the angles for which $q_\perp / q_\| \ll 1$ or $q_\|/q_\perp \ll 1$ and also, by the adjustment of the experimental results for the well-known approximations, increase the number of the viscosity coefficients measured by this method. The determination of the incidence and scattering angles in the liquid crystal medium (φ,θ) and in air (φ',θ') for the homeotropic and planar orientations is explained by Fig. 2.5.11 [149]. In both cases, the director is found in the scattering plane. Figure 2.5.12 shows the lines corresponding to the same value of the width of the scattering spectrum τ_α^{-1}, calculated for the homeotropic orientation of the nematic liquid crystal in the angle range $-90° \le \varphi \le 90°$ and $0 \le \theta \le 180\%$. The values of the refractive indexes, the ratios of the elasticity coefficients

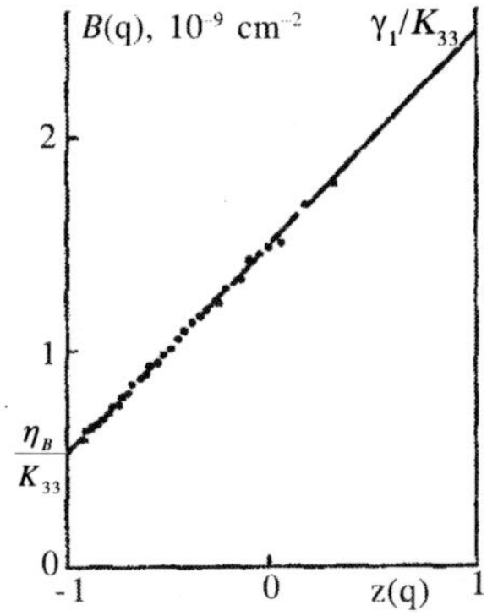

Fig.2.5.10. Dependence of **B** (q) associated with the half-width of the scattering line $\Delta\omega_{1/2}$ by the relationship (2.5.15), on $z(q)$ associated with the wave vector by (2.5.16); MB1OA; $\tau = T/T_{NI} = 0.9949$.

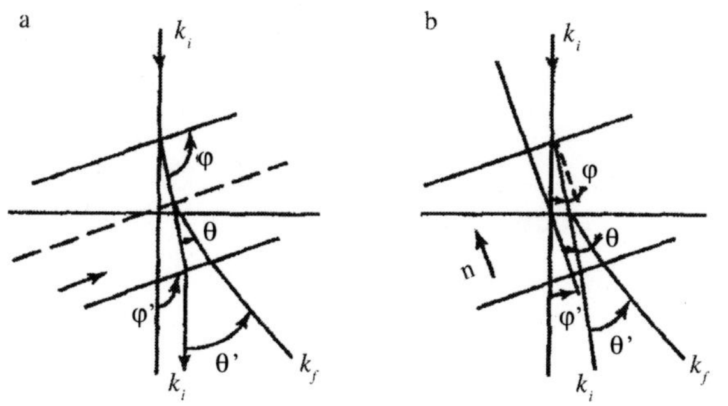

Fig.2.5.11. Determination of the incidence angles of the light beam in a liquid crystal and in air in relation to the direction of the director (φ and φ', respectively), the angles of scattering in the liquid crystal and in air (θ and θ') for planar (a) and homeotropic (b) orientations of the director on the substrate.

and viscoelastic ratios, taken from the calculations, are presented in [150]. The lines on the graph correspond to the fluctuations of the second mode, the extraordinary polarisation of the incident light and ordinary polarisation of scattering light $\left(\tau_2^{(e,0)}\right)$. The other relationships not equal to zero, i.e. $\tau_1^{(e,e)}$ and $\tau_2^{(o,e)}$ in relation to φ,θ, have the same form. These relationships have the centre of symmetry at the point with the coordinates $\varphi = 0$, $\theta = 0$ and the periodicity in respect of the angle equal to 180°. For the angles $|\varphi| \leq 30°$ and $\theta \sim 20$–$40°$, corresponding to the normal experimental conditions, the value τ_α^{-1} is equal to 10 kHz. For the constant scattering angle θ the dependence τ(φ) is non-monotonic. This is confirmed by the results, presented in Fig. 2.5.13 in [150], of the measurements of the inverse relaxation time and of the dependence on the incidence angle for different scattering angles in air for the homeotropic and planar orientations. The

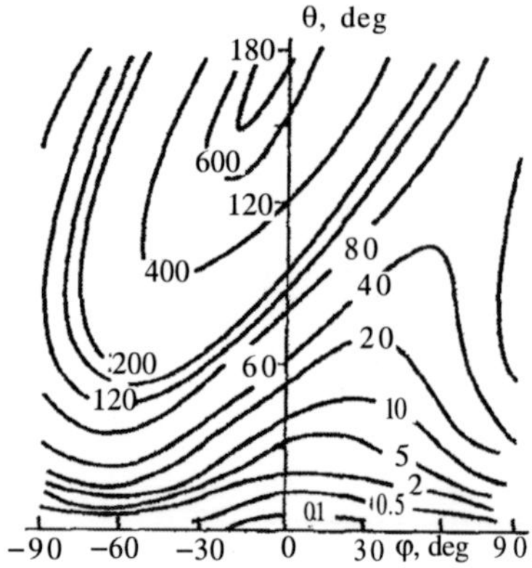

Fig.2.5.12. Isophots of scattering $I^{e,o}$ at different levels of incidence of the light φ and scattering θ in the liquid crystal medium. The director us in the scattering plane. The values of the physical parameters used for calculations were published in [150].

width of the lines for the geometries *e–o* and *o–e* is smaller than for the polarised scattering (*e–e*), i.e. it is smaller than the appropriate relaxation time and the recorded intensity of light scattering. This requires increasing the intensity of incident light for increasing the accuracy of measurement of the lines in the case of depolarisation of scattering.

The selection of the experiment geometry for the measurement of the viscosity coefficient is justified by Fig. 2.5.14 [140]. The dotted line indicates the lines corresponding to the zero intensity of Rayleigh scattering, i.e., not recommended for solving the given problem. They are described by the following equations:

$$a)\ \varphi'=\frac{\theta'}{2}(\text{hom. or.});\ b)\ \varphi'=\frac{\theta'}{2}+90^\circ(\text{plan. or.});\qquad b)\ \varphi'=0(\text{hom. or.});\ \ d)\ \varphi'=\theta'(\text{hom. or.}). \tag{2.5.17}$$

To measure η_S and γ_1 it is recommended to use the following selection: $\varphi' = -8^\circ$, $5^\circ \le \theta' \le 10^\circ$, the homeotropic orientation (in the first case, the polarised scattering of light (*e–e*), in the second case, depolarised scattering (*o–e*), lines *1* and *2* in Fig. 2.5.14). The value η_B can be measured at $\varphi' = 73^\circ$, $\theta' = 34^\circ$ in the planar-oriented cell, when the polarisation of incident light is normal to the director (*o–e*, point *3* in Fig. 2.5.14). Figure 2.5.15 [150] gives the dependence of τ^{-1} on the scattering angle for the fixed incidence angle and the values of K_{11}/η_S and K_{22}/γ_1 calculated from these values. In accordance with the data in Fig. 2.5.15, the width of the line of the scattering spectrum greatly increases with a small variation of the scattering angle. By selecting the different geometries of light scattering

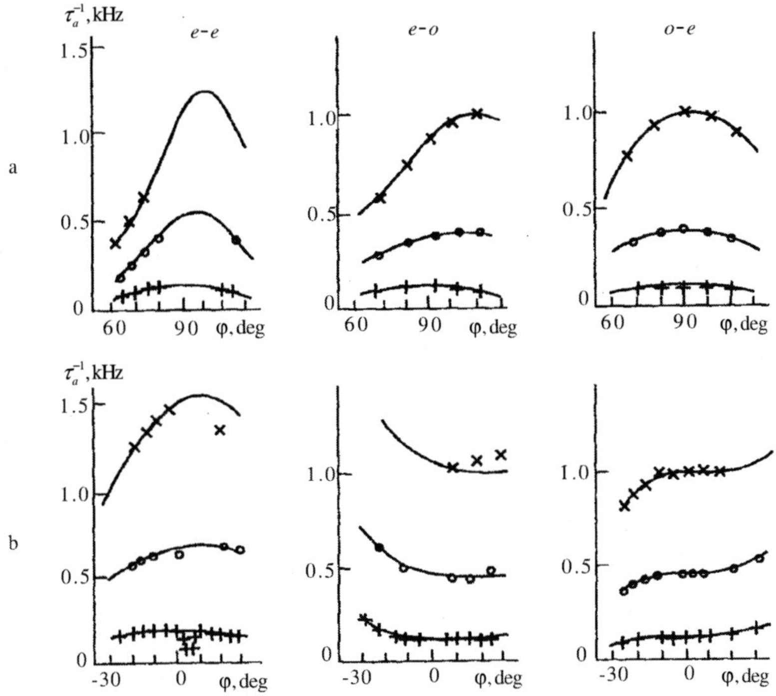

Fig.2.5.13. Dependencies of the width of the scattering spectrum τ_α^{-1} on the angle of incidence φ' for modes at changing values of the scattering angle θ': 8°(+), 16°(o), 24° (×); *a*) planar, *b*) homeotropic orientational director.

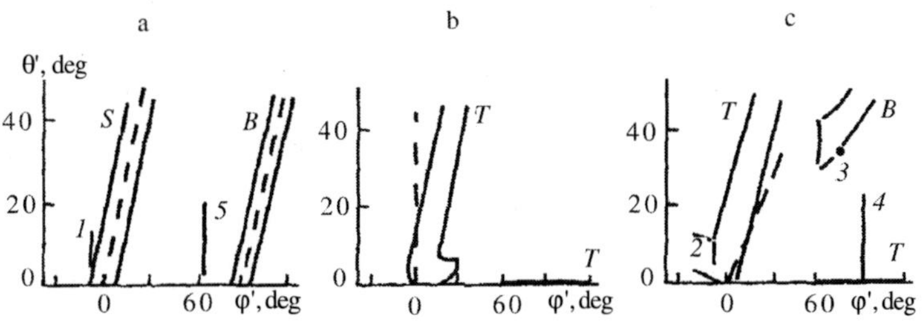

Fig.2.5.14. Ranges of the incidence φ' and scattering θ' angles for different modes in which the fluctuations of the director are determined by *S*-, *B*-, *T*-strains. The broken line – $I^\alpha = 0$; *1–5*) is measuring the combinations of the viscosity coefficients of Leslie geometry. Polarisation of light *e–e* (*a*), *e–o* (*b*), *o–e*(*c*).

and mathematical processing of the results of the measurements, it is possible to obtain the viscoelastic ratios for almost all Leslie viscosity coefficient. In the measurement of the relaxation time along line *4* in Fig. 2.5.14 (φ' = 90°, 0 ≤ θ' ≤ 28°), we estimated $\alpha_4 / 2\alpha_2^2$, and in the case of line *5* (φ' = 65°, 2° ≤ θ' ≤ 19.7°), i.e. coefficient α_1.

It was shown in [151] that when using the homeotropic orientation of

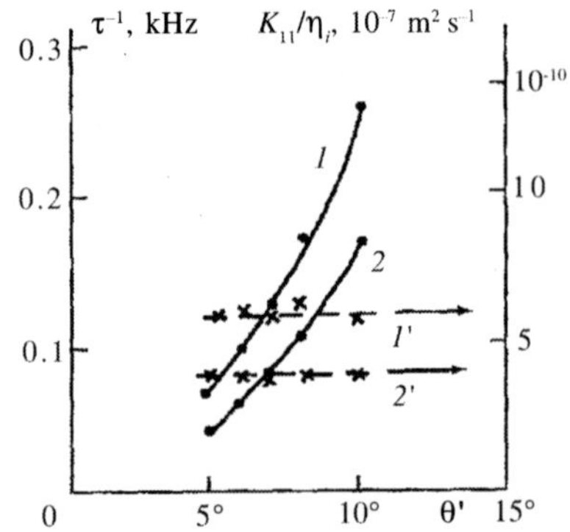

Fig.2.5.15. Dependence of the width of the scattering spectrum τ^{-1} on *S*- and *T*-modes (*1* and *2*, respectively) and of the ratio $K_{11}/\eta s$ and K_{22}/γ_1 calculated from the dependence (*1'* and *2'*, respectively) on scattering angle θ'. MBBA; φ' = –8°, homeotropic geometry.

the nematic liquid crystal and inclined incidence, the spectrum of the Lorenz type may also be obtained for very small values of the scattering angles (up to 0.015 rad ~ 1°) and is clearly visible in the background of static scattering by the homodyne method. The latter does not make it possible to determine the viscoelastic ratio, and is suitable for determining the ratio γ_1/χ_a or η_S/χ_a, η_B/χ_a, because in this case, due to the small value of the wave vector in the denominator of equation (2.5.4) there is a term describing the magnetic field $(K(q) + \chi_a H^2)$ which is controlling because of the small wave vector q.

When using the Raman scattering method in the specimens of nematic liquid crystals to which the electrical field is applied, it is possible to determine the relaxation time separately for the rigid core of the molecule and more flexible alkyl substitutes. This method is described in [152].

Although the measurements of the viscoelastic ratios by the methods of light scattering to require complicated experimental equipment and special spaces, they provide very valuable information because the experiments are carried out with the non-distorted specimen of the nematic liquid crystal and do not contain many errors typical of other methods. These measurements make it possible to examine efficiently the dependence of of rotational viscosity γ_1 on temperature and the degree of ordering [125].

Figure 2.5.16 shows the results of measurement of the temperature dependence of γ_1, obtained for BF-5 using three different methods [153]. The agreement of these results indicates that the given methods are adequate for the given purpose.

In conclusion, it is important to list the coefficients of viscosity of the nematic crystals indicating the possible methods of determination (Table 2.1) [154–156].

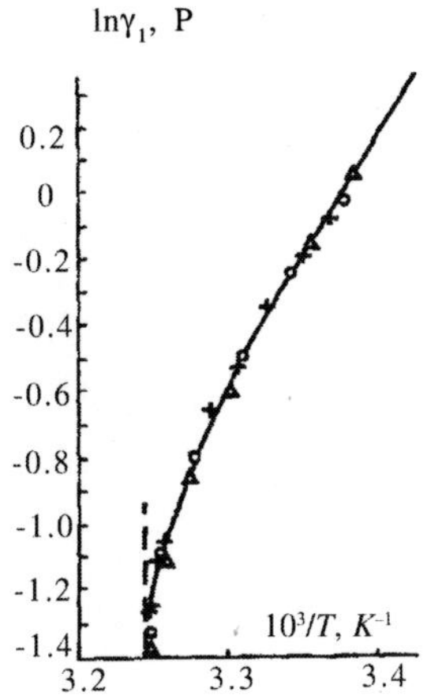

Fig.2.5.16. Temperature dependence of rotational viscosity γ_1, measured by the method of the spinning magnetic field (o), light scattering (Δ) and relaxation of the optical response of the Fredricks transition (+) [153].

Analysis of the methods described in this chapter indicates that for the formulation of the experiment it is very important to determine accurately the boundary conditions, the flow speed or the speed of rotation of the nematic liquid crystal, select the required strength of the external field and select suitable signal sensors. The largest amount of information on all coefficients is provided by the shear flow method which however requires a large number of the investigated matter. The total set of the viscosity coefficients can be obtained using the light scattering method on fluctuations of the director. The experiments require the minimum amount of material and at the same time special equipment. The simple procedure and the small amount of the liquid crystal are typical features of the methods of examination of relaxation of the Fredericks transition.

Optical measurements are more suitable for recording the dynamic response of the liquid crystal. Extensive possibilities are typical of the acoustic methods which are more suitable in operation with high pressures.

Organisations and groups using liquid crystal materials for practical purposes should use simple viscometeres of the Ostwald or Haak type for evaluating the kinematic or dynamic viscosity and the relaxation of the Fredericks transition in the planar-oriented cells for the evaluation of rotational viscosity.

Table 2.1. Viscosity of NLC and methods of measuring it

Notation	Relation to other viscosity coefficients	Parameter	measurement method	
α_i, i=1-6		Lesley viscosity coefficient		
η	$\approx\eta_2-\alpha_3\approx\eta_2$ $(-\alpha_2\gamma_2/\gamma_1+\alpha_4+\alpha_5)/2$	Dynamic viscosity	Kinematic viscosity	
ν	η/ρ	Mesovich viscosity coefficients	Rotational viscosity coefficient (Tsvetkov)	
η_1	$(-\alpha_2+\alpha_4+\alpha_5)/2$	Coefficient of twisting in speed gradient	viscosity coefficient of deformation of bi-transverse bending	$v\perp n$, $n\parallel\nabla v$
η_2	$(\alpha_3+\alpha_4+\alpha_6)/2$			$v\parallel n$, $n\perp\nabla v$
η_3	$\alpha_4/2$			$v\perp n$, $n\perp\nabla v$
η_{12}	α_1			$\angle(v,n)=45°$
γ_1	$\alpha_3-\alpha_2$	viscosity coefficient of deformation by longitudinal bending	Capillary or rotational viscometer, Ultra sound sheer	
γ_2	$\alpha_3+\alpha_2$	Capillary viscometer	Poisell Kuetta flow	
η_S	$\gamma_1-\alpha?/\eta_1$	Rotating magnetic field, Frederick's transition dynamics, scattering of light on thermal fluctuations of the director,	Kuetta flow	
η_B	$\gamma_1-\alpha?/\eta_2$	Frederick's transition dynamics scattering of light on thermal fluctuations of the director	Frederick's transition dynamicsscattering of light on thermal fluctuations of the director	

Chapter 3

Theory of the temperature dependence of the viscosity of nematic liquid crystals

Initial investigations of the viscosity of nematic liquid crystals show that it is of the activation nature. The initial theoretical studies described the dependence of the viscosity coefficient on the order parameter. Activation energy was represented by the height of the Maier–Saupe potential and this was reflected in the initial studies by Diogo and Martins.

In the development of liquid crystal substances and mixtures with the melting point below 0°C it was found that the activation mechanism does not describe the temperature dependence of viscosity in the entire nematic range. To explain this dependence, Schad and Zeller proposed an analogy between liquid crystals and polymers so that it was possible to use the free volume model and calculate the viscosity of any substance on the basis of increments.

Recently, several approaches were developed to constructing a molecular–statistical theory of the viscosity of nematic liquid crystals. Diogo and Martins described the temperature dependence of rotational viscosity on the basis of the theory of high-rate processes proposed by Eyring. A large part of the investigations was concerned with the derivation of a specific kinetic equation used for obtaining the specific form of the viscous stress tensor. Because of its simple form, the Fokker–Planck equation is used most frequently for the rotational function of the distribution, or the Landau–Khalatnikov equation is used for the order parameter.

In a number of studies by Nemtsov, investigations were carried out using the methods of advanced statistical thermodynamics – the Zwanzig–Mori projection operator method and the method of the Zubarev non-equilibrium statistical operator. In this case, the viscosity coefficient and other kinetic coefficients are expressed by time correlation functions whose accurate form is calculated using some kinetic equation or other methods. Although this approach is not connected with any type of kinetic equation, the Fokker–Planck equation is used in most cases for calculating the viscosity coefficient.

For the calculation simulation of the system, subjected to an arbitrary effect, Hess and Walls used the methods of non-equilibrium molecular dynamics which are simple and easy to understand.

These approaches to describing the temperature dependence of the viscosity efficient will be discussed.

3.1. Activation mechanism

The rotational frequency is a function of not only absolute temperature T but also of the degree of ordering S. The dependence of γ_1 on S was examined for the first time by Imura and Okano in [157]. They assumed that the strong temperature dependence, determined by the activation member exp (E/k_BT) (E is activation energy), is typical only of the viscosity of the isotropic phase and of the coefficient corresponding to this viscosity in the nematic phase, $\alpha_4 = 2\eta_3$. The temperature dependence of other coefficients of viscosity of the nematic liquid crystals is determined by the previously mentioned activation term and corrections for this term, proportional to different powers S. Expressing according to de Gennes [155] the order parameter in the form of a symmetric tensor who trace is equal to zero, and expanding the parameter written in this form in respect of the exponents (to the terms proportional to S^2) the symmetric and antisymmetric parts of the tensor of the viscous stresses, and performing the expression with the stress tensor according to Leslie (equation (1.2.1)), the authors of [157] obtained:

$$\gamma_1 = C_1S + C_2S^2, \tag{3.1.1}$$

Here C_1 and C_2 are the parameters which depend only slightly on temperature. Helfrich showed [89] that because the possibility $S < 0$ is not theoretically excluded (for example, conical deformation), the relationship $C_1 = 0$ should be fulfilled [159].

However, the temperature dependence of rotational viscosity is considerably stronger in comparison with that indicated by equation (3.1.1) (Fig. 1.2.2). The linear section of the dependence of ln γ_1 on reciprocal temperature T^{-1} was approximated according to Arrhenius [160]:

$$\gamma_1 = a\exp\frac{E}{k_BT}. \tag{3.1.2}$$

In order to take into account the deviation of the dependence $\gamma_1(T)$ from the exponential dependence on approach to clarification temperature T_{NI} and describe the behaviour of γ_1 in the entire nematic range, the degree of ordering S [161] was introduced into the dependence (3.1.2):

$$\gamma_1 = bS\exp\frac{E}{k_B T}. \tag{3.1.3}$$

The relationships (3.1.2) and (3.1.3) are purely empirical and do not have any strict physical justification, although (3.1.3), as shown in chapter 4, describes most efficiently the temperature dependence of the rotational viscosity of a number of substances.

The molecular–statistical theory of the nematic liquid crystal according to Maier and Saupe [162] was applied for the first time for the determination of the type of temperature dependence of viscosity in [163, 164]. In the reorientation of a molecule or a cluster consisting of m molecules, the following potential barrier is overcome:

$$E = D(\theta)_{max} - D(\theta)_{min} = D(\pi/2) - D(0) = \frac{3}{2}\frac{A}{mV_N^2}S = \varepsilon S, \tag{3.1.4}$$

Here $D(\theta)$ is the Maier–Saupe potential of the mean field, and is specified by the equation:

$$D(\theta) = -\frac{AS}{mV_N^2}\frac{3\cos^2\theta - 1}{2}, \tag{3.1.5}$$

Here A is a parameter which specifies the value of the potential and depends on the anisotropy of the polarisability of the molecules of the nematic liquid crystal, V_N is the volume of the molecule (or cluster) in the nematic phase, θ is the angle between the long axis of the molecule and the direction of the director. According to the Maier–Saupe theory [162]:

$$\frac{A}{mV_N^2} = 4.54 k_B T_{NI}. \tag{3.1.6}$$

The equations (3.1.4) and (3.1.6) yield two important circumstances: the activation energy in the nematic phase should be proportional to the degree of ordering of the nematic liquid crystal and clarification temperature.

The following form of the temperature dependence of γ_1 was determined in [163, 164]:

$$\gamma_1 = CS\exp\frac{\varepsilon S}{k_B T}. \tag{3.1.7}$$

3.2. The free volume model

In the development of low-temperature liquid crystal materials, the development of the technology of low temperatures in applications to liquid crystals and the improvement of the accuracy of measurement of γ_1 it was shown that the dependence $\gamma_1(T)$ deviates from the exponential dependence (3.1.2) not only the vicinity of the clarification temperature but also at low temperatures (of the order of 0°C and lower) [107].

This form of the temperature dependence of rotational viscosity is explained on the basis of the free volume models [165] which ignore all the components of the microscopic structure of the liquid and the free volume (the actual volume less the volume at absolute zero) which is a main factor determining the rate of the dynamic processes. During the dynamic process, for example reorientation, the volume occupied by the molecule (or the cluster) is greater in the intermediate conditions than in the initial or final condition. This means that these processes can take place only in cases in which the volume of the system is larger than the volume of closely packed solid rods. The size of the free volume determined by this procedure is not large and usually equals 2–3% of the total volume of the liquid.

The formation of the free volume is determined by two factors: the entropy factor S_g, associated with the internal and external degrees of freedom of the molecule which facilitate the formation of the free volume, and the energy factor E_g which prevents the formation of the free volume. At a specific temperature T_g, the interaction of these processes results in vitrification of the liquid. It may be assumed that below the glass transition temperature T_g all the dynamic processes are 'frozen' and the viscosity is equal to infinity. Of the liquids in which the temperature dependence of the viscosity is determined by the glass transition temperature, the polymers have been studied sufficiently. For the polymers, the viscosity is a universal function of the difference (T–T_g) [166]:

$$\frac{\eta(T)}{\eta(T_g)} = F(T - T_g) = \exp\left\{-\frac{P(T - T_g)}{[(T - T_g) + Q]}\right\} = G\exp\left\{\frac{PQ}{[T - (T_g - Q)]}\right\}, \quad (3.2.1)$$

$$P \cong 40\text{K},\ Q \cong 52\text{K}, G = \exp(-P).$$

Equation (3.2.1) is valid for the polymers in the temperature range $T_g < T < T_g + 100$ K. The previously described model of the free volume was also used by Schad and Zeller in [167] for describing the temperature dependence of the rotational viscosity of nematic liquid crystals which undergo glass transition in specific conditions (at cooling rates and at a temperature of approximately 200 K). In [167] it is assumed that the dependence for any liquid crystals is described by the equation identical to (3.2.1), i.e., it is the function of glass transition temperature T_g and has the form of the Fulcher–Tamman law:

$$\gamma_1 = b\exp\frac{B}{(T - T_0)}, \tag{3.2.2}$$

where b is a parameter which depends only slightly on temperature; $T_0 = T_g - Q$ is the so-called temperature of freezing of the movement of the director; $B = PQ$. It should be mentioned that in the free volume model, derived for the isotropic liquid, the value of V_{ff} is determined by the expression:

$$V_{ff} = \frac{T - T_0}{B}, \tag{3.2.3}$$

which is the functional free volume.

The value $1-V_{ff}$ shows the fraction of the volume of the nematic liquid crystal occupied by the molecules which are in the thermal motion state.

3.3. The Diogo–Martins theory

In 1981, Diogo and Martins published the molecular–statistical theory of rotational viscosity of nematic liquid crystals [168]. In their approach, they used the assumptions of the molecular–statistical theory of the nematic crystals proposed by Maier and Saupe [162], and the Eyring theory of high-rate processes [169] which makes it possible to combine both the activation mechanism and the free volume model for describing the dependence $\gamma_1(T, S)$. Special attention will be given to the assumptions of the Diogo–Martins theory and the main moments of derivation of the equation of the dependence $\gamma_1(T, S)$.

The main prerequisite of the theory is the relationship of γ_1 with frequency ν_0 of the reorientation of the particle (a molecule or a small group of m molecules) around the director direction (reorientation–rotations of the particles through π-radian from one minimum of the intermolecular potential of the mean field to another). When the particle is at the minimum of the potential ($\theta = 0$), it is in the ground state, when it is at the maximum ($\theta = \pi/2$) it is in the activated (excited) state. The quantities γ_1 and ν_0 are linked by the expression derived in [164]:

$$\gamma_1 = \frac{k_B T}{\pi^2 V^a}\nu_0^{-1}, \tag{3.3.1}$$

where k_B is the Boltzmann constant, T is temperature, V^a is the volume which the article should have in order to rotate through the π-radian (according to the assumptions by the authors of [168], the minimum value of the volume is obtained by the extrapolation of the molar volume from the isotropic phase to temperature T (Fig. 3.3.1)). Therefore, if β is

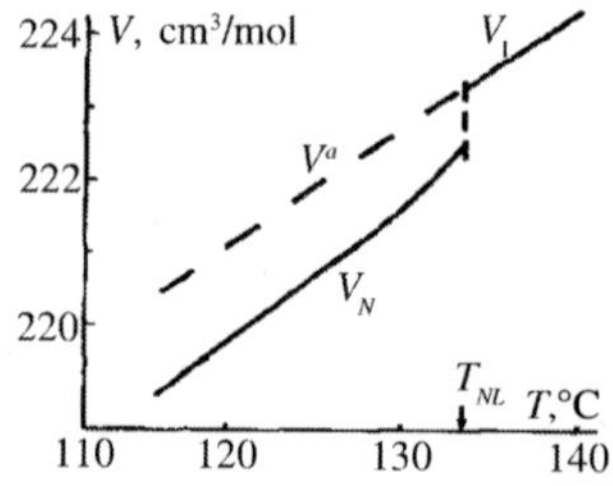

Fig.3.3.1. Molar volume of PAA in the isotropic phase V_I, in the nematic phase V_N, V^a– extrapolation of values of V_I to the nematic phase.

compressibility and V_N is the volume of the particle in the nematic phase, then

$$\frac{V^a - V_N}{V^a} = -\beta\Delta p = \frac{\beta A}{mV_N^3}S^2, \tag{3.3.2}$$

where $\Delta p = -(\partial\langle D\rangle/\partial V_N)_S$ is the variation of the internal pressure required for volume expansion, calculated from the potential of the Maier–Saupe mean field (equation (3.1.5)). Equation (3.3.2) yields an important conclusion on the proportionality of $\gamma_1 \sim S^2$.

The value of ν_0 is calculated on the basis of the theory of high-rate processes [169] according to which there is equilibrium between the number of the molecules in the activated condition N_a and the number of the molecules in the ground state N_g, determined by the relationship:

$$\frac{N_a}{N_g} = \frac{z_a}{z_g}\frac{Z_1^a Z_2^a Z_3^a}{Z_1^g Z_2^g Z_3^g}\exp\left(-\frac{\Delta\Sigma}{k_B}\right), \tag{3.3.3}$$

Here Z_i^a and Z_i^g are the functions of the rotational contribution of each of the three rotational degrees of freedom of the particle in the activated and ground states, respectively; z_a and z_g are the functions of the contributions for all other degrees of freedom, $\Delta\Sigma$ is the difference of the entropies of the activated and rground states. The functions Z_i may be written in the form:

$$Z_i = \left(\frac{2I_i k_B T}{h^2}\right)^{1/2}, \tag{3.3.4}$$

where h is the Planck constant, I_i are the appropriate moments of inertia, $I_2 = I_3$, if the molecules (or particles) are regarded as the ellipsoids of revolution.

The ratio N_a/N_g can also be determined using a different procedure. The number of the molecules 'jumping over' the potential barriers in unit time, is equal to $\nu_0 N_g$ and is also equal to N_a divided by the mean lifetime of the molecules in the activated state τ. Using the Boltzmann statistics, we determine the value of τ:

$$\tau = \frac{\pi}{\langle\omega\rangle} = \pi Z_2^{a}\left\{\int_0^{\infty} d\omega\,\omega \exp\left(-\frac{I_2^{a}\omega^2}{2k_BT}\right)\right\}^{-1} = \pi^{3/2}\frac{2I_2^{a}}{k_BT}. \tag{3.3.5}$$

From the last three equations we obtain the expression for ν_0:

$$\nu_0 = \frac{k_BT}{h}\frac{z_a}{z_g}\frac{I_2^{a}}{I_2^{g}}\left[\frac{I_1^{a}}{I_1^{g}}\frac{h^2}{2\pi^3 I_2^{a}k_BT}\right]^{1/2}\exp\left(-\frac{\Delta\Sigma}{k_B}\right). \tag{3.3.6}$$

The exponential member exp $(\Delta\Sigma/k_B)$ is the product of two co-multipliers: the probability of the particle having the energy sufficient for overcoming the potential barrier, and the probability of the particle having a free volume sufficiently large for reorientation. The derivation of the first co-multiplier exp $(-\varepsilon S/k_BT)$, described in [163], is described in section 3.1. The second co-multiplier has the following form:

$$\exp\left(-\gamma\frac{V^{a}-V_0}{V_N-V_0}\right) = e^{-\gamma}\exp\left(-\gamma\frac{V^{a}-V_N}{V_N-V_0}\right), \tag{3.3.7}$$

Here $V^a - V_0$ is the free volume which every particle should have for transition to the excited state, $V_N - V_0$ is the free volume per article, γ is the factor of overlapping of the free volumes whose value may vary from 0.5 to 1. If α is the difference of the thermal expansion coefficients of the nematic and crystalline phases, the following equation may be presented:

$$V_N(T) - V_0 \approx \alpha V_0(T - T_0), \tag{3.3.8}$$

where $V_0 = V_N(T_0)$ is the temperature at which the free volume disappears or, as mentioned previously, the temperature of freezing of motion of the director. Using equations (3.3.2), (3.3.7) and (3.3.8), we obtain:

$$\exp\left(-\frac{\Delta\Sigma}{k}\right) = e^{-\gamma}\exp\left(-\frac{\varepsilon S}{k_BT} - \frac{\theta S^2}{T-T_0}\right), \tag{3.3.9}$$

$$\theta = \frac{V^{a}\gamma}{V_0\alpha}\frac{\beta A}{mV_N^3}. \tag{3.3.10}$$

Consequently, the final equation of the temperature dependence has the form:

$$\gamma_1(T,S) = \left(\frac{2}{\pi}\frac{I_1^{g}}{I_1^{a}}k_BTI_2^{a}\right)^{1/2}\frac{z_g}{z_a}\frac{I_2^{g}}{I_2^{a}}\frac{\beta A e^{\gamma}}{mV_N^3(V^{a}-V_N)}\times$$
$$S^2\exp\left(\frac{\varepsilon S}{k_BT} + \frac{\theta S^2}{T-T_0}\right) = gS^2\exp\left(\frac{\varepsilon S}{k_BT}\right)\exp\left(\frac{\theta S^2}{T-T_0}\right). \tag{3.3.11}$$

Almost all quantities included in (3.3.11) can be obtained from other experiments or calculated from the geometry of the molecule, i.e., the equation is derived for predicting the value of γ_1 and its temperature dependence. Equation (3.3.11) shows that the dependence $\gamma_1(T)$ is determined mainly by two exponential co-multipliers, where the first co-multiplier is dominant in the high-temperature region (closer to T_{NI}) and the second one in the low-temperature region, if the liquid crystal is characterised by a low melting point or is capable of considerable supercooling.

An interesting example of the direct application of the Eyring theory [169] was proposed in [170, 171] where the correlation between the geometrical characteristics of the distribution function of the nematic elementary cells and the Mesovich coefficients [170] was determined. The validity of the relationship $0.5\,(\eta_3/\eta_2+\eta_3/\eta_1)\approx 1$ in the entire nematic range has been confirmed for a number of substances.

In [163, 164, 168, 169] special features of the short-range order were taken into account only by introducing the correlation multiplier m into the equation for the potential of the mean field (3.1.5) and the special features of the molecular structure of the nematic liquid crystals were not taken into account at all. Qualitative changes in clarification temperature T_{NI} and the activation energy E of rotational viscosity, which may result from the additional consideration of the local symmetry and the short-range order, associated with the special features of the structure of the molecules in specific nematic liquid crystals, are described in [172]. In this work, the author developed the approximation of the mean molecular field taking into account the symmetry of mesogenic molecules (for example, the molecule with the structural formula $R{-}C_6H_4{-}COO{-}C_6H_4{-}OR'$ is referred to as symmetric, and the molecule with the formula $R{-}C_6H_{10}{-}COO{-}C_6H_4{-}OR'$ as non-symmetric; C_6H_4 is the benzol fragment and C_6H_{10} the cyclohexane fragment). The modelling potential (pseudopotential) $D(\theta)$ is written in the form:

$$D(\theta)=-\sum_L\sum_M V\sum_L M\eta_L P_M(\cos\theta)=-V\eta[P_2(\cos\theta)+ \\ \delta_1 P_1(\cos\theta)+\delta_4 P_4(\cos\theta)], \tag{3.3.12}$$

where η_L are the parameters of the orientation order ($\eta_2=S$), P_M (cos θ) are Legendre polynomials of the M-th order, V_{LM} are the normalisation multipliers, δ_1,δ_4 are parameters. At $L = M = 2$, the pseudopotential (3.3.12) is reduced to the Maier–Saupe potential (3.1.5).

A different feature in comparison other approximations of the mean field, described in [2], is mainly the presence in (3.3.12) of crossed terms with $L\neq M$. Terms with P_2, P_4 and P_1 in the pseudopotential were used for actual calculations (this selection is justified in [2, 172]). The non-symmetric form of the structure of the molecules is reflected in an increase of the contribution to $D(\theta)$ of the term proportional to $P_1(\cos\theta)$ and, consequently, the parameter δ_1 reflects the value of the nonsymmetricity of the molecule. In accordance with the Diogo–Martins theory [168], the rotational viscosity coefficient is determined by the frequency of jumps ν_0 (equation (3.3.1)). For the pseudopotential of the type (3.3.12) which has for $\delta_1 > 0$ the absolute minimum at $\theta = 0$ and the local minimum at $\theta = 180°$ (Fig. 3.3.2 [172]), the value of ν_0 should increase with increasing δ_1 and, consequently, γ_1 should decrease. The probability of the jump from the position corresponding to the local minimum of the pseudopotential should also increase because of the smaller height of the potential barrier. Thus, the nematic liquid crystals consisting of non-symmetric molecules should have, according to [172], lower values of activation energy and rotational viscosity in comparison with the nematic liquid crystals consisting of symmetric molecules. For these crystals, the correlation between E and T_{NI}, following from the expressions (3.1.4) and (3.1.6), is not fulfilled.

3.4. Calculation of the viscosity coefficient of nematic liquid crystals using the kinetic equation

The expression for the temperature dependence of rotational viscosity γ_1, based on the solution of the relaxation equation of the order parameter, was derived by Hess [123, 124].

The initial equation in the thermodynamics non-equilibrium processes is the equation describing the time dependence of specific entropy:

$$\frac{dS}{dt} = T^{-1}\left(\frac{du}{dt} + p\frac{d\rho^{-1}}{dt}\right) - T^{-1}\frac{dG_a}{a_{\mu\nu}}\frac{da_{\mu\nu}}{dt}. \qquad (3.4.1)$$

Here ρ is density, u is the internal energy of the unit volume, G_a is the part of the specific chemical potential (Gibbs potential), associated with the orientation, a_{ik} is the symmetric traceless anisotropy (orientation) tensor having the uniaxial form in equilibrium:

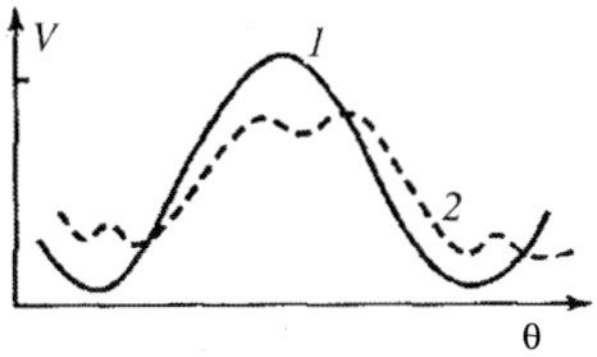

Fig.3.3.2. Comparison of the type of dependence of the pseudopotential D on the angle of orientation of molecules θ in relation to the director for symmetric (*1*) and non-symmetric (*2*) molecules.

$$a_{ik}^{eq} = a^{eq}\left(n_i n_k - \frac{1}{3}\delta_{ik}\right), \tag{3.4.2}$$

where a^{eq} is the value of anisotropy or scalar order parameter in equilibrium, i.e. $a^{eq} = S$, n_j are the components of the director.

The Gibbs potential is described by the equation:

$$G_a = \frac{k_B T}{m}\left(\Sigma + 0.5\xi_0^2 (\nabla_\lambda a_{\mu\nu})^2\right) \tag{3.4.3}$$

with potential Σ identical with the Landau potential:

$$\Sigma = \frac{1}{2} A a_{\mu\nu} a_{\mu\nu} + \frac{\sqrt{6}}{3} B a_{\mu\nu} a_{\nu\lambda} a_{\lambda\mu} + \frac{1}{4} C (a_{\mu\nu} a_{\mu\nu})^2, \tag{3.4.4}$$

where m is the mass of the molecule, ξ_0 is correlation length, A, B, C are the dimensionless coefficients which are the function of temperature T and pressure P (summation is carried out with respect to the repeating indexes). The temperature dependence of A is proposed in the form:

$$A = A_0\left(1 - \frac{T^*}{T}\right), \tag{3.4.5}$$

where the characteristic temperature T^* is slightly lower than T_{NI}. The thermodynamic equilibrium and relaxation of the order parameter to the equilibrium value described by the Landau–Khalatnikov equations:

$$\frac{\partial F}{\partial a_{ik}} = 0, \quad \frac{\partial a_{ik}}{\partial t} + \Omega_{iklm} \frac{\partial F}{\partial a_{lm}}, \tag{3.4.6}$$

where Ω_{iklm} is the tensor kinetic coefficient.

According to [124], in the absence of flows ($v = 0$):

$$\frac{\partial a_{ik}}{\partial t} + \frac{\Sigma_{ik}}{\tau_a} = 0. \tag{3.4.7}$$

Here, τ_a is the duration of orientational relaxation of the molecule which is considerably longer than the duration of rotational relaxation:

$$\Sigma_{ik} = -\frac{M}{k_B}\left(\frac{\partial \Sigma_a}{\partial a_{ik}} + \frac{\varepsilon a_{ik}}{T}\right), \tag{3.4.8}$$

where Σ_a is the specific orientational entropy, M is the molecular mass, ε is a quantity characterising the ordering energy. Using the expansion of Σ_a in respect of the degrees a_{ik}, it may be shown that $\Sigma_{ik} = dF/da_{ik}$ and that (3.4.7) is identical to (3.4.6). Consequently, we can determine the relationship of the kinetic coefficient Ω_{iklm} with time τ_a. In [124], potential Σ_{ik} is expressed in terms of the molecular field h_i by the equation:

$$h_i = -2\sqrt{\frac{3}{2}} a^{eq} P_c \Sigma_{ik} n_k, \tag{3.4.9}$$

where $a^{eq} = S$, $p_c = \rho k_B T/M$ is the so-called kinetic pressure, ρ is the mass density. This gives the final equation for the viscosity coefficient:

$$\gamma_1 = 3(a^{eq})^2 P_c \tau_a = \frac{3\rho S^2 k_B T}{M} \tau_a,$$
$$\gamma_2 = 2\sqrt{3} a^{eq} P_c \tau_{ap}, \tag{3.4.10}$$

where τ_{ap} is the relaxation time describing torsion in the speed gradient. Comparing (3.4.10) with (3.3.1), it can be seen that the quantity γ_1 in these equations differs by a multiplier of the order of 30.

If the director orientation is spatially nonuniform, the value τ_a^{ef} decreases [123] (see (2.4.32)).

To describe the non-equilibrium processes in the nematic liquid crystals, it is necessary to derive the equation for the function of distribution of the molecules in respect of the orientation and not the equation for the order parameter. In most cases, the kinetic equation is derived using a one-particle approximation, i.e. rotational diffusion of asymmetric rigid particles in the anisotropic viscous liquid is investigated. This approximation is referred to as the suspension model although, taking into account the dimensions of the molecules and particles characteristic of the nematic crystals, it would be more accurate to refer to the system as a colloid. In this case, the natural timescale is the time to establishment of thermal equilibrium in the statistical ensemble of molecules as a result of Brownian motion [173–175]:

$$\tau_B \sim \frac{\eta V_0}{T}, \tag{3.4.11}$$

where V_0 is the molecule volume. The molecule (particle) is subjected to the effect of the mean molecular field and the external (usually magnetic) field. The frequency of the external influences ω is small ($\omega\tau_B \ll 1$). The distribution of the molecules with respect to the orientations (angle of rotation) is described by the function W and its dependence by the Fokker–Planck equation whose simplest form is:

$$\frac{\partial W}{\partial t} + \hat{R}(\omega W) = 0, \tag{3.4.12}$$

where $\hat{R} = \mathbf{e} \times \partial/\partial \mathbf{e}$ is the operator of an infinitely small rotation, **e** is the unit director along the axis of symmetry of the particle, ω is the angular speed of the molecule or particle proportional to the momentum of forces L_{ik} acting on the particle and rotational mobility M_{ik} or rotational diffusion D. In most cases, a small perturbation of the distribution function W is considered. The solution of the kinetic equation yields the tensor of viscous stresses of the nematic liquid crystal σ_{ik} or the equation of motion of the director and after comparison of the results with (1.1.7) or (1.2.3), we obtain the expressions for Leslie coefficient α_i and rotational viscosity γ_1 or γ_2.

This is the procedure used by all the authors who derived and solved the kinetic equation for the distribution function. The difference in the results obtained by these authors is caused mainly by the selection of the solution method and also by the initial assumptions. Therefore, some of the expressions for the viscosity coefficient, obtained by individual answers, are identical.

The function of orientation distribution $W(\mathbf{e},t)$ is determined as the probability of the direction of the longitudinal axis of an arbitrarily selected molecule coinciding with the direction of the unit vector $\mathbf{e}$ at the moment of time t. According to [176, 177], the variation of function $W(\mathbf{e}, t)$ with time is described by the kinetic equation:

$$\frac{\partial W}{\partial t} = D\hat{R}\left[\hat{R}W + \frac{W}{k_B T}\hat{R}V\right] - \hat{R}(\omega W). \tag{3.4.13}$$

The first member in the right-hand part of (3.4.13) describes the rotational Brownian motion with the rotational diffusion coefficient D. The second term describes the rotation induced by potential $V(\mathbf{e})$ which is the sum of the potential of the mean field V_{MF} and the potential of the external (magnetic) field V_H. Since the exact form of V_{MF} depends on the model applied in this case, for further calculations we use only the dependence of V_{MF} on the cosine of the angle between the adjacent molecules identical with the Maier–Saupe potential. The last term in (3.4.13) appears as a result of the presence of the speed gradient which leads to the rotation of the molecule with angular speed ω for the spheroid of revolution with the ratio of the axes p equal to:

$$\omega = \kappa\left\{\left[\mathbf{e}\times\left(\overline{\overline{A}}\mathbf{e}\right)\right] - \left[\mathbf{e}\times\left(\overline{\overline{\Omega}}\mathbf{e}\right)\right]\right\}; \tag{3.4.14}$$

$$\kappa = \frac{p^2 - 1}{p^2 + 1}. \tag{3.4.15}$$

The first and second terms in the right-hand part of (3.4.14) are the symmetric and and asymmetric parts of the tensor of the velocity gradient $\overline{\overline{K}} = (\nabla\mathbf{V})$. The components of the tensors $\overline{\overline{A}}$ and $\overline{\overline{\Omega}}$ are described by the equations (1.1.4). We now examine the hypothetical low strain $\overline{\overline{\varepsilon}}$, applied over a very short period of time δt to the system with the tensor of the velocity gradient $\overline{\overline{K}}$. The quantities $\overline{\overline{\varepsilon}}$ and $\overline{\overline{K}}$ are linked by a simple relationship:

$$\overline{\overline{\varepsilon}} = \overline{\overline{K}}\delta t. \qquad (3.4.16)$$

This deformation results in the rotation of the molecule through a small angle $\delta\theta$:

$$\delta\theta = \omega\delta t = \kappa\left[\mathbf{e}\times\left(\overline{\overline{A}}\delta t\mathbf{e}\right)\right] - \kappa\left[\mathbf{e}\times\left(\overline{\overline{\Omega}}\delta t\mathbf{e}\right)\right] = \mathbf{e}\times\left(p^2\overline{\overline{\varepsilon}} - \overline{\overline{\varepsilon}}^*\right)\frac{\mathbf{e}}{p^2+1} \qquad (3.4.17)$$

and in the variation of the distribution function

$$W' = W + \delta W = W(\mathbf{e},T) - \hat{R}[\delta\theta\times f(\mathbf{e},t)]. \qquad (3.4.18)$$

From the difference of the free energies of the nematic liquid crystals corresponding to the deformed and non-deformed conditions, it is possible to determine the tensor of viscous stresses $\overline{\overline{\sigma}}$:

$$F[W'] - F[W] = \overline{\overline{\sigma}}:\varepsilon^+ + O(\varepsilon^2). \qquad (3.4.19)$$

Then, reducing the tensor to the form obtained by Leslie and Ericksen (equation (1.1.2)), we derive equations for the Leslie viscosity coefficients:

$$\alpha_1 = -\frac{nk_BT}{D}\kappa^2 S_4;\quad \alpha_2 = -\frac{nk_BT}{2D}(1+\lambda^{-1})S_2;$$
$$\alpha_3 = -\frac{nk_BT}{2D}(1-\lambda^{-1})S_2;\quad \alpha_4 = -\frac{2nk_BT}{70D}\kappa^2(7-5S_2-2S_4);$$
$$\alpha_5 = -\frac{nk_BT}{2D}\kappa\left[\frac{\kappa}{7}(3S_2+4S_4)+S_2\right]; \qquad (3.4.20)$$
$$\alpha_6 = -\frac{nk_BT}{2D}\kappa\left[\frac{\kappa}{7}(3S_2+4S_4)-S_2\right];$$

where $n = N_A/V$ is the number of the molecules in the unit volume, V is the molar volume, $\lambda^{-1} = -\gamma_1/\gamma_2$, S_2, S_4 are the order parameters equal to the Legendre polynomials of the second and fourth degree. Equation (3.4.20) gives:

$$\gamma_1 = \frac{nk_BTS_2}{D\lambda}, \qquad (3.4.21)$$

i.e. to calculate viscosity γ_1, it is necessary to know the coefficient of rotational diffusion, in addition to the density and the order parameter.

The identical derivation of the temperature dependence of the viscosity coefficient in comparison with [177] was proposed in [175, 178, 179] and also in [180]. The appropriate Fokker–Planck equation:

$$\frac{\partial W}{\partial t}+\hat{R}_{\alpha}(\omega_{\alpha}W)-D\hat{R}_{\alpha}\left\{W\hat{R}\left[\frac{E}{T}-\frac{\kappa}{2D}A_{\mu\nu}e_{\mu}e_{\nu}+\ln W\right]\right\}=0, \qquad (3.4.22)$$

is almost identical with (3.4.13). Writing this equation with the random component proportional to ln W, we can find an equilibrium solution, having the Gibbs form:

$$W\sim\exp\left(-\frac{E}{T}\right). \qquad (3.4.23)$$

Consequently, when fulfilling the condition $\omega\tau_{B}\ll 1$, where ω is the frequency of external effects, and examining the variation of the distribution function with time, it can be assumed, in accordance with [175]. that the liquid crystal passes through a number of equilibrium states, i.e., the distribution function at every moment of time has the Gibbs form: $W \sim \exp[-E(t)/T]$, corresponding to the instantaneous value of the energy of the potential barrier $E(t)$. In other words, it is assumed that at every moment of time the non-equilibrium distribution function in the true field coincides with the equilibrium function of distribution in the effective field. As in the case of (3.4.18), the nonequilibrium distribution function is presented in the form:

$$W=W_{0}[1+b_{ik}(t)(e_{i}e_{k}-\langle e_{i}e_{k}\rangle)]. \qquad (3.4.24)$$

If (3.4.24) is regarded as linearisation:

$$W\sim W_{0}\exp[b_{ik}(t)(e_{i}e_{k}-\langle e_{i}e_{k}\rangle)]\sim\exp\left[-\frac{E}{T}+b_{ik}(t)(e_{i}e_{k}-\langle e_{i}e_{k}\rangle)\right], \qquad (3.4.25)$$

then the difference of $E(t)$ from the equilibrium form can be associated with the small tensor in b_{ik}. In this case, the moments of the distribution function are expressed through the tensor parameters of the order a_{lm} proportional to S_2 and S_4, and for the deviations of these functions from the equilibrium values:

$$\varepsilon_{lm}=a_{lm}-a_{lm}^{eq} \qquad (3.4.26)$$

we write a system of relaxation equations:

$$\frac{\partial\varepsilon_{lm}}{\partial t}+\sum_{l'}\Lambda_{ll'}^{(m)}\varepsilon_{l'm}=0,\quad \bar{\bar{\Lambda}}=\bar{\bar{\Gamma}}\,\bar{\bar{\chi}}^{-1}, \qquad (3.4.27)$$

in which the components of the tensor of the kinetic coefficients $\Gamma_{ll'}^{(m)}$ are a linear combination of S_2 and S_4, and the dynamic susceptibility $\chi_{ll'}^{(m)}$ contains the relaxation time spectrum. The difference between equation (3.4.27) and (3.4.6) is the presence of the coefficient of the tensor parameter of the fourth rank which increases the number of the investigated relaxation times. On the basis of the previous considerations, the authors of [178] derived an equation for the movement of the director in the external field **H** in the vector form:

$$S_2[\mathbf{n}\,(\mathbf{n}-\Omega\times\mathbf{n})]=[(16S_4+5S_2+14)/35]\times \left\{b\chi_a[\mathbf{n}\times\mathbf{H}](\mathbf{nH})+\kappa\left[\overline{\overline{\mathbf{n}}}\left(\overline{\overline{\mathbf{dn}}}\right)\right]\right\}, \quad (3.4.28)$$

from which

$$\gamma_1=\frac{n}{b}\left[\frac{35S_2^2}{16S_4+5S_2+14}\right]; \qquad \gamma_2=-\frac{n}{b}\kappa S_2, \quad (3.4.29)$$

where $b=(6\eta V_0\delta)^{-1}$ is the rotational mobility coefficient, δ is the form factor, $\delta=1$ for a spherical particle, $\delta\cong 3$ for an ellipsoid with the ratio of the axes 4. Using the Maier–Saupe potential as the mean field, the expression for γ_1 may be simplified, because the following relationship should be fulfilled:

$$S_4=\frac{5S_2-35T/U+7}{12}. \quad (3.4.30)$$

The variation of the effective field during the relaxation of the distribution function (see equation (3.4.25) was taken into account for the first time in the problem of the relaxation of magnetisation of a suspension of magnetic particles [181]. Consequently, it should be mentioned that as regards the form and method of derivation, the conclusions made in [175, 178, 179] are identical to the previously made conclusions [181, 182]. In [181] the Leslie–Ericksen equation is obtained in the zero approximation of expansion with respect to the characteristic duration of orientational relaxation and the movement of the director. This examination is identical with the model of the magnetic liquid with internal rotation [183] which makes it possible, by means of the appropriate definition of the function of state – the chemical potential in the internal orientation space – to determine the dependence of the rotational viscosity of the magnetic liquid on the strength of the magnetic field **H**. Without presenting the equation of motion of the director, with the exception of the small components identical with (3.4.28), the equations for Leslie α_i and Tsvetkov γ_1 coefficients, taking into account integration with respect to orientations **e** have the following form:

$$\alpha_1 = \left(\frac{\beta_1}{4} - \alpha n\gamma_1^2\right)\left[\frac{7}{3} + \frac{5S}{3} - \frac{35S}{2X}\right],$$

$$\alpha_2 = -\alpha n\gamma_1 S - \frac{1}{2}\alpha n S^2\left[\frac{S+2}{3} - \frac{2S}{X}\right]^{-1},$$

$$\alpha_3 = -\alpha n\gamma_1 S + \frac{1}{2}\alpha n S^2\left[\frac{S+2}{3} - \frac{2S}{X}\right]^{-1},$$

$$\alpha_4 = 2\eta + \alpha n\gamma_1^2\left[\frac{2-2S}{3} + \frac{S}{X}\right] + \frac{2}{3}\beta_2(1-S) - \frac{1}{4}\beta_1\left[(2-2S) - \frac{S}{X}\right],$$

$$\alpha_5 = 2\alpha n\gamma_1^2\left[\frac{2S+1}{3} - \frac{5S}{2X}\right] + \frac{\beta_1}{2}\left[\frac{5S}{2X} + \frac{S-1}{3}\right] + \beta_2 S + \alpha n\gamma_1 S,$$

$$\alpha_6 = 2\alpha n\gamma_1^2\left[\frac{2S+1}{3} - \frac{5S}{2X}\right] + \frac{\beta_1}{2}\left[\frac{5S}{2X} + \frac{S-1}{3}\right] + \beta_2 S - \alpha n\gamma_1 S, \qquad (3.4.31)$$

$$\gamma_1 = \alpha n S^2\left[\frac{S+2}{3} - \frac{2S}{X}\right]^{-1} = \alpha n S^2\left[\frac{S+2}{3} - \frac{0.2936T}{T_{NI}}\right]^{-1}, \qquad (3.4.32)$$

where $X = \sigma S/k_B T$ is a coefficient σ linked in the Maier–Saupe theory, with temperature by the relationship $k_B T_{NI} = 0.1468\sigma$.

The previously presented equations for the coefficients α_i,α_j, obtained by different methods, contain several 'seeding' quantities, τ_a, D, η, α, which must be taken from the results of measurement of other physical parameters. This complicates the preliminary calculation of viscosity from the available characteristics of the molecules (dimensions, polarisability, dipole moment, etc) and the comparison of the results with the results of independent experiments. The authors of [184–187] attempted to link the value of the microscopic friction coefficient ς with the dimensions of the molecules and the energy of intermolecular interaction. Since the approach of the authors of the above studies is identical with that used in [177] and this resulted in the agreement of the equations for all viscosity coefficients with the exception of α_2 and α_3, we examine in greater detail the factors supplementing the conclusions of the study [177]. The coefficient ς is linked with rotational diffusion D in [177] by the relationship:

$$\varsigma = \frac{k_B T}{D}. \qquad (3.4.33)$$

Microscopic friction forms when the actual angular velocity of rotation of the molecule $\psi = \dot{\varphi}\mathbf{e} + [\mathbf{e}\times\dot{\mathbf{e}}]$ differs from the mean velocity ψ_0 = rot $\mathbf{v}$ + $\kappa[\mathbf{e}\times A_{\mu\rho}\mathbf{e}]$, $A_{\mu\rho}$ (see (1.1.4)). It is important to note the small perturbation of the distribution function in proportion to $A_{\mu\rho}$. Consequently, the variation of the momentum of the forces acting on the molecule is proportional to the tensor of microscopic friction:

$$\Gamma = \Gamma_{\rm el} - \hat{\varsigma}(\psi - \psi_0), \qquad (3.4.34)$$

where $\Gamma^{\alpha}_{\rm el} = -\varepsilon_{\alpha\beta\gamma} e_{\beta}(\partial U/\partial e_{\gamma})$, i.e., the elastic momentum is the derivative of the potential of the mean field in respect of orientation. The fluctuation–dissipative theorem indicates that the microscopic friction coefficient is the time correlation function of the momentum of elastic forces:

$$\varsigma = (3k_{\rm B}T)^{-1}\int_0^{\infty}\langle\Gamma(t)\Gamma(0)\rangle_{eq}\,dt, \text{ where } \langle\Gamma\rangle_{eq} = 0. \qquad (3.4.35)$$

if the relaxation time of the correlator of the stochastic force is very short. This integral is calculated from the Fokker–Planck approximation. The quantity τ is evaluated as the ratio of the fraction of the free angular volume

$$\Delta\theta \sim \left(\sqrt{4/(\pi nL)} - d\right)L^{-1} \qquad (3.4.36)$$

($n = N_{\rm A}/N$ is the density of the number of particles, L and d are the dimensions of the molecule) to the mean velocity of thermal rotation proportional to $\tau_a^{-1} = (k_{\rm B}T/I_{\perp})^{1/2}$ ($I_{\perp}$ is the moment of inertia of the molecule, measured around the transverse axis). Consequently, the friction coefficient is:

$$\varsigma \cong \tau\langle\Gamma^2(0)\rangle_{eq} \cong \Delta\theta\left(\frac{I_{\perp}}{k_{\rm B}T}\right)^{1/2}\langle\Gamma^2(0)\rangle_{eq}, \qquad (3.4.37)$$

where $\langle\Gamma\rangle = 0$.

To calculate $\langle\Gamma^2(0)\rangle_{eq}$, we use the paired intermolecular interaction potential [188]:

$$U(1,2) \cong -\frac{G}{r_{12}^6} - \frac{1}{r_{12}^6}\left[J_0(\mathbf{e}_1,\mathbf{e}_2)^2 + J_2(\mathbf{e}_1,\mathbf{e}_2,\mathbf{r}_{12})\right], \qquad (3.4.38)$$

which is divided into the isotropic part, corresponding to the attraction of the molecules, and the anisotropic part, described by J_0 and J_2 ($J_0 \sim J_2 \ll G$).

The one-particle equilibrium distribution function is given by the mean field potential ($W \sim \exp(-U(\mathbf{en})/k_{\rm B}T)$) which is the average value of the paired potential (3.4.38) with respect to the coordinates and the angles of the neighbouring molecule:

$$U(\mathbf{e}_1\mathbf{n}) = n\int U(\mathbf{e}_1,\mathbf{e}_2,\mathbf{r}_{12})W_1(\mathbf{e}_2\mathbf{n})\,d\mathbf{r}_{12}\,d\mathbf{e}_2 \cong$$
$$\cong \text{const} - 2\pi d^3 n\left(J_2 + \frac{2J_0}{3}\right)S(\mathbf{e}_1\mathbf{n}). \tag{3.4.39}$$

The final form of the microscopic friction coefficient in [186] is:

$$\varsigma \cong 100 V_{fg} n^2 d^6 \left(\frac{d}{L}\right)^2 (k_B T)^5 G^3 \left(\frac{I_\perp}{k_B T}\right)^{1/2} \exp\left[\frac{3(G+J_0)}{k_B T}\right], \tag{3.4.40}$$

where $V_{fg}=1-k_p$ is the geometrical free volume, $k_p=N_A V_0\rho/M$ is the molecular packing coefficient (for more details see section 4.2).

The expressions for the viscosity coefficient can be determined using the macroscopic stress tensor which is calculated by averaging the microscopic stress tensor. It is convenient to use the microscopic tensor in the form of the sum of its symmetric and antisymmetric parts. The Leslie coefficient α_1, α_4, $\alpha_5 + \alpha_6$, $\alpha_2 + \alpha_3$ are included in the symmetric part, which may be calculated without solving the Fokker–Planck equation. To average the antisymmetric part of the microscopic tensor, it is necessary to know the appropriate nonequilibrium function which is the solution of the following equation:

$$\dot{W} + \varepsilon\partial_k(\Omega_k W) = \varepsilon^2\partial_k\left[\partial_k W - \frac{\Gamma^k_{y??}}{k_B T}W\right], \tag{3.4.41}$$

where $\varepsilon = \tau_\omega/\tau_a = (I_\perp/\lambda)(I_\perp/k_B T)^{1/2} = (I_\perp/k_B T)^{1/2}/\lambda \ll 1$ is the small parameter describing the two-stage nature of relaxation of the Maxwell distribution function with respect of speeds (τ_ω) and the slow establishment of the Boltzmann distribution with respect to the angular coordinates or thermal rotation (τ_a).

Averaging over the angles gives the following equation:

$$\gamma_1 \cong -\frac{n}{2\varepsilon Z}\int_0^\pi \frac{\partial U(\theta)}{\partial\theta}\exp\left[-\frac{U(\theta)}{k_B T}\right]\sin\theta \int_0^\theta \exp\left[\frac{U(x)}{k_B T}\right]\frac{1}{\sin x}\times$$
$$\int_0^x \exp\left[-\frac{U(Z)}{k_B T}\right]\frac{\partial U(Z)}{\partial Z}\frac{\sin Z}{k_B T}\,dZ\,d\theta, \tag{3.4.42}$$

where the normalisation multiplier is:

$$Z = 2\pi\int_0^\pi \exp\left(-\frac{U}{k_B T}\right)\sin\theta\,d\theta \cong \frac{2\pi}{q}.$$

If $q = J/k_B T \gg 1$ (J is the coefficient in front of $(\mathbf{an}) = \cos\theta$ in (3.4.39)), for the calculation of the integral (3.4.42) we may use the saddle method. Consequently

$$\gamma_1 \cong \frac{n\varsigma}{6}\left(\frac{J}{k_B T}\right)^{1/2} \exp\frac{J}{k_B T}. \tag{3.4.43}$$

The equations (3.4.43) and (3.4.40) show that quantity γ_1 depends on the geometrical free volume V_{fg}, the transverse size of the molecule, the ratio of its axes and its activation depends temperature with the activation energy $[3\,(G + J_0) + J]$.

Until now we studied in this section the case of suspension of non-spherical particles in an isotropic liquid. On the basis of the approach proposed in [173, 174], the authors of [189–195] constructed a self-consistent rheological model which takes into account the anisotropy of the liquid. The special features of the model and of the derivation of the model are presented below.

The anisotropic properties of the nematic liquid crystals are determined by investigating the movement of a separate particle having the form of an ellipsoid of rotation with the ratio of the half-axes $p = L/d = 1+\beta$ with the orientation vector e_j formed by the potential of some mean field. The evolution of the orientation function of distribution of the separated particles is given by the Fokker–Planck diffusion equation of type (3.4.12) and is regarded as the result of rotation or diffusion of this particle in an effective anisotropic viscous environment with the angular velocity:

$$\Omega_i = M_{ik} L_k. \tag{3.4.44}$$

The tensor of rotational mobility M_{ik} taking into account self-consistency arguments in the linear approximation with respect to the anisotropy tensor a_{ik} (see (3.4.2) (and with respect to β has the following form:

$$M_{ik} = (8\pi\eta R)^{-1}\,[(1-1.2\beta)\delta_{ik} + 0.6\beta a_{ik} + 0.6 e_i e_k]. \tag{3.4.45}$$

The equation for the momentum of the forces L_k orienting the separated particle is presented in the following form:

$$L_{js} = \int_S \left(\sigma^m_{jk} m_k x_s - \sigma^m_{sk} m_k x_j\right) dS + Tv(a_{ji} e_s - a_{si} e_j) e_i +$$
$$T\sigma h_i h_k\,(\delta_{jk} e_s e_i - \delta_{sk} e_j e_i) + T\left(e_j \frac{\partial \ln W}{\partial e_s} - e_s \frac{\partial \ln W}{\partial e_j}\right). \tag{3.4.46}$$

Its value is determined by the viscous hydrodynamic friction (σ^m_{ik} is the tensor of the microstresses of the environment calculated on the basis of the fields of pressure and speed of the liquid, m_k is the vector of the normal to the surface S of the separated particle); the mean field of environment of the particle with the energy constant v, the external magnetic field $H_i/H_c = \sigma h_i$; the effect of Brownian motion. These four factors determine the terms in (3.4.46).

To obtain the complete equation of the tensor coefficient of resistance to the rotational and translational movement of the separated particles, the authors of [190, 191] solved the important problem of the theory of dynamics of non-Newtonian liquid – the flow of an anisotropic liquid around a particle. It should be mentioned that the relationship between the translational and rotational motion of a spherical particle in the isotropic liquid was described in [196–198]. The authors of [192, 193] derived equations for the tensor of rotational mobility in the case of an arbitrary axisymmetric particle and an anisotropic medium. Thus, the one-particle Fokker–Planck equation is completely defined and is solved using the main principle of self-consistency which in this problem is equivalent to the claim of the identical structure of the tensors for micro- and macrostresses. This yields phenomenological transfer coefficients in dependence on β, i.e., the parameter of the non-spherical form of the particle and, consequently, with the accuracy to the terms of the order β^2, S^3, the equations for the Leslie viscosity coefficient α_i:

$$\begin{gathered}\alpha_1 = 0;\quad \alpha_2 = -3(1+0.2\beta)\,\eta_0 S^2 - 1.2\beta S\eta_0;\\ \alpha_3 = 3(1+0.2\beta)\eta_0 S^2 - 1.2\beta\eta_0 S;\quad \alpha_4 = \left[2-\frac{8}{7}\beta S\right]\eta_0;\\ \alpha_5 = \frac{102}{35}\beta S\eta_0;\quad \alpha_6 = \frac{18}{35}\beta S\eta_0;\end{gathered} \tag{3.4.47}$$

and for rotational viscosity γ_1:

$$\gamma_1 = 6(1+0.2\beta)\,\eta_0 S^2. \tag{3.4.48}$$

The authors of [193, 194] also presented equations for the Mesovich viscosity coefficient η_i:

$$\begin{gathered}\eta_1 = \eta_0\left[1+\frac{103}{35}\beta S\right];\quad \eta_2 = \eta_0\left[1-\frac{23}{35}\beta S\right];\\ \eta_3 = \eta_0\left[1-\frac{4}{7}\beta S\right]\end{gathered} \tag{3.4.49}$$

and for the difference $\Delta\eta = \eta_1 - \eta_2$ in relation to the strength of the magnetic field which was not previously examined:

$$\Delta\eta = 0.04\ \sigma\eta_0\beta\left(1+\frac{\nu S}{21}\right) \text{ or } \sigma \ll 1,$$

$$\Delta\eta = \frac{12}{35}\frac{\sigma}{(\sigma+\nu S)} \text{ or } \sigma \ll 1, \tag{3.4.50}$$

Here ν is the constant of interaction of the particle with the 'mean' field.

As indicated by (3.4.48) and (3.4.49), the quantities γ_1, η_i depend explicitly only on the geometrical parameter of the particle and are independent of the type of mean field. This dependence is evident only in the presence of a weak external field (3.4.50).

The derivation of the equations for the viscosity coefficients using the Fokker–Planck kinetic equations is also described in [199, 200], where the concepts proposed in [123, 124] and also in [201] are developed further, and the angular gradients are taken into account in the rotational operator $\hat{R}$.

It should also be mentioned that the authors of [202] solved the Smolukhovskii equations in the form identical to the equations (3.4.13), (3.4.22) which was derived for the paired correlation function (a suspension of interacting colloidal particles (charged monodispersed polymer spheres)). It was shown in [203] how the relaxation time of the orientation, described for the nematic crystals by the equation (3.4.10), depends on the concentration of matter in a liotropic syste,.

3.5. Calculation of the internal friction and rotational viscosity coefficients by the method of the nonuniform statistical operator

As already mentioned in the beginning of this chapter, the kinetic coefficient can be expressed through the time correlation functions (TCF) using the Zwanzig–Tori projection operators [204] or the Zubarev operator [205]. In [206, 207], the authors developed the relaxation hydrodynamics of nematic liquid crystals on the basis of the Zubarev nonuniform statistical operator (the non-equilibrium distribution function in classic examination). The principal result of this theory is the statistical description of the final orientational deformation and relaxation of orientational ordering [206].

In [208], the method of non-equilibrium averaging of the appropriate microscopic equations was used to derive equations for the stress tensor and the equations of motion for the director and the tensor order parameter. The average parameters have the form of integrals in respect of time for the TCF flows and sources expressed by means of the characteristics of motion and interaction of the molecules [208]. The time correlation functions were calculated using the Fokker–Planck equation for the rotational distribution function derived in examination of the Brownian motion in external fields [209], and the equilibrium averaging in the approximation of the mean field determined by the effective potential energy of the molecule:

$$U = \frac{3}{2} k_B T b \cos^2 \theta, \tag{3.5.1}$$

where $b = 4.5415$ was determined on the basis of the Maier–Saupe theory. The following equation was obtained in [208] for rotational viscosity γ_1:

$$\gamma_1 = 3\varsigma n f_1(S) = \frac{3nS^2(1+bq/6)}{q+(2+S)S^2(1+bq/6)+6S^2-1.5bS^4}. \tag{3.5.2}$$

The dependence of function $f(S)$ on the order parameter S is slightly weaker than the first degree dependence (Fig. 3.5.1).

The internal friction coefficient is $\varsigma = k_B T/D_\perp$, where $D_\perp$ is the component of the rotational diffusion tensor transverse to the director. The quantity $D_\perp$ is expressed by means of the TCF of the angular speed of the molecule $\omega = d\theta/dt$, using the equation $D_\perp = \int_0^\infty \langle\omega(0)\omega(t)\rangle dt$, which, being expressed through the distribution functions of the second and fourth order taking (3.5.1) into account, has the following form:

$$D_\perp = \frac{2\sqrt{2}}{Z}\frac{(k_B T)^{1/2}}{(3\pi b S I_\perp)^{1/2}}\frac{(\langle\cos^4\theta\rangle - \langle\cos^2\theta\rangle)^{1/2}}{\langle\cos 2\theta\rangle^{3/2}} = \frac{2\sqrt{2}}{m}\left(\frac{k_B T}{3\pi b I_\perp}\right)^{1/2} f_2(S), \tag{3.5.3}$$

where

$$\langle\cos^4\theta\rangle = \frac{2S+1}{3} - \frac{2}{3b};\ \langle\cos^2\theta\rangle = \frac{2S+1}{3};\ \langle\cos 2\theta\rangle = \frac{4S-1}{3}. \tag{3.5.4}$$

Here Z is the fitting parameter containing information on the distribution functions, $I_\perp$ is the moment of inertia of the molecules in relation to the short axis. Taking into account (3.5.3), (3.5.4) at $b = 4.5418$, the dependence is presented in the following form:

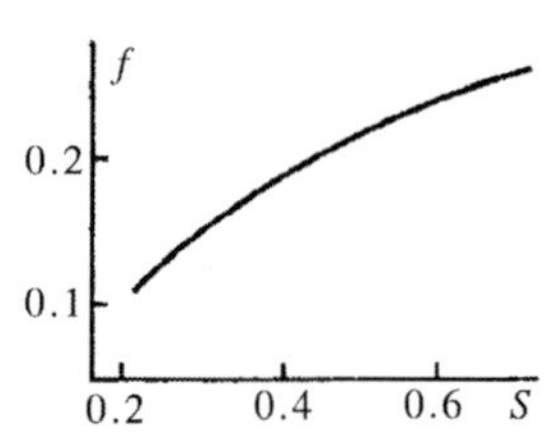

Fig. 3.5.1. Dependence of function $f(S)$ on the order parameter S.

$$\gamma_1(T,S)=\frac{3m(\pi b I_\perp k_B T)^{1/2}}{2\sqrt{2}}\times$$

$$\times\frac{S^2(3.180+0.757S)}{(-5.243S^4+4.694S^3+12.360S^2+S+2.881)}\times$$

$$\frac{S^{1/2}(4S-1)^{3/2}}{(-4S^2+2S+0.679)}\sim f_1(S)f_2(S). \quad (3.5.5)$$

In [210] this approach was developed for determination of the relationship of microscopic friction with the dimensions and form of the molecule, and also for improving the accuracy of the temperature dependence. For the coefficient ς the authors obtained an expression having the thermal activation form with the activation energy E equal to the value of the binary potential of the mean force for two contacting molecules:

$$\varsigma=\frac{I_\perp d^2 n}{2(k+1)}\left(\frac{\pi k_b T}{m}\right)^{1/2}\left(\frac{8}{9}+\frac{84\kappa-40\kappa S/3}{105}\right)\exp\frac{E}{k_B T}, \quad (3.5.6)$$

where m is the mass of the molecule, d is its transverse dimension, $k = 4I_\perp/md^2$, κ is described by equation (3.4.15). In [211], the Enskog methods for the model of the molecule in the form of a solid convex body (ellipsoid of rotation or spherical cylinder) was used to improve the accuracy of the form of the rotational self-diffusion tensor expressed through the TCF of the angular speed of the molecule:

$$D_\perp=\frac{3\gamma D}{1+2\gamma}, \quad (3.5.7)$$

and

$$\gamma=\frac{1-S+(2+S)/\gamma_0}{1+2S+2\gamma_0(1-S)}, \quad \gamma_0=\frac{D_\perp^0}{D_\parallel^0}, \quad (3.5.8)$$

$$D=3\,(4I_\perp)^{-1}(k+1)(m\pi k_B T)^{1/2}(nA\bar{g})^{-1}, \quad (3.5.9)$$

where $D_\perp^0$ and $D_\parallel^0$ are the transverse and longitudinal (in relation to the director) rotational self-diffusion coefficients in the ideally ordered state of the medium, $\bar{g}=\exp(E/k_BT)$ is the mean value of the binary distribution function in respect of the orientations of two contacting molecules, A is the surface area of the excluded volume of the two molecules, i.e., the surface area formed by the centre of gravity of the second molecule at different orientations of this molecule in relation to the first molecule in their contact. It should be mentioned that this approach takes into account the effect of the nearest neighbours of the molecule on the values ς and γ_1. The final equation for the rotational friction coefficient is:

$$\varsigma = \frac{1+2\gamma}{9\gamma}\frac{4I_\perp}{k+1}(\pi m k_B T)^{-1/2} nA \exp\frac{E}{k_B T}. \tag{3.5.10}$$

If the molecule is regarded as the ellipsoid of rotation with the imensions of the axes L and d, and taking into account $I_\perp = m\,(L^2+d^2)/20$ and $p = L/d$, the surface area of the excluding volume is equal to:

$$A = 2\pi d^2\left\{1+\left[p^2(p^2-1)^{1/2}\right] \arcsin\,(1-p^{-2})^{1/2}\right\} \tag{3.5.11}$$

and equation (3.5.10) is converted to the form:

$$\varsigma_{mol} = \frac{2}{9}\left[\frac{1+2\gamma}{\gamma}\right](1+p^2)(p^2-1)^{1/2} \arcsin\,(1-p^{-2})^{1/2} \times$$

$$\left[1-\frac{5}{p^2+6}\right](\pi k_B T m)^{1/2} d^4 n \exp\frac{E}{k_B T}. \tag{3.5.12}$$

For a spherical cylinder with length L and radius d:

$$A = 2\pi d^2(1+L/d), \tag{3.5.13}$$

$$\varsigma_{mol} = \frac{2}{9}\frac{(1+2\gamma)}{\gamma}(1+p)\left[1-\frac{5}{p^2+6}\right]\times$$

$$d^4(\pi k_B T m)^{1/2} n \exp\frac{E}{k_B T}. \tag{3.5.14}$$

The form of the molecules and the mutual distribution of the molecules determine the parameters γ and p included in the equations (3.5.12) and (3.5.14). Quantity $\alpha = (1+2\gamma)/\gamma$ is almost completely independent of p and S. If $S = 0.6$, the value of α changes from 3.15 to 3.44 with the value of p changing from 0 to 10. If $p = 3$, the value of α changes from 3.25 to 3.34 with S changing from 0.6 to 0.7. Therefore, to estimate the values of the coefficient ς, we can use the value $\alpha = 3.25$. The temperature dependence of rotational viscosity γ_1 can be used to calculate the molecular friction coefficient ς and to compare the value with that calculated from equations (3.5.12) or (3.5.14) (see section 4.3).

The theory of the kinetic processes in nematic liquid crystals can be improved further using the results obtained in describing the temperature and density dependence of the viscosity of simple liquids [212–215].

Further improvement of Nemtsov's theory, using the results of the Osipov–Terent'ev theory [184], can be found in [216–218] where calculations were carried out of the rotational diffusion coefficient and the order parameters of the second and fourth degrees using the results of analysis of molecular–dynamic examination taking the real interatomic interactions into account. In the final analysis, the calculated and experimental temperature dependences of rotational viscosity for BF-5 were in satisfactory agreement (up to ±7%). In this case, the ratio $J_0/k_B T$, which usually corresponds to the height of the mean field potential, is assumed to be the adjustable parameter [218]. All the parameters can be calculated from the molecular characteristics, using the Gay–Berne potential [219, 220].

The frequency dependence of the viscosity coefficient was investigated in [221]. The equilibrium state of the nematic liquid crystal is the non-degenerate state with respect to the orientations of the director because of the presence of external fields. In calculation of the viscosity coefficient measured on the basis of the reflection of ultrasound shear waves from the interface of the nematic liquid crystal and a piezocrystal, for example, quartz, the size of the molecular field h_i included in the dynamics equation of the nematic liquid crystal contains not only the traditional contribution for the degenerate nematic crystals $h_i^0 = -\delta f/\delta n_i$, proportional to the second spatial derivatives of the director, but also the additional contribution $h_1' = -B\delta n_i$, where $B = 2S^2 b_3/V$ (V is the volume of the system) are the function of fluctuations of the transverse components of the order parameters $\langle \delta a_{13}^2 \rangle = \langle \delta a_{23}^2 \rangle = T\langle 2b_3 \rangle^{-1}$, $b_3 = \partial F/\partial \upsilon - S^{-1}\, \partial F/\partial u$ is the derivative of free energy in respect of the invariants $u = n_i a_{ij} n_j$ and $v = n_i a_{ik} \times a_{kl} n_l - n_i a_{ik} n_k n_j a_{ji} n_l$. For the three characteristic mutual orientations of the equilibrium director, the speed and the wave vector (see Fig. 2.2.1), the viscosity coefficients are presented in the following form:

$$\eta_A = \eta_A(\omega = 0), \eta_B = \eta_B(\omega = 0) + \gamma_1(\lambda - 1)^2[4(1 + i\omega\tau)]^{-1},$$
$$\eta_C = \eta_C(\omega = 0) + \gamma_1(\lambda + 1)^2[4(1 + i\omega\tau)]^{-1}, \tag{3.5.15}$$

where $\tau = \gamma_1/B$ is the relaxation time of the director, $\lambda = -\gamma_2/\gamma_1$, ω is the frequency of the shear wave. For a degenerate nematic crystal $\tau \to \infty$ and $\eta_B = \eta_C$, i.e. Rapini equality is fulfilled. For non-degenerate nematic liquid crystals this equality is not fulfilled which, however, does not indicate that the Parodi relationship has been violated [7]. The latter is fulfilled in the initial equations describing the speed of reorientation of the director.

3.6. Description of the viscosity of nematic liquid crystals by the molecular dynamics methods

In a number of cases, it is useful to describe a nematic liquid crystal by a simulation system consisting of a relatively small number of particles, interacting together in accordance with a specific law. Consequently, it is possible to examine the behaviour of the system in the extreme conditions, for the arbitrary given dimensions of the specified parameters and the form of the particles, which is not always possible in the experiments. The algorithm for the calculation of these characteristics is presented in [222]. The symmetric tensor of the pressure of the liquid decomposes to the isotropic part (scalar hydrostatic pressure p) and the symmetric anisotropic part with the zero trace $\mathbf{p}$ whose relaxation from the nonequilibrium state $\mathbf{p}(0)$ at time $t = 0$ is described by the correlation function $c(t)$ in accordance with the equation:

$$\mathbf{p}(t) = c(t)\mathbf{p}(0). \tag{3.6.1}$$

The anisotropic part of the pressure tensor is the consequence of the application, to the system at the initial moment of time, of a strain characterised by tensor $\mathbf{e}$ with the strain connected with $\mathbf{p}(0)$ through the shear modulus $G(0)$ by the relationship:

$$\mathbf{p}(0) = -2G(0)\mathbf{e}. \tag{3.6.2}$$

From (3.6.1) and (3.6.2), we obtain:

$$\mathbf{p}(t) = -2G(t)\mathbf{e}, \tag{3.6.3}$$

Here $G(t) = G(0)\,c(t)$ is the relaxation modulus. Consequently, shear viscosity (or, more accurately, its exponential part, see below) is given by the equation:

$$\eta = \int_0^\infty G(t)dt = G(0)\tau. \tag{3.6.4}$$

The effective relaxation time in (3.6.4) is determined by the correlation function:

$$\tau = \int_0^\infty c(t)dt. \tag{3.6.5}$$

In [223] an identical procedure is used for an ellipsoid of rotation described (taking into account the orientation of the longer axis of the ellipsoid parallel to director u_α) by the equation:

$$A_{\mu\nu} = \left[\left(1-\frac{1}{3}A\right)^2\left(1+\frac{2}{3}A\right)^{-1/3}(\delta_{\mu\nu} + A\overline{u_\mu u_\nu})\right], \tag{3.6.6}$$

where A $(-3/2 < A < 3)$ is the non-sphericity parameter. The dimensions of the half-axes of the ellipsoid a and b are linked with A by the expressions:

$$a=\left[\left(1-\frac{1}{3}A\right)\left(1+\frac{2}{3}A\right)^{-1}\right]^{1/3} r_A,$$

$$b=c=\left[\left(1-\frac{1}{3}A\right)^{-1}\left(1+\frac{2}{3}A\right)\right]^{1/6} r_A. \tag{3.6.7}$$

Consequently, the ratio of the half-axes Q and the parameter of non-sphericity A may be expressed through each other:

$$Q=\frac{a}{b}=\left[\left(1-\frac{1}{3}A\right)\left(1+\frac{2}{3}A\right)^{-1}\right]^{1/2};\ A=[3(1-Q^2)\,(1+2Q^2)]^{-1}. \tag{3.6.8}$$

For elongated (rod-shaped) particles $Q > 1$, $A < 0$, and for flattened (disc-shaped) particles $Q < 1$, $A > 0$.

For the nematic liquid crystals, the molecular–dynamic method of calculating the viscosity coefficients η_i and γ_1 in relation to the shape of the molecule in the form of the ellipsoid of rotation with the ratio of the axes $A^{\perp}/A^{\parallel} = Q^2$, $(A^{\perp})^2 A^{\parallel} = 1$, is presented in [224]. Let there be an ensemble of $N = 128$ non-spherical particles with ideal orientation ($S = 1$) enclosed in a right-angled parallelepiped with volume V since the unit volume contains $n = N/V$ particles (in the calculations it is assumed that $n = 0.6$). The interaction potential $\varphi_A(\mathbf{r})$ of these particles in the Kuetta flow, situated at the distance $\mathbf{r}$, is described by the potential of 'soft ellipsoids' $\varphi_A=\varphi_{SE}\equiv r_A^{-12}$ or in the form of the Lenard–Jones potential $\varphi_A = \varphi_{LJP}\equiv 4\,(r_A^{-12} - r_A^{-6})$. The vector $\mathbf{r}_A$ is the affine transformation of vector $\mathbf{r}$ in which the potential φ_A becomes spherical $\varphi_A(\mathbf{r}) = \varphi(\mathbf{r}_A)$. The equipotential surfaces of the oriented non-spherical molecules are ellipsoids, determined by the equation [224]:

$$A_1 r_x^2 + A_2 r_y^2 + A_3 r_z^2 = r_A^2. \tag{3.6.9}$$

In both studies in the calculations it was assumed that $r_A \leqslant 2.5$. For the geometries of the Kuetta flows, denoted by I, II, III and corresponding to the Mesovich experiment (Fig. 1.2.1), with the flow speed $\mathbf{v}$

$$\upsilon_x = \gamma y,\ \upsilon_y = 0, \upsilon_z = 0, \tag{3.6.10}$$

and also for the spherical particles, the calculations are carried out at the shear rate $\gamma = \partial\upsilon_x/\partial_y$ equal to 1/4, 1/8, 1/2 and 1, respectively. Viscosity η_i is determined by dividing the component p_{yz} of the pressure tensor $\overline{\overline{p}}$ by $(-\gamma)$. The pressure tensor $\overline{\overline{p}}$ is the sum of the kinetic and potential contributions calculated from the relationships:

$$V\overline{\overline{p}}^{\,\mathrm{kin}} = m\mathbf{c}_i\mathbf{c}_i,\ V\overline{\overline{p}}^{\,\mathrm{pot}} = \sum_{i,j,i<j} r_{ij}F_{ij}, \tag{3.6.11}$$

where $\mathbf{c}_i$ is the velocity of the particle i, $\mathbf{r}_{ij} = \mathbf{r}_i - \mathbf{r}_j$, F_{ij} is the force with which the particle j acts on the particle i. The motion equations are solved using the uniform shear algorithm used for the examination of the flows in the conventional liquids [225]. For the orientations I and II, the following relationships are fulfilled:

$$\frac{P_{yx} - P_{xy}}{\gamma_I} = \frac{\gamma_1 + \gamma_2}{2},\ \frac{P_{yx} - P_{xy}}{\gamma_{II}} = \frac{\gamma_1 - \gamma_2}{2}, \tag{3.6.12}$$

so that it is possible to calculate the values of γ_1 and γ_2. Table 3.1 gives the kinetic and potential contributions to viscosity η_i (at $Q = 7/3$), and γ_1, γ_2 for different types of potential φ. The notation Sp denotes spherical particles ($Q = 1$).

For all the viscosity coefficients with the exception of η_2, the kinetic contribution is small in comparison with the potential contribution. Instead of the complicated procedure of calculating viscosity on the basis of the interaction of the particles, we can use a simpler method based on the affine transformation used in the conversion of the potential. The potential contributions to the anisotropic viscosity coefficients are proportional to the potential contribution to the viscosity of the system of the spherical particles η^{pot} with coefficient H_i which depends on the ratio of the axes Q [226]:

$$\begin{gathered}
\eta_i^{\mathrm{pot}} = H_i\eta^{\mathrm{pot}};\ H_1 = Q^{-2},\ H_2 = Q^2,\ H_3 = 1, \\
\gamma_1 = G_i\eta^{\mathrm{pot}};\ G_1 = (Q - Q^{-1})^2,\ G_2 = Q^{-2} - Q^2, \\
\alpha_4 = 2\eta^{\mathrm{pot}},\ \alpha_1 = -0.5\,(Q - Q^{-1})\ \alpha_4, \\
\alpha_2 = -\alpha_5 = 0.5\,(1 - Q^2)\ \alpha_4, \\
\alpha_3 = \alpha_6 = 0.5\,(Q^{-2} - 1)\ \alpha_4
\end{gathered} \tag{3.6.13}$$

This is in agreement with the earlier conclusions by Helfrich [227]. The dependences of the coefficients η_i, γ_1, γ_2 on Q for the rod- and disc-shaped

Table 3.1

φ	I		II		III		Sp		γ_1	γ_2
	kin	pot	kin	pot	kin	pot	kin	pot		
E	0.03	0.09	0.04	2.8	0.04	0.52	0.04	0.52	1.9	–2.7
LDP	0.08	0.15	0.15	3.5	0.12	0.70	0.15	0.73	2.2	–3.5

particles are presented in Fig. 3.6.1 [123]. For the former, the inequalities $\eta_1 > \eta_3 > \eta_2$, $\gamma_1 > |\gamma_2|$ are fulfilled and for the latter $\eta_2 > \eta_3 > \eta_1$, $\gamma_1 < |\gamma_2|$. The identical relationship of the Mesovich coefficient η_i with the viscosity corresponding to the isotropic phase is also shown in (3.4 .49). In accordance with the data in Table 3.2, where the viscosity η_i^*, γ_i^* values, normalised for the mean viscosity $\bar{\eta} = (\eta_1+\eta_2+\eta_3)/3$ are presented, this calculation method gives the results which differ only slightly from those obtained using the Lenard–Jones potential or r^{-12}.

Errors of the calculations using the Mesovich and Lenard–Jones potentials are associated with the fluctuations of the correlation function from zero at relatively high relaxation times (Fig. 3.6.2) [223]. The accuracy of calculation of viscosity η_i by this method is 10–50%. In the calculation of the viscosity coefficients, the shear rate Γ was assumed to be very low. It was shown in [128] that in a relatively wide range of the values of Γ, the viscosity of the disc-shaped nematic liquid crystals is constant (Fig. 3.6.3).

The extent to which the experimental data are in agreement with the dependences of the viscosity coefficients on temperature and the molecular structure, presented in this chapter, is discussed in the following chapter.

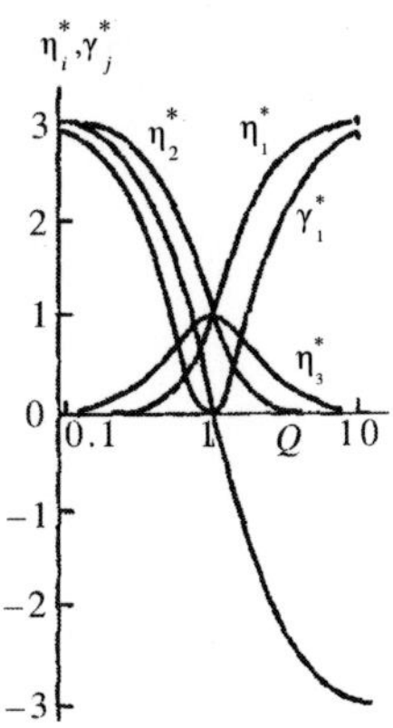

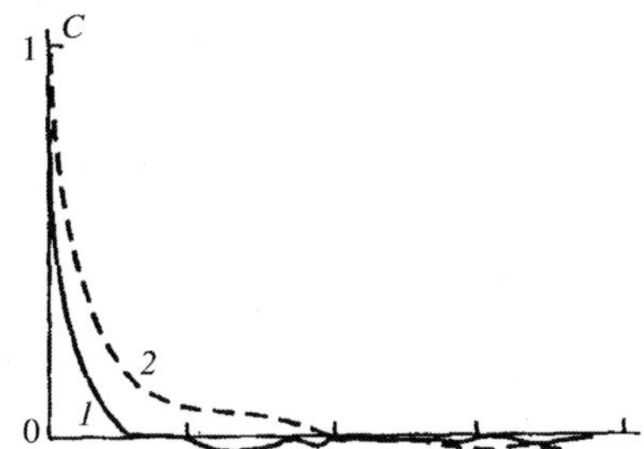

Fig. 3.6.1. Dependence of the viscosity coefficients of viscosity on the ratio of the length of the axes of the particles Q obtained by the method of nonequilibrium molecular dynamics. The values of viscosity were normalised for $\eta^{pot} = (\eta_1^{pot} + \eta_2^{pot} + \eta_3^{pot})/3)$.

Fig. 3.6.2 (right). The time correlation function C (t) for the liquid with the potential of international the particles in the form of the Lenard–Jones potential (l) or the potential of rigid ellipsoids (2) [224, 228].

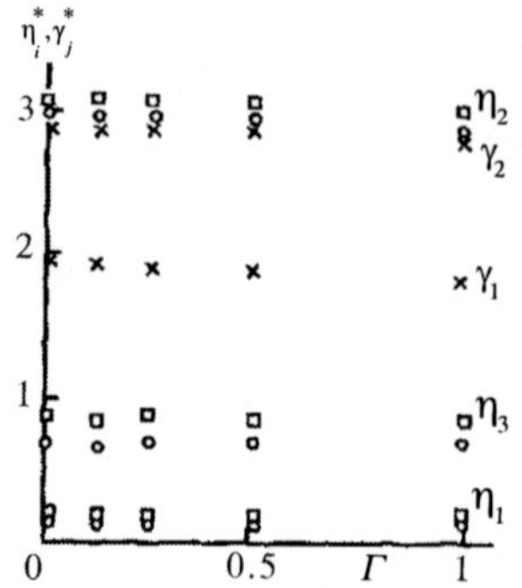

Fig. 3.6.3. Dependence of the rotational viscosities γ_1 and γ_2 (×), tMesovich viscosity η_i (□) and of their potential part η_i^{pot} (o) on the shear rate $\Gamma = dV_x/dy$ for disc-shaped particles with the ratio of the axes Q = 3/7 [228].

Table 3.2

	η_1*	η_2*	η_3*	γ_1	γ_2
Sp POT	2.47	0.08	0.45	1.65	2.37
ME	2.6±0.3	0.08±0.02	0.48±0.1	1.79±0.4	2.5±0.3
LDP	2.33±0.3	0.015±0.07	0.52±0.1	1.40±0.2	2.08±0.3
N4 (50°C)	2.14	0.32	0.54	1.75	-
Esters (40°C)	2.08	0.37	0.56	1.62	-
MBBA	1.99	0.38	0.64	1.55	-

Chapter 4

Dependence of the viscosity of nematic liquid crystals on temperature and pressure

This chapter contains the experimental results, mainly for rotational viscosity, obtained by the methods described in chapter 2. It was shown in section 4.1 that the experimental data can be approximated by different functions; it has been attempted to determine the physical meaning of the parameters present in these approximations, especially energy parameters. In sectioN 4.2, restrictions are imposed on the applicability of the activation model, and it is shown that when using the increment method for calculating the glass transition temperature of the nematic liquid crystals, good experimental results are obtained using the free volume model. In section 4.3 the physical principles of the series, described in chapter 3, are discussed more extensively than in 4.1. The results are used for the evaluation of the duration of reorientation of the molecules or clusters of nematic liquid crystals, rotational diffusion, molecular friction. Although special attention is given to the temperature dependence of viscosity and parameters proportional to it, to explain many processes it is necessary to use the special features of the molecular structure of the nematic liquid crystal, concentration dependences of the characteristics of the mixtures of the nematic liquid crystals which, generally speaking, is the subject of chapters 5 and 6. In section 4.4, attention is given to the dependence of viscosity on pressure. Because of the small number of studies in this area associated with the experimental difficulties, in this section special attention is given to the experimental procedures, the results and theoretical discussion. In section 4.5, the dependence of the Mesovich and Leslie coefficients on temperature and molecular structure is reviewed, and the discussion of the results is supported by the conclusions in chapter

4.1. Activation nature of the temperature dependence of rotational viscosity

The typical temperature dependences of the measured viscosity coefficients η_i, γ_1 are presented in Fig. 1.2.2. The majority of these relationships have a distinctive activation nature in the entire nematic range, with the exception of the areas in the vicinity of clarification temperature T_{NI} and, in some cases, the region of reduced temperatures (below 20°C). In [153] the method of least squares is used to examine the accuracy of the description of the experimental data by the approximations of the following type:

$$a)\ln\frac{\gamma_1}{\Delta n} = A_1 + \frac{B_1}{T}$$

$$b)\ln\frac{\gamma_1}{(\Delta n)^2} = A_2 + \frac{B_2}{T} \qquad (4.1.1)$$

$$a)\ln\frac{\gamma_1}{\Delta n} = A_3 + \frac{B_3}{T - T_{03}}.$$

$$b)\ln\frac{\gamma_1}{(\Delta n)^2} = A_4 + \frac{B_4}{T - T_{04}}. \qquad (4.1.2)$$

In these approximations, S was replaced by the birefringence Δn measured in the same experiment as γ, because the ratio $S/\Delta n$ is constant in almost the entire nematic range and only slightly increases (by 2–3%) in the vicinity of T_{NI}. The parameters of approximations and the smallest sums of the square for every pair of the approximation are presented in Tables 4.1 and 4.2. It should be mentioned that in a number of cases the approximations (4.1.2) proved to be unsuitable because the values of T_{03} or T_{04} were located in the examined nematic range or above this range.

The data presented in Tables 4.1 and 4.2 show that the simple activation approximations (4.1.1) give the smallest sum of the squares for the majority of the investigated substances. For the substances with molecules characterised by elongated bond chains, $\gamma_1 \sim \Delta n \sim S^1$. For the substances whose molecules contain a fragment with saturated bonds, $\gamma_1 \sim (\Delta n)^2 \sim S^2$ (Fig. 4.1.1). The replacement of the fragment with non-saturated bonds results in a different type of the potential of intermolecular interaction and, consequently, in higher values of T_{NI} and lower values of activation energy E. This was reported in [172, 229]. For the first group of the substances, with the exception of azoxy compounds and alkyl benzoic acids, $B_1 = E = 5100 \div 5800$ K, for the second group, $B_2 = E = 2700 \div 3700$ K.

For the mixtures of the substances containing and not containing the

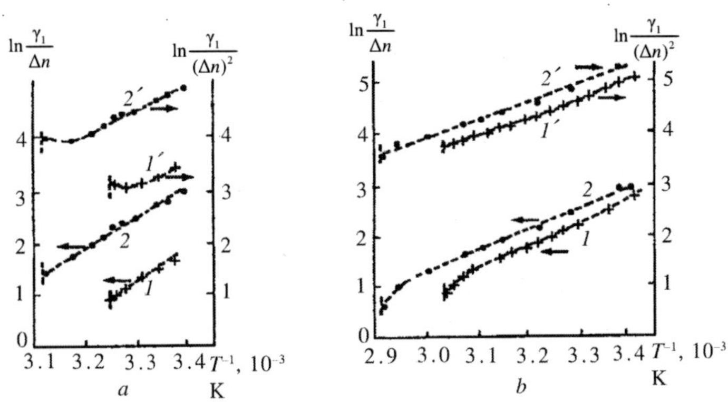

Fig.4.1.1. Dependence of $\ln(\gamma_1/\Delta n)$ (*1,2*) and $\ln[\gamma_1/(\Delta n)^2]$ (*1', 2'*) on reciprocal temperature: *a*) BF-5 (*1, 1'*) and mixtures of phenyl esters of alkylbenzoic acid (*2,2'*); *b*) FTsG-7 (*1,1'*) and the mixture of phenyl esters of alkyl cyclohexane carbonic acid (*2,2'*) [25]

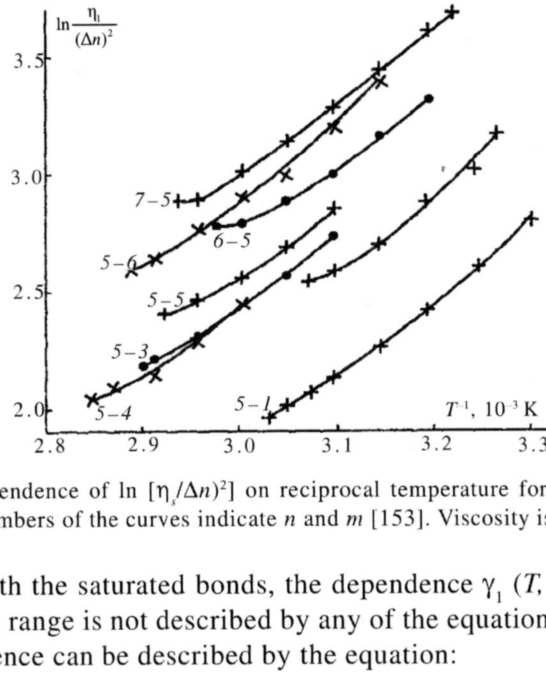

Fig.4.1.2. Dependence of $\ln[\eta_s/\Delta n)^2]$ on reciprocal temperature for tolane derivatives *n*TO*m*. The numbers of the curves indicate *n* and *m* [153]. Viscosity is in P.

fragment with the saturated bonds, the dependence γ_1 (T, S) in a specific concetration range is not described by any of the equations (4.1.1) so that this dependence can be described by the equation:

$$\gamma_1(T,S) = x_1 b_1 S \exp\frac{E_1}{k_b T} + x_2 b_2 S_2 \exp\frac{E_2}{k_b T}, \qquad (4.1.3)$$

Here x_1, x_2 are the molar fractions of the components. It is important to note the agreement between (4.1.3) and (3.1.1) and it will be shown that there is a different approach to describing the concentration dependence of viscosity.

Figure 4.1.2 shows the dependences of the logarithm of the ratio $\gamma_1/(\Delta n)^2$ on the reciprocal temperature for the members of the homologous

Table 4.1. Approximation parameters (4.1.1.) – (4.1.2.) for LNC with the elongated molecular bonding chain

Substance	Structural formula	Molar fraction	T_{ni}, K	Δ_n, $T = 0.95T_{ni}$	γ_1, P_3 $T = 25°C$	A_1	$B_1 = E/k$, K
Alkyl cyanobiphenyls (BF)	C_5H_{11}-B-B-CN C_7H_{15}-B-B-CN	0.4 0.6	312.2	0.185	1.10	-17.1	5640
a) BF-5			308.2	0.195	0.82	-20.0	5840
b) BF-8			313.7	0.153	1.11	-15.8	5310
			$T = T_{SN} = 307.2$K				
Alcoxy cyanobiphenyls (OBP)	C_5H_{11}O-B-B-CN C_7H_{15}O-B-B-CN	0.4 0.6	345.2	0.177	2.22 (304.2 K)	-14.6	5150
Phenyl pyridines (PPr)	C_5H_{11}-Pr-B-B-CN C_7H_{15}-Pr-B-B-CN	0.4 0.6	317.4	0.173	1.84	-16.0	5400
Phenyl pyrimidines (PPm)	C_5H_{11}-Pm-B-B-CN C_7H_{15}-Pm-B-B-CN	0.4 0.6	324.4	0.174	1.91	-14.9	5110
Phenyl benzoates (PEBK)	C_4H_9-B-B-B-OC_2H_5 C_6H_{13}-B-B-B-OC_2H_5 C_6H_{13}-B-B-B-CN	0.4545 0.4545 0.091	320.8	0.126	2.20	-16.7	5820
Azoxy compounds (AB)	C_4H_9-B-N(O)=N-B-- OCH_3 C_4H_9-B-N(O)=N-B-- OCOC_6H_{13}	0.667 0.333	345.2	0.189	1.95	-11.6	4160
Alkyl benzoic acids (BA)	C_4H_9-B-COOH C_6H_{13}-B-COOH	0.5 0.5	381.8	0.120	0.36 (345 K)	-11.9	4460

Table 4.1. *(continued)*

Substance	Structural formula	CHK_1	CHK_2	CHK_3	T_{03}, K	CHK_4	T_{04}, K
Alkyl cyanobiphenyls (BF)	C_5H_{11}-B-B-CN C_7H_{15}-B-B-CN	0.0097	0.024	1.09	323.2	0.019	314.9
a) BF-5		0.038	0.060	0.745	311.5	0.320	289.7
b) BF-8							
Alcoxy cyanobiphenyls (OBP)	C_5H_{11}O-B-B-CN C_7H_{15}O-B-B-CN	0.00054	0.0024	0.00055	99.5	0.00015	205.8
Phenyl pyridines (PPr)	$C_5H_1$1-Pr-B-B-CN $C_7H_1$5-Pr-B-B-CN	0.0026	0.011	0.0013	393.5	0.0314	291.9
Phenyl pyrimidines (PPm)	C_5H_{11}-Pm-B-B-CN C_7H_{15}-Pm-B-B-CN	0.0035	0.097	103	301.2	0.0043	280.7
Phenyl benzoates (PEBK)	C_4H_9-B-COO-B-OC_2H_5 C_6H_{13}-B-COO-B-OC_2H_5 C_6H_{13}-B-COO-B-CN	0.027	0.060	0.519	331.5	10.3	296.0
Azoxycompounds (AB)	C_4H_9-B-N(O)=N-B-OC- H_3 C_4H_9-B-N(O)=N-B-OC- OC6H_{13}	0.016		0.010	249.4		
Alkyl benzoic acids (BA)	C_4H_9-B-COOH C_6H_{13}-B-COOH	0.127		0.033	395.8		

Table 4.2. Approximation parameters (4.1.1.) – (4.1.2.) for NLC containing a fragment of saturated bonds

Substance	Structural formula	Molar fraction	T_{ni}, K	Δn, $T = 0.95T_{ni}$	γ_1, P_3 $T = 25$ °C	A_1	$B_1 = E/k$, K
Phenyl cyclohexanes (PCH)	C_5H_{11}-Cr-B-CN C_7H_{15}-Cr-B-CN	0.4 0.6	329.6	0.121	1.02	-8.0	3680
a) PCH-7			330.2	0.102	1.40 (1.09)**	-10.05	3250 (2920)**
b) PCH-5			328.1	0.149	0.97	-5.8	2760
Phenyl cyclohexyl carboxylates (PECHCC)	C_4H_9-Cr-COO-B-O-C_2H_5 C_6H_{13}-Cr-COO-B--OC_2H_5 C_6H_{13}-Cr-COO-B--CN	0.4545 0.4545 0.091	346.8	0.073	1.34	-6.4	3450
Alkyl cyclohexane carbonic acids (CHCA)	C_4H_9-Cr-B-COOH C_6H_{13}-Cr-B-COOH	0.5 0.5	366.2	0.042	1.85	-3.4	2970
Phenyl bicycloctanes (BCO)	C_3H_7-PBO-B-CN C_5H_{11}-PBO-B-CN	0.4 0.6	367.7	0.118	2.96	-7.8	3750

Table 4.2 *(continued)*

Substance	Structural formula	CHK	CHK	CHK	T_0K	CHK	T_0,K
Phenyl cyclohexanes (PCH)	C_5H_{11}-Cr-B-CN C_7H_{15}-Cr-B-CN	0.238	0.049	0.175	334.9	112	322.2
PCH-7			0.010*			0.0038	207.8
PCH-5			0.0092	0.0035	247.4	0.0102	286.6
Phenyl cyclohexyl carboxylates (PECHCC)	C_4H_9-Cr-COO-B-OC_2H_5 C_6H_{13}-Cr-COO-B-OC_2H_5 C_6H_{13}-B-COO-B-CN	0.044	0.038			0.466	273.9
Alkyl cyclohexane carbonic acids (CHCA)	C_4H_9-Cr-B-COOH C_6H_{13}-Cr-B-COOH	0.103	0.098	5.45	291.6	8.04	293.5
Phenyl bicycloctanes (BCO)	C_3H_7-PBO-B-CN C_5H_{11}-PBO-B-CN		0.041			0.0030	198.8

* Sum of the least squares calculated for approximation ln $(\gamma 1/(\Delta n)^2) = A + B/T + D/(T - T_0)$

* Two measurement series

series of the derivatives of (C_nH_{2n+1}–C_6H_4–C≡C–C_6H_4–OC_mH_{2m+1} (nTOm, $n - m$)). For the homologs with the sum $n + m \leqslant 8$ the relationship $\gamma_1 \sim S^2$ is fulfilled, for the homologs with the higher value of the alkyl substituents of the exponent at S decreases [153].

The difference in the values of exponent S in front of the exponent for different groups of substances is also explained by the fact that the approximations, discussed in chapter 3, are in fact expansions with respect of the sizeable parameter ($S > 0.3$). This was reported in [230] in explaining the difference between the theoretical and experimental dependences of the flexocoefficient on the order parameter. In [116, 186] $\alpha_i \sim S^2$, if the potential $U = JP_2(\cos\theta) \sim \varepsilon S$, but the situation in which $\alpha_i \sim S$ is also possible, if $U \sim P_1(\cos\theta)$ which is observed in the case of dimerisation of the cyano derivative molecules with the saturated bonds [2] or for the non-symmetric molecules (see (3.3.12)).

For a number of substances from the classes of the azoxy compounds and Shiff bases, investigations were carried out to determine the correlation of the approximations corresponding to theoretically described dependences $\gamma_1(T, S)$, presented in chapter 3 [231] with the experimental results:

$$\ln\frac{\gamma_1}{S} = A_1 + \frac{B_1}{T}, \tag{4. 1.4}$$

$$\ln\frac{\gamma_1}{S^2} = A_2 + \frac{B_2}{T}, \tag{4.1.5}$$

$$\ln\frac{\gamma_1}{S} = A_3 + \frac{B_3}{T} + (x_3 - 1)\ln S, \tag{4.1.6}$$

$$\ln\frac{\gamma_1}{S} = A_4 + \frac{B_4 S}{T}, \tag{4.1.7}$$

$$\ln\frac{\gamma_1}{S^2} = A_5 + \frac{B_5 S}{T}, \tag{4.1.8}$$

$$\ln\frac{\gamma_1}{S^2} = A_6 + \ln\frac{f_1(S)}{S^2} + \ln\rho + \frac{B_6}{T}, \tag{4.1.9}$$

$$\ln\frac{\gamma_1}{S^2} = A_7 + 4\ln T - 1.5\ln\, S + \ln,\, \rho + \frac{B_7}{T}, \tag{4.1.10}$$

$$\ln\frac{\gamma_1}{S^2} = A_8 + \ln\rho\text{-}\ln\left[5S + 16\langle P_4(\cos\theta)\rangle + 14\right] + \frac{B_8}{T}. \tag{4.1.11}$$

For MBBA, OH-MBBA and the members of the homologous series 4-methoxy benzylidene-4'-alkanoiloxy aniline (MBnOA, denoted in the literature by APAPA), the date for ρ and S were taken from [232], and for the ratio γ_1/K_{22} from [233], for the elasticity coefficients from [234], for 4, 4'-dialkyl azoxy benzols (nAB) ρ and S from [235], γ_1 from [236], for

4-pentyl-4'-cyano biphenyl (BF-5) S from [237], ρ from [238] and γ_1 from [24].

The approximations (4.1.4) and (4.1.5) are identical to (4.1.1); equation (4.1.6) is derived for calculating the exponent at the order parameter giving the smallest sum of the squares; (4.1.7) and (4.1.8) are the variants of description of γ_1 (T,S) according to Diogo and Martins (see 3.1.7); (4.1.9) according to Nemtsov (see 3.5.2) taking into account (3.5.7); (4.1.10) according to Osipov and Terent'ev (see 3.4.42) and (3.4.43), the identical approach by Kudzu and Doy (3.4.21); (4.1.11) according to Stepanov (see 3.4.29), with the approach used by Tsebers (3.4.32) identical to this one taking into account (3.4.30). The comparative processing of the experimental results for a number of substances according to Diogo and Martins, Osipov and Terent'ev was carried out in [239].

Table 4.3 gives the parameters of the approximations (4.1.4)–(4.1.11) A_i and B_i, and also the values of the sum of the least-squares $S_i = \sum (y_j - A_i - B_i x_j)^2$, characterising their linearity, and the correlation coefficient R_i, $y = \ln (\gamma_1/S)$ or $\ln (\gamma_1/S^2)$, $x = T^{-1}$ or S/T.

For the majority of the parameters A_i, B_i, the deviations are specified in the ranges corresponding to the confidence range $P > 0.95$. The same order of magnitude B as the one detected in previous approximations is also found in the case of (4.1.9) and (4.1.11). The largest deviation from linearity is obtained for the relationship (4.1.10). This is associated with the only slight dependence on S in the formula (3.4.43) ($\gamma_1 \sim S^{1/2}$). Processing of the data using equation (4.1.10) gives the highest value of activation energy ($B_7 \sim 10^4$ K) which corresponds to the actual value of the isotropic part of the energy of intermolecular interaction $G \approx (1/3)\, B_7 \approx 3000$ K.

Thus, almost any of the theoretical dependences described in chapter 3 is suitable for the mathematical description of the experimental results obtained for rotational viscosity. Therefore, to calculate the temperature dependence of rotational viscosity, it is convenient to use the simple approximations (4.1.4) or (4.1.5) which are also suitable when there are no data on the order parameter S and there are only the data on the quantities ($\Delta\chi$, Δn, etc) proportional to the order parameter. This approach is used in compiling the tables of viscosity of the nematic liquid crystals [240].

In [168, 186] the activation energy is interpreted as the height of the potential barrier preventing the rotation of the molecular particles of the angle of 180° (see (3.1.4)). Which physical quantity is this barrier?

It was shown in [241] that the activation energy of *para*-azoxyanisole and some derivatives of cholesterine is proportional to surface tension $\sigma = H - sT$, where H is the surface part of the potential energy, s is surface entropy.

In rapid cooling to low temperatures (approximately 200 K), the nematic liquid crystals undergo glass transition. It is well known that in the glass-forming liquids the activation energy of the viscous flow is comparable with the melting enthalpy [242]. Therefore, this approach may also be used for the nematic liquid crystals. If we compare the data on activation energy E and melting enthalpy ΔH_m for a number of nematic liquid crystals (Table 4.4) it may be seen that for the substances whose molecules have a fragment with saturated bonds (cyclohexane), $E \approx \Delta H_m$. For the nematic liquid crystals in which the molecules are characterised by the elongated chain of connection, the relationship $E \approx 1.5 \div 2\Delta H_m$ is fulfilled, i.e., in the nematic liquid crystals differing in the type of intermolecular interaction, the activation energy is characterised by a different relationship with melting enthalpy. However, this is not valid for the entire nematic range. Figure 4.1.3 shows that for the substances of the first type the temperature dependence of γ_1 can be divided into two sections with different activation energies: in the case of FCG-7 $E_1 = 3250$ K, $E_2 = 4300$ K, in bicyclooctanes $E_1 = 3750$ K, $E_2 = 6060$ K, i.e., in the vicinity of the melting point for the majority of the nematic liquid crystals $E \approx 1.5 \div 2\Delta H_m$.

The solution of the equation of state of the nematic liquid crystals shows that the activation energy of rotational viscosity is close in value to the orientation part of interaction of the molecules. According to [243] the

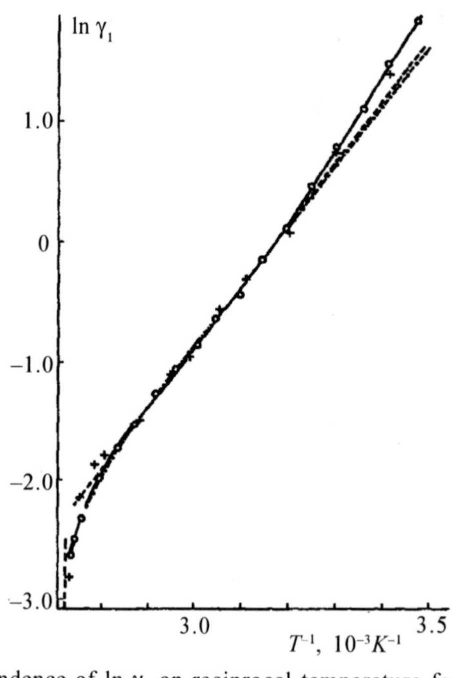

Fig.4.1.3. Dependence of ln γ_1 on reciprocal temperature for BCO-3 and BCO-5; γ_1 is taken from P.

Table 4.3. Approximation parameters (4.1.3.)–(4.1.11) of the temperature dependence of the rotational viscosity of azoxy compounds and Schiff basis in the confidence range P = 0.95 (S_i–the sum of least squares, R_i – correlation coefficient, γ_1 expressed in P, B_i in K)

No	Substance	A_1	B_1	S_1	R_1	A_2	B_2	S_2	R_2
1	MBBA	-18.92±2.79	5819±854	0.0315	0.9919	-12.66±0.88	4108±270	0.00315	0.9984
2	N4(BMAOB)	-14.46±0.77	4550±275	0.0349	0.9978	-10.68±0.55	3511±177	0.0178	0.9981
3	AB-4	-21.80±7.55	6503±2254	0.0133	0.9703	-12.5±6.8	3980±2039	0.0109	0.9382
4	AB-5	-13.65±1.87	4256±611	0.0691	0.9849	-9.12±0.98	2951±318	0.0188	0.9914
5	AB-6	-15.52±1.90	4758±601	0.0167	0.9921	-9.26±1.06	2951±336	0.00522	0.9936
6	AB-7	-17.18±5.90	5498±1983	0.0678	0.9406	-8.78±5.85	2821±1966	0.0667	0.8202
7	MB10A	-17.61±3.07	6261±1143	0.0638	0.9719	-10.64±2.01	3910±749	0.0274	0.9693
8	MB20A	-17.14±2.56	5965±940	0.243	0.9729	-10.98±1.26	3958±461	0.0582	0.9853
9	MB30A	-14.58±0.82	5032±298	0.0434	0.9960	-9.41±1.68	3358±615	0.185	0.9639
10	MB40A	-15.31±0.91	5280±324	0.0238	0.9962	-9.09±1.04	3276±372	0.0313	0.9873
11	MB50A	-17.44±1.52	6116±558	0.0104	0.9928	-7.73±0.88	2768±325	0.00354	0.9881
12	MB90A	-23.71±3.81	8417±1381	0.0613	0.9803	-12.21±1.80	4497±652	0.0137	0.9845
13	OH-MMB	-22.37±2.37	7349±774	0.0380	0.9917	-14.38±1.87	4947±609	0.0235	0.9888
14	BP-5	-18.48	5617	0.0048	0.9943	-4.649	1645	0.0091	0.8923

Table 4.3. *(continued)*

substance	$-A_3$	B_3	X_3	S_3	R_3	$-A_4$	B_4	S_4	R_4
MBBA	11.52±3.13	3795±866	2.183±0.477	0.00243	0.9994	2.797±0.344	1690±200	0.0205	0.9948
N4(BMAOB)	12.18±0.88	3922±248	2.604±0.218	0.0049	0.9997	3.671±0.495	1801±261	0.241	0.9846
AB-4						2.301±0.716	1549±482	0.0108	0.9758
AB-5	7.77±3.05	2566±891	2.296±0.636	0.0160	0.9965	3.123±0.247	1383±133	0.0317	0.9931
AB-6	9.50±4.82	3020±1401	1.962±0.745	0.00521	0.9987	2.666±0.222	1183±120	0.0108	0.9949
AB-7	12.47	3996	2.56	0.0648	0.9434	2.602±0.657	917±335	0.0693	0.9393
MB10A						3.106±0.240	1652±170	0.0210	0.9909
MB20A						3.493±0.152	1862±107	0.0336	0.9963
MB30A						3.159±0.482	1479±294	0.453	0.9578
MB40A						2.743±0.286	1423±175	0.0934	0.9851
MB50A						2.514±0.140	1138±95	0.0088	0.9939
MB90A						2.433±0.236	1365±164	0.0336	0.9893
OH-MMB						2.606±0.428	1665±258	0.0802	0.9825
BP-5						1.245	797	0.0082	0.9902

Table 4.3. (*continued*)

No	substance	R_4	$-A_5$	B_5	S_5	R_5	$-A_6$	B_6	S_6	R_6
1	MBBA	0.9948	1.241±0.460	1169±267	0.0367	0.9808				
2	N4(BMAOB)	0.9846	2.289±0.623	1352±328	0.382	0.9584				
3	AB-4	0.9758	0.593	950±457	0.0097	0.9450	19.97	6353	0.0104	0.9973
4	AB-5	0.9931	1.773±0.312	937±169	0.0507	0.9765	11.75	4146	0.0523	0.9878
5	AB-6	0.9949	1.258±0.295	719±159	0.0191	0.9763	13.77	4676	0.0135	0.9934
6	AB-7	0.9393	1.266±0.689	455±351	0.0762	0.7913	12.61	4650	0.0676	0.9198
7	MB10A	0.9909	1.556±0.252	1017±178	0.0232	0.9740	16.05	6178	0.0551	0.9750
8	MB20A	0.9963	1.885±0.189	1209±132	0.0515	0.9867	15.29	5783	0.1793	0.9785
9	MB30A	0.9578	1.682±0.591	915±361	0.682	0.8535	13.22	5026	0.0453	0.9959
10	MB40A	0.9851	1.218±0.389	839±238	0.173	0.9277	13.81	5229	0.0158	0.9974
11	MB50A	0.9939	0.949	479	0.0202	0.9299	15.89	6339	0.0061	0.9957
12	MB90A	0.9893	0.772±0.327	679±228	0.0645	0.9246	20.81	8211	0.0042	0.9857
13	OH-MMB	0.9825	1.001±0.525	1077±315	0.120	0.8595	20.50	7757	0.0269	0.9939
14	BP-5	0.9902	0.4495	203	0.0179	0.7740	16.02	5278	0.0012	0.9983

Table 4.3. (*continued*)

No	$-A_7$	B_7	S_7	R_7	$-A_8$	B_8	C_8	S_8	R_8	$-A_9$	B_9	S_9
1												
2												
3	53.43	9019	0.0136	0.9837	25.32	-443	2245	0.1070	0.991	11.52	4483	0.0085
4	42.92	6199	0.1221	0.9837	32.95	2155	1649	0.0177	0.986			
5	45.60	6927	0.0322	0.9927	34.47	2797	1212	0.0031	0.991			
6	48.39	8202	0.0765	0.9684	37.74	4168	1304	0.0665	0.968			
7	13.17	8920	0.0947	0.9792	30.43	1004	2094	0.0211	0.979			
8	13.06	8405	0.6959	0.9776	31.65	1365	2205	0.0321	0.978			
9					41.05	5667	522	0.0425	0.949			
10	45.90	7696	0.0751	0.9944	38.75	4672	836	0.0120	0.976			
11	49.95	9267	0.0327	0.9901	36.17	3604	1067	0.00044	0.990			
12	56.99	11812	0.1368	0.9778	37.24	3970	1295	0.0106	0.978			
13	51.34	9999	0.0818	0.9900	40.21	4886	1150	0.0132	0.982			
14	52.25	8811	0.0202	0.9903						4.142	2341	0.0244

Table 4.4. Activation energy E of rotational viscosity and melting enthalpy ΔH_m of a number of NLC (in kJ/mol)

Substance	E	ΔH_m
4-alkyl-4'-cyano biphenyls (nBP)		
5BP 40% mol, 7BP 60% mol	46.9	22.41
5BP	48.5	17.14
8BP	44.1	22.15
4-alkoxy-4'-cyano biphenyls (nOBP)		
50BP 40% mol, 70BP 60% mol	42.8	28.84
5-(4'-cyano phenyl)-2-alkylpyridines (nPPr)		
5PPr 40% mol, 7PPr 60% mol	36.4	23.66
5PPr	29.1	20.92
6PPr	47.8	25.73
7PPr	38.2	25.50
4-alkyl-(4'-cyano phenyll)-cyclohexanes (nPCH)		
5PCH 40% mol, 7PCH 60% mol	22.9	19.6
5PCH	27.1	25.5
4-ethoxyphenylester of 4'-butyl cyclo hexa carbonic acid	28.7	28.9
4- alkyl-(4'-cyano phenyl)-bicyclo [2.2.2] octanes	31.2	19.36

interaction energy of the molecules with the environment in the unit volume is given by the relationship:

$$V = -J_1 n - J_2 n P_2(\mathbf{e}\cdot \mathrm{n}), \qquad (4.1.12)$$

where **e** is the direction of the longer axis of the molecule, n is the direction of the director, $n = k_p/V_0$ is the number of molecules in the unit volume, $k_p = N_a V_0/V$ is the coefficient of molecular packing of the nematic liquid crystals, V_0 is the van den Waals volume of the molecule, N_a is the Avogadro number, $V = M/\rho$ is the molar volume of the nematic liquid crystal, M is the molecular mass, ρ is the density of the nematic liquid crystal. Quantity J_1 is the isotropic part of intermolecular interaction, J_2 is the anisotropic part, responsible for the orientation of the molecules of the nematic liquid crystals. The values of J_1, J_2 can be determined by solving the equations of state of the nematic liquid crystals [243]:

$$p_N = p_I, g_N = g_I, S = \int P_2(\mathbf{e}, n) f(\mathbf{e})\, d\mathbf{e}, \qquad (4.1.13)$$

where p is the pressure of the NLC, g is the chemical potential, $f(\mathbf{e})$ is the distribution function, the symbols N and I relate to the nematic and isotropic phases, respectively. The values p, g and S are taken at the temperature T_{NI}. In [243], the pressure p and the chemical potential g are expressed by the packing coefficient k_p and the ratio $x = L/d$, where L and d is the length and the width of the spheroidal cylinder, approximating the form of the molecule:

$$pv_0^2 = -\frac{1}{2}J_1k_p^2 - \frac{1}{2}J_2k_p^2 + k_pV_0k_BT\frac{1+\tau+\tau^2-\tau^3}{(1-\tau)^3}, \qquad (4.1.14)$$

$$\frac{g}{k_BT} = 3\ln\left[h(2\pi mk_BT)^{1/2}\right] + \ln\frac{k_p}{V_0} - 4 - \frac{J_1k_p}{V_0k_BT} + \\ + \frac{15\pi}{16}k_p\gamma\frac{2-\tau}{(1-\tau)^3}S^2 + \frac{4-4\tau+3\tau^2-\tau^3}{(1-\tau)^3} - \ln Z, \qquad (4.1.15)$$

where $\tau = k_p + (3/2)\,\gamma k_p\,[\pi/4 - (5\pi/32)\,S^2]$, $\gamma = 2(x-1)^2 / [\pi(3x-1)]$, h is the Planck constant, k_B is the Boltzmann constant, m is the mass of the molecule, V_0 is the van der Waals volume of the molecule, Z is the normalisation constant of the distribution function determined from the condition $\int f(\mathbf{e})d\mathbf{e}=1$. Since the distribution function in the nematic phase depends on S:

$$f(\mathbf{e}) = \frac{1}{4\pi Z}\exp\left[\left[\frac{15\pi k_p\gamma(2-\tau)}{16(1-\tau)^3} + \frac{J_2k_p}{V_0k_BT}\right]SP_2(\mathbf{en})\right], \qquad (4.1.16)$$

then in contrast to [243] it is convenient to solve the system of equations of state of the nematic liquid crystal without the equation for the order parameter S, i.e. define its value at temperature T_{NI} in the nematic phase (evidently, in the isotropic phase $S = 0$) determined in the experiments. The solution of the system of equations are the values J_1 and J_2 determined by varying the parameter x and they correspond to the minimum of the potential (4.1.12). Table 4.5 gives the values of J_1/k_BV_0 and J_2/k_BV_0 for a number of nematic liquid crystals, the values x corresponding to the solution, and also the values of k_N, k_I and S ($T = T_{NI}$) used in the calculations [244]. For comparison, the table also gives the results for the PAA, obtained in solving the complete system of equations (4.1.13) [243]. For all substances $J_1 \times (k_BV_0)^{-1} \approx 5{\div}9{\cdot}10^4$ K. This value is in agreement with the energy of the paired intermolecular interaction $E = 4.7{\cdot}10^5$ J/mole $= 5.7{\cdot}10^4$ K, obtained for 4-propyl-4'-cyano biphenyl (BF-3) by quantum-chemical calculations [245, 246]. The value of the order of $2{\cdot}10^4$ is also obtained if we add up the increments of the energy of attraction for different fragments of the molecules (for more details see later). Values of the ionisation potential

Table 4.5. Comparison of activation and energy values E of rotational viscosity with isotropic and anisotropic parts of the paired intermolecular interaction (J_0 and J_2) calculated from the equation of state at the clarification temperatures of NLC

Substance	T_{ni}, K	$S(T = T_{ni})$	J_0, K	J_2, K	E, K
BP-5	308.5	0.373	91 950	2 922	5 840
PCH-5	330.2	0.358	78 950	3 545	3 600
BCL-5	374.2	0.386	89 760	3 435	3 680
BMAOB [N4]	346.2	0.337	-19 800	4 085	4160
10 AB (PAA)	409	0.437	42 540	2 918	-

$I \sim 7 \div 8$ eV $\approx 8 \div 9 \cdot 10^4$ K with the order of magnitude similar to J_1/k_BV_0 were published in [247]. Thus, it may be assumed that J_1 corresponds to the total energy of the molecule.

Comparing the values of J_2/k_BV_0 and E it can be seen that for the majority of substances these values are similar, i.e., energy J_2 is responsible for the short-range order in the liquid. Therefore, it may be assumed that the energy of the orientation interaction determines the activation energy of viscosity, and also melting enthalpy, surface energy, etc.

Intermolecular interaction determines the molecular packing and the density of the nematic liquid crystal. Therefore, the latter characteristics can be used for calculating the activation energy. In the following section, it is shown that, knowing the density of the nematic liquid crystal, it is also possible to determine approximately the value of viscosity.

4.2. Relationship of viscosity with molecular packing and free volume

As indicated by the Tables 4.1, 4.2, for most substances (azoxy compounds, phenyl cyclohexanes and phenyl bicyclooctanes) the experimental results are approximateed most efficiently by equation (4.1.2) indicating that the free volume model (section 3.2) can be used to describe the temperature dependence of the viscosity of these substances. For these substances, the activation energy and, consequently, the energy of intermolecular interaction is considerably lower (by approximately 2500 K) in comparison with the nematic liquid crystal in which the molecules contain only aromatic rings. Thus, one of the criteria for the applicability of the model of the free volume for describing the temperature dependence of rotational viscosity is the low value of the activation energy (less than 0.45 eV $\approx$ 4900 K in the determination of E using equation (3.1.2)). Other criteria include the width of the nematic range (for a large number of the nematic liquid crystals, the effects of the free volume are not evident because of the small width of the nematic range, 20÷30°C) and also low temperature (0°C and lower) at which the compressed nematic liquid crystals prevent the displacement and rotation of the molecules. From this viewpoint, the effects of the free volume are quite evident in the derivatives of phenyl cyclohexane and phenyl bicyclooctane and mixtures based on them, characterised by low melting point and high clarification temperatures (Fig. 4.1.3). The dependence of viscosity on temperature for these substances is given by the expression:

$$\gamma_1(T,S) = bS^2 \exp\frac{B}{T-T_0}, \quad (4.2.1)$$

so that the equation (3.2.3) can be used to determine the value of the fluctuation free volume for these substances and compare the result with the geometrical free volume $V_{fg} = 1 - k_p = 1 - N_a \rho V_0/M$, obtained in the density measurements (Fig. 4.2.1). Comparison of these relationships shows that the small increase of the packing coefficient with the decrease of temperature is accompanied by a larger decrease of V_{ff} ($\delta V_{ff} \sim 4\, \delta V_{fg}$), and the reorientation of the molecule becomes less probable. We introduce the concept of transition temperature T_a above which the dependence $\gamma_1(T, S)$ is described by the simple activation approximation of the type (4.1.1) or (4.1.4). Below temperature T_a it is necessary to take into account the effect of the decreasing free volume resulting in a deviation from the simple activation dependence. For FTsG-7 $T_a \approx 35$°C, and for the mixtures of bicyclooctanes $T_a \approx 45$°C; at $T = T_a$ the following relationship is fulfilled:

$$V_{fg} - V_{ff} \approx 0.10 \div 0.13. \quad (4.2.2)$$

The free volume (this concept was described in section § 3.2) can be determined directly from the measurements of the molar volume (density) in the solid crystal–nematic liquid crystal transition (see, for example [248]). However, for certain substances or mixtures the coefficient of thermal expansion in the solid phase is higher in comparison with the coefficient in the nematic phase and this indicates the zero free volume in the nematic phase.

It will be attempted to estimate the free volume from other considerations, examining the packing of the molecules with different types of interaction and with similar geometrical dimensions. Figure 4.2.2 [272] compares the temperature dependences of the coefficient of molecular packing k_p obtained for derivatives of biphenyl, phenyl cyclohexane and bicyclooctane. The length of the molecules of the five homologs of these nematic liquid crystals is equal to 18 Å and their thickness is determined by the thickness of the

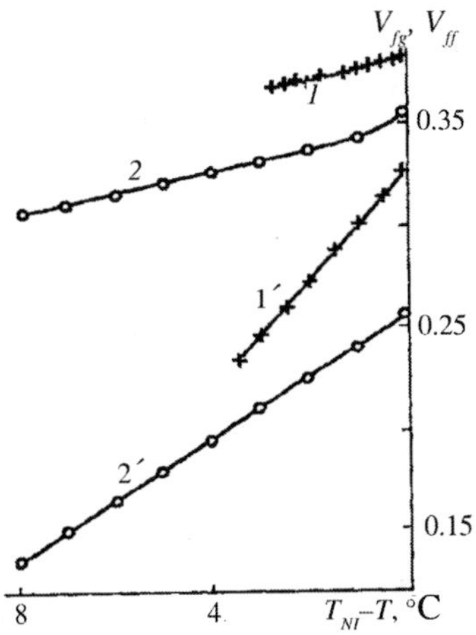

Fig.4.2.1. Temperature dependence of geometrical and fluctuation free volumes [V_{fg}(*1*,*2*) and V_{ff}(*1'*,*2'*) respectively for PCH-7 (*1*,*1'*) and the mixture BCL-3/5 (*2*,*2'*).

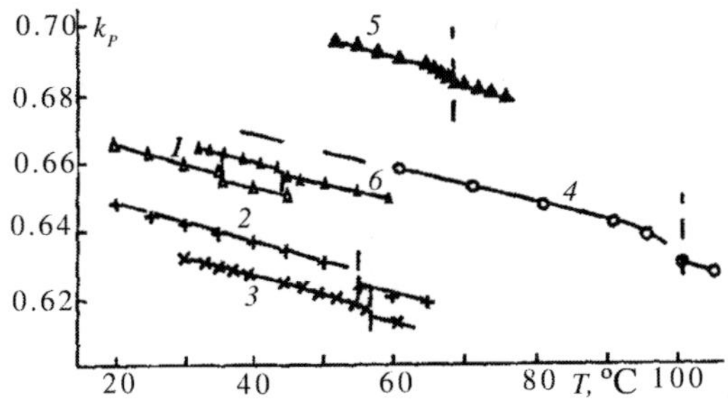

Fig.4.2.2. Temperature dependence of the coefficient of molecular packing k_p for bicyclic cyano derivatives of NLC: *1*) BF-5, *2*) PCH-5, *3*) PCH-7, *4*) BCL-5, *5*) OBP-5, *6*) PPr-5.

aromatic fragments and is equal to 4.3, 4.7 and 5.0 Å for BP, PCH and BCO, respectively [249]. The high value of the aliphatic fragments in comparison with the phenyl fragments should result in denser packing of the nematic liquid crystal and, correspondingly, in a high value of γ_1. However, the weakening of intermolecular interaction in the replacement of the phenyl ring in BP-5 by the cyclohexane fragment in PCH-5 results in a decrease of the value of k_p by 1.7% with the value of γ_1 remaining on the previously obtained level. The value of the variation of the free volume δV_{fg} is comparable with the value of the free volume (2–3%) in the appropriate model (section §3.2). In bicyclooctanes, the denser packing of the molecules of the nematic liquid crystal, resulting from the larger thickness of the bicyclooctane fragment and is special form, is accompanied by the large (by a factor of 3) increase of γ_1. However, the lower energy of the intermolecular interaction (the activation energy of bicyclooctanes $E = 3600$

K in comparison with $E = 5800$ K for BP-5) ensures the validity of the model of the free volume when describing the dependence γ_1 (T, S) also for the derivatives of bicyclooctane, regardless of the smaller fraction of υ_{fg} in this case.

Figure 4.2.3 shows the relationship of the quantities γ_1 and V_{fg} for several cyano derivatives of the nematic liquid crystals. On the whole, the graph confirms the validity of the previously made considerations. At the same time, care must be taken when using the results of density measurements for predicting viscosity. For example, comparison of the packing coefficient of the 5 and 7 homologs of PCH (Fig. 4.2.2) shows that the value of the free volume for PCH-7 is higher. At the same time, the viscosity γ_1 of the substance is also higher which seems to contradict the conclusions made and appears in all likelihood to be the consequence of the higher anisotropy of polarisability (section 5.2) or higher molecular friction coefficient (section 4.3).

In the free volume model, viscosity depends on the glass transition temperature T_g or the related temperature of 'freezing' of movement of the director T_0, or more accurately on the difference $\Delta T_g = T - T_g$ or $\Delta T_0 = T - T_0$, irrespective of the chemical structure of the molecules of the nematic liquid crystal. This is confirmed by the measurements of the value for low-temperature liquid crystals [167] or the relaxation time of the twist effect ($T_s \sim \gamma_1/K_T$) of different nematic crystals at low temperatures [250] (Fig. 4.2.4). The method of calculating the glass transition temperature on the basis of the chemical structure of the molecules is described below [251].

It was shown in [167] that the temperature dependence $\gamma_1(T)$ of different nematic liquid crystals characterised by a wide nematic range can be be

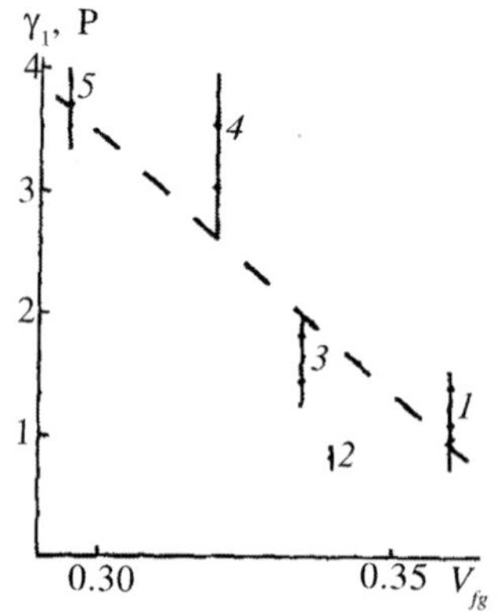

Fig.4.2.3. Relationship of the rotational viscosity γ_1 of bicyclic cyano derivative compounds with the geometrical free volume V_{fg}: (*1*) C_nH_{2n+1}–CH–B–CN (PCH) (*2*) C_nH_{2n+1}–B–B–CN (BP), (*3*) C_nH_{2n+1}–C_5H_3–B–CN (PPr), (*4*) C_nH_{2n+1}–$C_4H_2N_2$–B–CN (PPm), (*5*) C_nH_{2n+1}–O–B–B–CN (OBP).

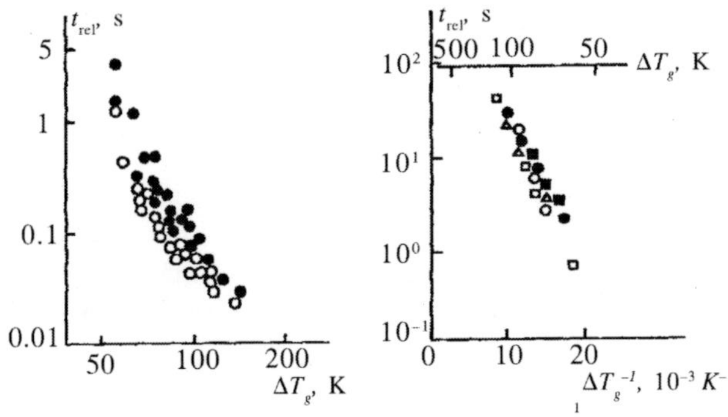

Fig.4.2.4. Dependence of the realisation time of the optical response of the twist effect t_{rel} on the difference $\Delta T_g = T - T_g$ for different NLM.

described by the equation (3.2.2), in which B = 1225 K, $T = T$ +50K.

For the liquid crystal materials consisting of the substances belonging to different chemical classes, the value b is almost identical and equals $6.03 \cdot 10^{-4}$ P. Equation (3.2.2) is identical with the equation describing the temperature dependence of the viscosity of linear polymers. For these substances, the value of T_g and, consequently, the viscosity can be calculated by the increment method using the equation:

$$T_g^{\text{lin}} = \alpha \sum_i E_i^{\text{coh}} z_i + \beta, \tag{4.2.3}$$

where E_i^{coh} is the energy of cohesion (binding) of the i-th member (atom) with their neighbours, z_i is the number describing the freedom of rotation of the i-th member [167]. The values of E_i^{coh} and z_i for different atoms and molecular fragments are presented in Table 4.6. For linear polymers α = 0.061 K mol/cal, β = 120 K. The equation (4.2.3) is also valid for the nematic liquid crystals at α = 0.0286 K mol/cal, β = 141 K, i.e. the actual glass

Table 4.6. Increments of the individual growth of molecules of mesogens and the binding energy of molecules E^{coh} and the number of degrees of freedon Z.

Fragment	Z_i	E_i^{coh}, cal/mol
Phenyl	1	7630
Cyclohexane	1	6310
Pyrimidine	1	11170
CN	0	6100
CO_2 (ester bridge)	1	4300
CH_2	1	1180
CH_3	1	1125
O (oxygen)	0	800

transition temperature of the nematic liquid crystal is linked with T_g^{lin} by the relationship:

$$T_g = 0.469T_g^{\text{lin}} + 84.7. \qquad (4.2.4)$$

The authors of [252] described a method of calculating the glass transition temperature of linear polymers based on the summation of the increments of van der Waals volumes of the individual atoms ΔV_i taking into account the coefficients describing the interaction of the atoms with each other and the specific interaction:

$$T_g^{\text{lin}} = \frac{\Sigma_i \Delta \upsilon_i}{\Sigma_i (a_i \Delta \upsilon_i + b_i)}. \qquad (4.2.5)$$

Values of T_g^{lin}, calculated using this procedure, are in good agreement with the experimental values. It will be shown that equation (4.2.5) may be used to calculate the rotational viscosity of the nematic liquid crystal.

Figure 4.2.5 shows the structural forms of the molecules of typical nematic liquid crystals indicating the increments of the volumes of the atoms ΔV_i taken from [240]. The largest contribution to the value of T_g^{lin} is provided by the vibrations of the light hydrogen atoms (for hydrogen $a = 19.98 \cdot 10^{-3}$ K, and for carbon $a = 0.021 \cdot 10^{-3}$ K). Therefore, the lowest values of the glass transition temperature are shown by the derivatives of phenyl cyclohexanes. The presence of the longitudinal dipoles reduces the denominator of the fraction (4.2.5) (for example, the dipole–dipole interaction is described by the constant $b = -55.4 \cdot 10^{-3}$ K·Å^3), and therefore the polar compounds are characterised by relatively high values of T_g. The values of a and b for different atoms and types of interaction are presented in Table 4.7 which is identical with Table 3.3 [241].

The abscissa of Fig. 4.2.6 gives the values of T_g^{lin} calculated using equation (4.2.5), and the ordinate gives the experimental values of T_g taken from [167, 250]. The plotted points fit with sufficient accuracy the straight line, and described by the equation:

$$T_g = mT_g^{\text{lin}} + n = 0.303T_g^{\text{lin}} + 112.3. \qquad (4.2.6)$$

The coefficients in the equations (4.2.4) and (4.2.6) slightly differ, i.e. the method of calculating T_g on the basis of the increments of the cohesion energy E_i^{coh} or atomic volumes ΔV_i and coefficients a_i, taking into account the anharmonism of the vibrations of the atoms, leads to different values of T_g^{lin}. It should be mentioned that the relationship (4.2.6) gives relatively accurate values of T_g. The maximum difference is 20 K (6%). The difference between T_g and T_g^{lin} is associated in all likelihood with the difference between the dimensions of the mesogenic and polymer molecules. Consequently, the value of constant a_i for carbon in polymers is extremely

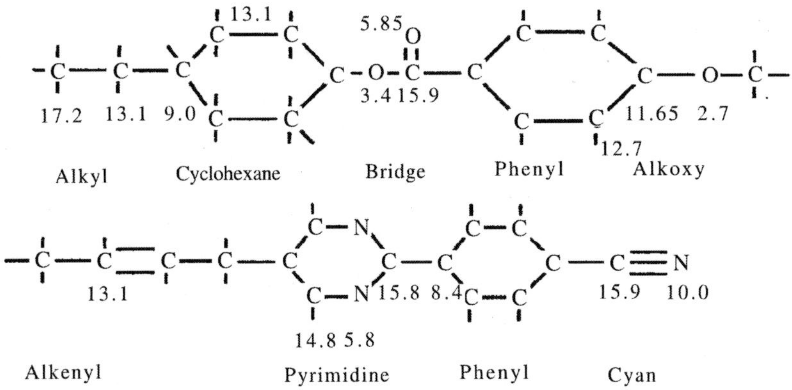

Fig.4.2.5. Structural formulae of mesogenic substances indicating the increments of volumes of different atoms; ΔV_H=2Å.

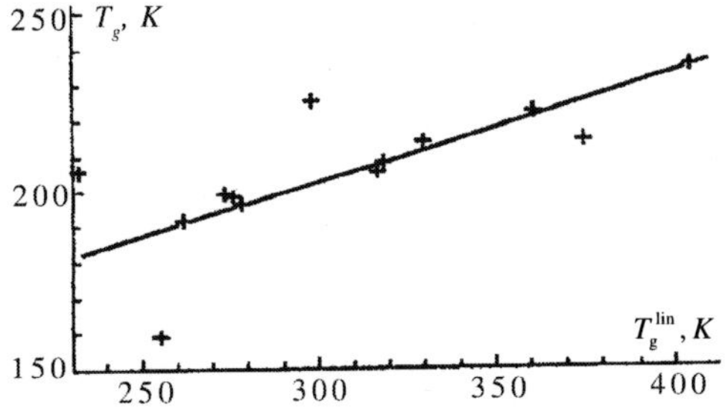

Fig.4.2.6. Relationship of the experimental values of T_g for different LC substance with values of T_g calculated from equation (4.2.5). The direct line is described by the realtionship (4.2.6).

low. If it is assumed for carbon that $a = (1.93\pm0.18)\cdot10^{-3}$ K, then the values of the glass transition temperature calculated from equation (4.2.5) and those determined in the experiments coincide. Since the values of the coefficients a_i determined by the intermolecular interaction and the dimensions of the atoms, the higher values of a_c is more suitable for calculating the characteristics of low-molecular compounds.

Table 4.8 gives the experimental values and the values calculated from (3.2.2), (4.2.5) and (4.2.6) for rotational viscosity and its activation energy E for different nematic liquid crystals. The value of E was determined by the equation:

$$\frac{E}{k_b} = \frac{\partial(\ln\gamma_1)}{\partial(1/T)} = \frac{BT^2}{(T-T_0)^2}. \qquad (4.2.7)$$

Equation (4.2.7) is the binding term describing the temperature dependence of the viscosity of the nematic liquid crystals by two models based on

Table 4.7. Numerical values of increments a_i and b_i for different atoms and types of intermolecular interactions [252]

Atom	a_i, 10^{-3} K^{-1}	Type of intermolecular interaction	b_i, $10^{-3} \times K^{-1}$
Carbon in low-molecular compounds	1.93 [251]	Dipole-dipole	−55.4
Carbon in polymers	0.021	Hydrogen bond	206.78
Hydrogen	19.98	Symmetry coefficient in *ortho*- and *para*-substitution in the aromatic core	16.0
Oxygen in the main chain	22.95		−25.6
Oxygen in the lateral chain	7.03	Coefficient for polydienes, taking into account the high degree of freedom of rotation in chains with double bonds	92.22
Nitrogen in the main chain	8.62		
Nitrogen in the lateral chain	6.35		
Chlorine	4.01		
Sulphur	3.53		
Silicon	8.3		

Table 4.8. Numerical values of increments a_i and b_i for different atoms and types of intermolecular interaction [252]

Structural formula	T_g, K[167]	γ_1, P		$\frac{\Delta\gamma_1}{\gamma_1}$, %	E, K		$\frac{\Delta E}{E}$, %
		Calc.	Exp.		Calc.	Exp.	
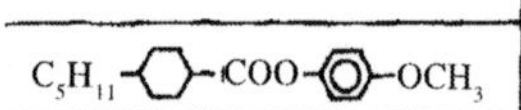	197	1.62	1.34	20	5020	4680	7.0
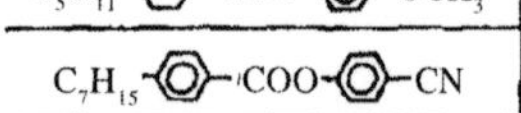	214	4.77	4.10	15	6370	6480	−1.4
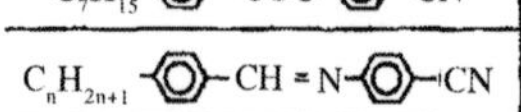	212	4.06	3.60	11	6190	6480	15.5
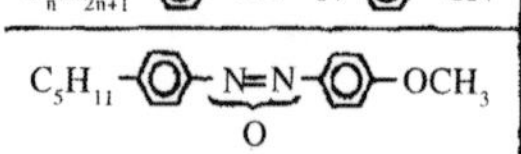	199	1.78	1.89	−6	5160	4530	−15.0
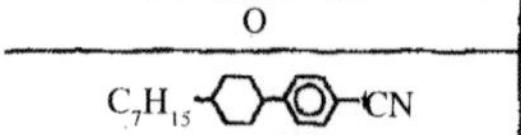	200	1.92	1.40	37	5230	4670	12.0
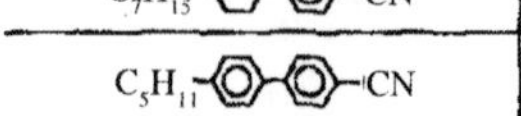	207	2.06	0.82		5760	5840	−1.3
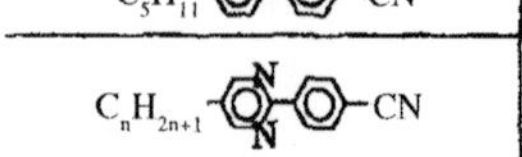	215	5.13	1.91		6470	5640	15.0

different physical principles – activation and free volumes. The same equation includes both the activation energy and the temperature of 'freezing' of movement of the director.

For the majority of of substances, the relative difference of rotational viscosity $\Delta\gamma/\gamma_1$ is 6÷20%, activation energy $\Delta E/E$ = 1÷15%. For certain bicyclic compounds whose molecules do not contain bridge bonds (–CH=N–, –COO– etc.), the calculated values are approximately three times higher than the experimental values and it may be taken into account by introducing lower values of pre-exponential multiplier b for these substances.

The case often encountered in the development of new nematic liquid crystals will now be discussed. To produce or intensify the required property of the initial molecule, one of the atoms is replaced by another atom or a group of atoms. If the initial substance has the glass transition temperature T_g determined from (4.2.5) and (4.2.6), the volume of the molecule $v_0 = \Sigma_i \Delta v_i$, the substituted atom – the increment of the volume Δv_j, the atom group of atoms replacing it–the increment of the volume Δv_x, and the coefficients a_x, b_x, the variation of the glass transition temperature for the new substance may be described by the equation:

$$T_g^{\text{lin}} = T_g^{(0)} + \Delta T_g^{\text{lin}} = \frac{v_0 - \Delta v_j + \Sigma\Delta v_x}{\Sigma(a_i\Delta v_i + b_i) - a_j\Delta v_j + \Sigma(a_x v_x + b_x)} =$$

$$\left(T_g^{(0)} + T_g^{(0)}\frac{\Sigma\Delta v_x - \Delta v_j}{v_0}\right)\left[1 + T_g^{(0)}\frac{a_j\Delta v_j - \Sigma(a_x v_x + b_x)}{v_0}\right] = \tag{4.2.8}$$

$$T_g^{(0)} + T_g^{(0)}\frac{\Delta v}{v_0} + \frac{\left(T_g^{(0)}\right)^2}{v_0}\left[a_j\Delta v_j - \Sigma(a_x v_x + b_x)\right].$$

In (4.2.8) it is taken into account that $\Delta v_j \ll v_0$, $\Delta v = \Sigma\Delta v_x - v_j \ll v_0$ and $\Delta T_g \ll T_g^{(0)}$. Equation (4.2.6) yields the equality:

$$\Delta T_0 = \Delta T_g = m\Delta T_g^{\text{lin}}, \tag{4.2.9}$$

from which for the new substance:

$$\gamma_1 = A\exp\left\{\frac{B}{(T - T_0) - \Delta T_g}\right\} = A\exp\left\{\left[\frac{B}{T - T_0}\right]\left[1 + \frac{\Delta T_g}{T - T_0}\right]\right\} =$$

$$\gamma_1^{(0)}\exp\left[\frac{B\Delta T_g}{(T - T_0)^2}\right]. \tag{4.2.10}$$

At $\Delta T_g < 15$ K, the relationship (4.2.10) can be reduced to a simpler form (with the accuracy of up to 5%):

$$\gamma_1 = \gamma_1^{(0)}\left[1+\frac{B\Delta T_g}{(T-T_0)^2}\right]. \qquad (4.2.11)$$

The relationship (4.2.10) obtained for the condition $\Delta T_g \ll (T-T_0)$, taking into account (4.2.8) and (4.2.9), makes it possible to estimate the correction to the value of rotational viscosity resulting from some fragment. The applicability of the relationships (4.2 .0), (4.2.11) for the evaluation of the variation of viscosity in the replacement of the molecular fragment is indicated by the data in Table 4.9. The table gives the values of ΔT_g^{lin}, calculated using equation (4.2.8) for different derivatives of phenyl cyclohexane in which the group $(CH_2)_2$ is replaced by other fragments. The substance whose viscosity is compared with the viscosity of other substances – 4-propyl-(4'-ethyl phenyl) cyclohexane (C_3H_7–C_6H_{10}–C_6H_4-C_2H_5) has at t = 20°C, the kinematic viscosity of ν = 0.04 St. Because of the absence of the data on rotational viscosity, the values of $\gamma_1/\gamma_1^{(0)}$, calculated from the equations (4.2.10), (4.2.11) are compared with the ratio

Table 4.9. Variation of viscosity of derivatives of phenylcyclohexane C_3H_7 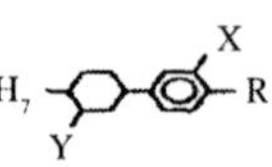 at substitution of the substituent

R	X	Y	ΔT_g	$\gamma_1/\gamma_1^{(0)}$	$\eta/\eta^{(0)}$, $\nu/\nu^{(0)}$
C_2H_5	H	H	0	1.00	1.00
C_2H_5	H	CH_3	1.0	1.03	1.50
OC_2H_5	H	H	7.9	1.20	1.0; 1.5
F	H	H	9.1	1.23	1.25
F	F	H	5.3	1.13	1.75
CN	H	H	44.6	3.07	2.95
CN	F	H	25.6	1.91	1.90
NCS	H	H	26.3	1.94	1.93
$CH{=}CF_2$	H	H	12.4	1.31	1.0; 1.25
$OOCC_3H_7$	H	H	26.5	1.95	2.14
$COOCH_3$	H	H	18.4	1.53	8.00

$\eta/\eta^{(0)}$ or $\nu/\nu^{(0)}$, where η and ν is the dynamic and kinematic viscosity, respectively. According to [26], for the substances with a similar molecular structure the quantity γ_1/η may be regarded as constant. The values of η, ν were taken from [240].

The data presented in Table 4.9 show that the equations (4.2.10), (4.2.11) describe quite accurately the variation of the viscosity of the substance in the replacement of the molecular fragment.

However, it should be mentioned that the equations (4.2.3) or (4.2.5) with the coefficients given for the nematic liquid crystals are not always suitable if the form of the molecule becomes more complicated. It was shown in [253] that for the nematic liquid crystals whose molecules are tricyclic and contain naphthalene fragments, benzo fragments with one or two lateral alkyl, alkoxy- or alkanoyloxy-substituents, the branched terminal substituents, and so on, glass transition temperature T_g is 40–60 K higher than the value given by equation (4.2.3) and, correspondingly, several orders of magnitude higher than viscosity.

The data presented in [253] show that for the bicyclic and tricyclic compounds whose molecules not contain lateral substituents

$$\ln \eta \sim 0.108 \frac{E^{\mathrm{coh}}}{z} - 2.9. \qquad (4.2.12)$$

In this case, the homologous series is characterised by a decrease of viscosity with the increase of E^{coh}/z, i.e. the viscosity given by (4.2.12) characterises the class of the compound. For the substances containing fragments with the lateral substituents and naphthalene fragments:

$$\ln \eta \sim 0.614 \frac{E^{\mathrm{coh}}}{z} - 28.45. \qquad (4.2.13)$$

Evidently, the constants α and β in (4.2.3) or coefficient a in (4.2.5) depend on the degree of approximation of the form and dimensions of the molecule or of the supermolecular structure, defined by them, in relation to the polymer structure. In particular, a decrease of a_i in (4.2.5) results in an increase of T_g and viscosity.

The authors of [153] calculated the values of T_g for tolane derivatives nTOm. With the increase of the sum $n+m$ the value of T_g decreases from 191.9 K for 5TO1 to 187.1 K for 7TO5. Figure 4.2.7 shows the dependence $\ln [\eta_S/(\Delta n)^2]$ on $(T-T_0)^{-1}$ for the homologous series of the tolane derivatives at the calculated values of T_g. The mean slope of the dependence was equal to $B = 949$ K, which differs from $B = 1225$ K in equation (3.2.2). Thus, the constants in the equations (4.2.1) and (3.2.2) differ. For 5TO1 the resultant value was $b = 5 \cdot 10^{-5}$ P taking into account the temperature dependence of the order parameter S determined in [254]. For the given

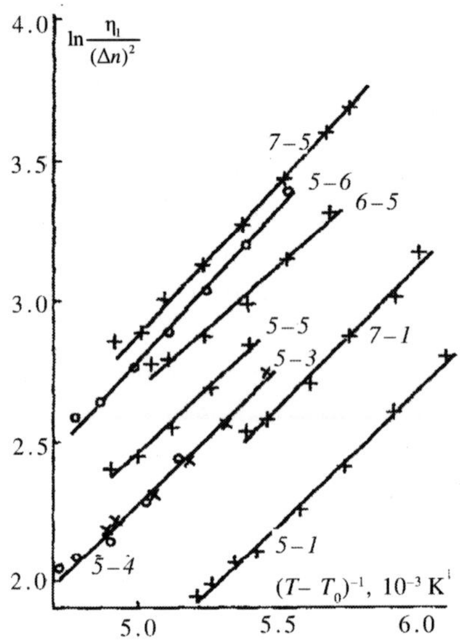

Fig.4.2.7. Dependence of ln $[\eta_s/(\Delta n)^2]$ or $(T-T_0)^{-1}$ for toluene derivatives *n*TO*m*. Values of T_0 were calculated from equation (4.2.5), viscosity is taken in P.

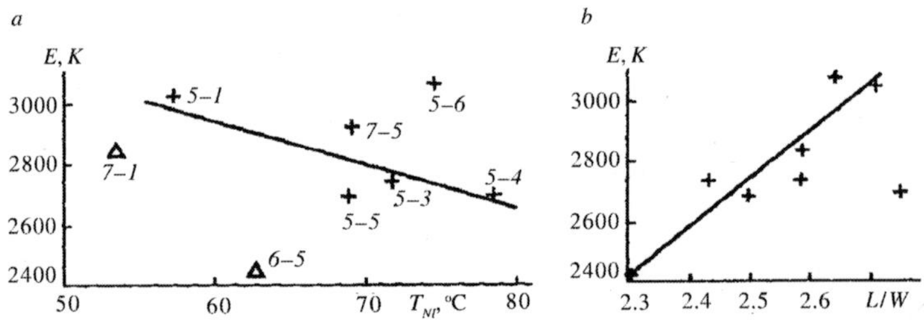

Fig.4.2.8. Relationship of the activation energy E_1 determined by the ratio (4.2.14), with the clarification temperature T_{NI} of tolene derivatives *n*TO*m* (*a*) and the form factor of the excluded volume *L*/*W* (*b*).

homologous series, the free volume model gives a decrease of the activation energy determined from (4.2.7) with the increase of the alkyl substituent: at $T = 60°$, the value

$$E_2 = \frac{\partial \ln\left[\eta s/(\Delta n)^2\right]}{\partial(1/T)} = \frac{BT^2}{(T-T_0)^2} \tag{4.2.14}$$

decreases from 2849 (5TO1) to 2711 K (7TO5). Figure 4.2.8 shows the dependence of the activation energy E_2, calculated at the reduced temperature $T = T_{NI}-10$K Assuming a small change δT_{NI} for different materials and simplifying the equation (4.2.14) at $T = T_{NI}$, gives the variation of activation energy:

$$\delta E_2 = -\frac{2E_2T}{T(T-T_0)}\delta T, \qquad (4.2.15)$$

from which at $T_g \sim 200$ K, $T_{NI} \sim 340$ K, and $E_2^{(0)} = 2500$ K we obtain $\delta E/\delta T_{NI} \sim -20$. The linear form of dependence (4.2.15) in the temperature range 300÷360 K and the value of the slope are in agreement with the experimental value $\delta E/\delta T_{NI} \sim -15$ (Fig. 4.2.8) and with the value –13.2 (Fig. 4.2.9) obtained from the identical dependence $E = BT^2/(T - T_0)^2$ for the nematic liquid crystals belonging to different chemical classes [26].

4.3. Comparison of the experimental data with the results of the molecular–statistical approach to describing the viscosity of the nematic liquid crystal

In this section, attention is given to the applicability of the results presented in section §3.3–3.5 for describing different dynamic processes taking place in the nematic liquid crystals on the molecular level and associated in some manner with viscosity: self-diffusion, rotational diffusion, orientational relaxation, molecular friction, etc. The specific features of the study are taken into account. For example, the theory proposed by Diogo and Martins is used for evaluating the orientational relaxation time. The theories proposed by Osipov and Terent'ev, Nemtsov and similar investigations enable conclusions to be drawn on the value of the coefficient of internal friction and rotational diffusion for which the number of experimental data is unfortunately very small.

In [25] in comparing the sum of the least-squares obtained in the approximation of the experimental dependences $\gamma_1(T)$ by different equations, the sum in the approximation by equations (3.1.7) and (3.3.11), obtained by Diogo and Martins [168] was the highest. However, this does not indicate that the theory proposed by Diogo and Martins or the individual assumptions of the theory cannot be used to describe the rotational viscosity of the nematic liquid crystals.

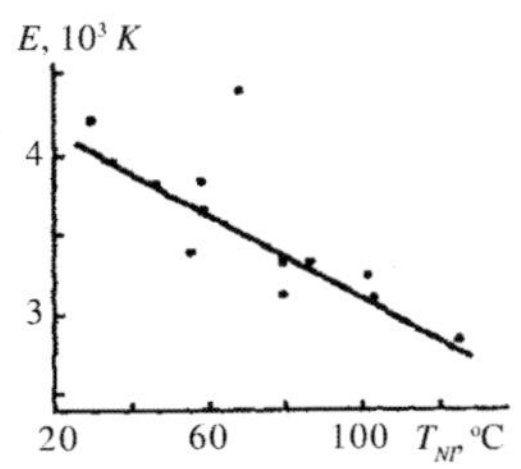

Fig.4.2.9. Relationship of the activation energy E of viscosity γ_1/S^2 with clarification temperature T_{ni} for different bicyclic LC substances.

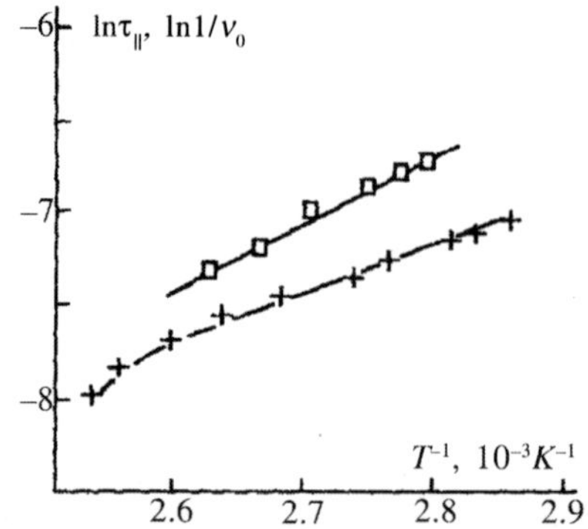

Fig.4.3.1. Temperature dependence of the dielectric relaxation time of 4,4'-dibutoxy azoxybenzol (4OAB) $\tau_{\|}$ (+) and the duration of the orientation of the molecule (particle) $1/\nu_0$ ( ), calculated in the accordance with (3.3.1) using the data for rotational viscosity γ_1.

We examine the initial assumptions of the theory. The value of γ_1 is associated with the reorientation time of the particle $\tau_\gamma = \nu_0^{-1}$ by the relationship (3.3.1). It was assumed in [164] that τ_γ is the duration of dielectric orientational relaxation of the molecules of the nematic liquid crystal $\tau_{\|}$. Figure 4.3.1 shows the temperature dependences of the reorientation time of the molecules of butyloxy azoxybenzol (4OAB), determined in dielectric measurements, and the reorientation time ν_0^{-1} converted using (3.3.1) on the basis of the results of measurements of γ_1 and the molar volume ($V^a \approx V_N$) [164]. These times differ only very slightly (by approximately a factor of 1.5) and the temperature dependences of these times are of the activation nature with similar values of activation energy.

If we compare the values of $\tau_{\|}$, obtained for *n*OAB in [255], and the calculated values of τ_γ, then these values coincide for all members of the homologous series, with the exception of $n = 3$ and $n = 5$ (Fig. 4.3.2). The activation energy is $E \sim 80 \div 100$ kJ/mole $\sim 9.6 \div 12 \cdot 10^3$ K, which is 2÷2.5 times higher than the value of E. The same figure shows the values of the duration of rotation around the longer axis $\tau_\perp$ which is approximately 500 times shorter than the duration of rotation around the short axis $\tau_{\|}$.

In [256], the value of viscosity γ_1 (or η_S) is used for calculating of the relaxation time using the classic Debye equation:

$$\tau = \frac{\zeta}{2k_b T}, \tag{4.3.1}$$

where $z = 8\pi\gamma_1 L^3 f(L/d)$ is the friction coefficient, L and d are respectively the long and short half-axis of the ellipsoid of rotation, approximating the

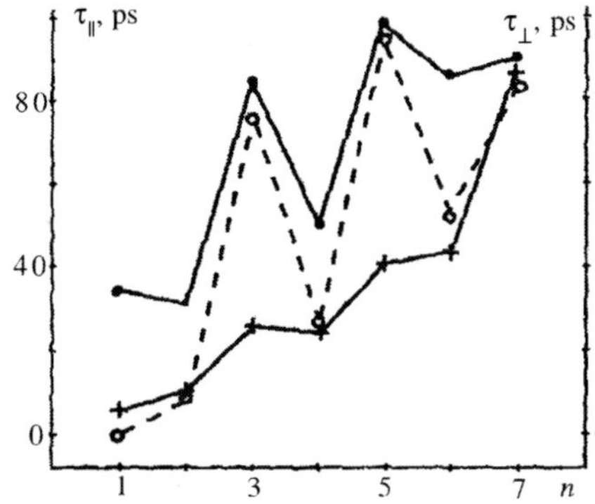

Fig.4.3.2. Variation of the values of $\tau_{\|}$(o), τ_{γ}(+) and $t_{\perp}$(•) in the homological series of 4,4'-dialkoxyazoxybenzols (*n*OAB) the data for $\tau_{\|}$, $\tau_{\perp}$ were taken from [255].

form of the molecule, *f(x)* is the Perren equation, given by the expression

$$f(x) \approx \frac{2}{3}\left[2\ln\left[1+\sqrt{(1-x^2)}/x\right]-1\right]^{-1}. \tag{4.3.2}$$

Table 4.10 compares the values of, calculated from equation (4.3.1) and the experimental values of τ for the nematic liquid crystals, belonging in different chemical classes. With the exception of dichloroderivative of phenyl benzoyl benzoate, the calculated relaxation times for substances are in good agreement with the experimental values. The values of the activation energy of rotational viscosity E_{γ} and the relaxation time E_{τ} for the azoxy compounds are almost identical. For the tricyclic compound $E_{\tau} \approx 2E_{\gamma}$. This is caused by the difference in the nature of the investigated processes. In the former case, rotation is performed by a single molecule, in the second case by the director of the liquid crystal (continuum). In this case, the height of the barrier decreases by 0.1 eV because in the rotation of the single molecule a free volume forms for the rotation of the adjacent molecules. The additional decrease of activation energy by 0.2 eV possibly takes place as a result of the rotation of not only around the short but also the long axis of the molecule [256].

On the other hand, in [168] it is claimed that it is not possible to compare directly the relaxation times of orientational the molecules of the liquid crystals calculated on the basis of the rotational viscosity with the duration of dielectric relaxation because the values of these parameters are greatly influenced by the orientation potential. However, the difference in the calculated and measured values of the relaxation time may be associated with the fact that $\tau_a = \nu_0^{-1}$ was determined from equation (3.4.10) derived by Hess. If we compare the relaxation times calculated using equations (3.3.1) and (3.4.10), they differ by the multiplier $3\pi^2S^2 \sim 20$ which generally

Table 4.10. Comparison of the experimental values of the duration of orientation relaxation $\tau_{\parallel}$ and calculated from viscosity γ_1 using (4.3.1.) (τ) [256]

Substance	L	D	$f(L/D)$	T, K	γ_1, P	τ_{γ}, 10^{-8} s	$\tau_{\parallel}$, 10^{-8} s
CH_3O-C_6H_4- -CH=N- -C_6H_4-C_4H_9	10	3.1	0.25	295	1.25	9.6	14.5
CH_3O-C_6H_4- -N(O)=N- -C_6H_4-OCH_3	8.9	2.9	0.26	398	0.06	0.25	0.4
C_4H_9-C_6H_4- -N(O)=N- -C_6H_4-C_4H_9	11	3.1	0.23	299	0.57	5.38	4.2
C_6H_{13}-C_6H_4- -N(O)=N- -C_6H_4-C_6H_{13}	13.5	3	0.20	315	0.64	9.1	10
C_4H_9-C_6H_4- -COO-C_6H_4- -COO-C_6H_4- -C_4H_9	14.5	3.2	0.20	397	1.3	18	28
Cl-C_6H_4- -COO-C_6H_4- -COO-C_6H_4- -Cl	11.3	3.2	0.23	408	0.263	1.94	30

eliminates the contradiction in the order of τ [168]. However, in comparison of v_0^{-1} and $\tau_{\parallel}$ it is necessary to take care, especially for substances characterised by similar smectic ordering. On the other hand, since the values of $\tau_{\parallel}$ and v_0^{-1} are similar, it is possible to compare the rate of orientation processes in different liquid crystal phases. In the calculation of v_0^{-1}, γ_1 should be replaced by η_S because this viscosity coefficient does not show any critical difference in the vicinity of the nematic liquid crystal–smectic liquid–crystal transition A, so that the values of η_S differ only slightly in both cases [257]. In the case of the smectic crystals C it is necessary to use viscosity coefficient γ_θ describing the dynamics of the variation of the angle of inclination of the director in relation to the normal of the smectic layer. The values of $\gamma_1(T^{-1})$ in the nematic phase and $\gamma_\theta(T^{-1})$ in the smectic phase C can be determined by the extrapolation of the values of viscosity from one phase to another [258]. The temperature dependences of $\tau_{\parallel}$ in the nematic and various smectic phases have been examined in [259]. In transition from one phase to another the value of $\tau_{\|}$ shows small 'jumps' which may be compared with the changes of γ_1 and V_N, η_S and V_A, γ_1 and V_C in the nematic, smectic A, smectic C and other phases.

Previously, it has been mentioned that the height of the potential barrier in rotation of the molecule decreases because a free volume forms around the rotating molecule or particle. The free volume was determined in [168] and is described by expression (3.3.2). From the expression (3.3.2), the value reciprocal to volume V^a included in the expression (3.3.1) changes to the final expression (3.3.11), describing the temperature dependence of rotational viscosity.

Thus, the characteristic reorientation time of the molecules of the nematic liquid crystal which determines γ_1 is in the range $10^{-8}\div10^{-7}$ s (or $10\div100$ ns). The relaxation time values of similar magnitude were obtained from the measurement of the frequency dependence of the complex shear modulus [69]. The comparatively high value of τ_{rel} (10^{-6} s) and its critical dependence on both sides of T_{NI} support the assumption that the relaxation time measured by this method is determined by the value of the kinetic coefficient of fluctuations with the same dimension as viscosity (see, for example, [260]).

It was shown in [261, 262] that the rotational viscosity of a nematic liquid crystal in the vicinity of transition to the isotropic liquid is determined by the orientation relaxation of the dielectrically anisotropic super molecular fluctuation formation. De Gennes determined the relationship of rotational viscosity γ_1 with viscosity η_0 in the isotropic phase [87]:

$$\gamma_1 = \frac{9}{2}\eta_0 S^2. \qquad (4.3.3)$$

The temperature dependence of the relaxation time τ, characterising the dispersion mechanism of the Kerr effect, is described by the expression:

$$\tau^{-1} = \frac{A}{\eta_0 a} = \frac{A_0(T-T^*)}{\eta_0 a}, \qquad (4.3.4)$$

Where $A = A_0\,(T - T^*)$ is the coefficient of expansion of the free energy in respect of the degrees of the order parameter (see 3.4.5), $T^* < T_{NI}$ is the temperature of fictitious phase transition of the second kind, a is a parameter characterising the regime of controlling the layer of the liquid-crystal (in free relaxation $a = 1$, in relaxation in a sinusoidal field $a = 2$). The coefficient A_0 is included in the expression for the heat of transition 'the nematic liquid-crystal–isotropic liquid':

$$L = 0.5aT_{NI}Q_{NI}^2, \qquad (4.3.5)$$

from which

$$\gamma_1 = 9\tau(T - T^*)\frac{L}{aT_{NI}}. \qquad (4.3.6)$$

In [261, 262] for the substances with molecules characterised by the longitudinal or transverse dipole momentum measurements were taken of the time t on the basis of the transmission of light by a Kerr cell in the measurement of the frequency of the electrical field and the value of γ_1 on the basis of the variation of the phase delay when the voltage is switched off (paragraph 2.4). Table 4.11 gives the values of γ_1, calculated from the equation (4.3.6) and determined in the experiments. Within limits of the accuracy of measurements, these values coincide, indicating the fact that the mechanisms of rotational relaxation in the nematic and isotropic phases are similar.

It was reported in [262, 263] that the relaxation time τ which determines the value of γ_1 in the vicinity of T_{NI} is 1–2 orders of magnitude longer than the time, determined by the dipole orientation of the molecules. The value τ in the isotropic phase is determined by the supermolecular process of establishment of the macroscopic parameter of the order from the fluctuation nuclei of the mesophase. Thus, it is not completely justified to compare the value of ν_0^{-1}, determined on the basis of γ_1 and $\tau_{\|}$, although interesting information may be obtained by comparing the temperature dependences of these quantities. It should also be mentioned that both the values of τ or $\tau_{\|}$, and also the values of ν_0^{-1} depends strongly on the calculation method or the methods used for processing the measurements. It was mentioned previously that the application of a specific equation at the same values of the parameter is included in the equation may give the results differing by a large amount.

Although the activation dependence of $\tau_{\|}$ on temperature was considered in the previous cases, it was shown in [264] that this nature of the temperature variation of the relaxation time is detected only in a relatively narrow range. The measurements of the frequency $f_0 = \tau_{\|}^{-1}$ in the range from 370 to 200K from the isotropic to glassy phase showed out the variation of f_0 by 10 orders of magnitude is described quite accurately by the Fogel–Fulcher equation within the framework of the free volume model (Fig. 4.3.3):

$$f_0 = cT^{1/2}\exp\frac{B}{T - T_0}. \qquad (4.3.7)$$

The values of T_0 greatly differ for the rotation around the long and short axes of the molecule (255 K and 231K, respectively, see [264])

Figure 3.3.1 explains the concepts of the extrapolated volume V^a and the free volume ($V^a - V_N$). Consequently, for the substances with similar chemical constitution, for example, for members of the homologous series,

Table 4.11. Comparison of the experimental values of ${}_1$ at $T = T_{ni}$ with the values calculated using the Kerr effect relation time [261, 262]

Substance	T_{ni}, K	L, 10^6 J/m^3	τ $(T-T^*)$, 10^{-6} s·K	γ_1, P Experiments	γ_1, P Calculated from (4.3.6)
C_5H_{11}-C_6H_4--C_6H_4-CN	308	1.41	0.84±0.0-7	0.21±0.02	0.17±0.02
C_4H_9O-C_6H_4--CH=N--C_6H_4-C_4H_9	317	2.18	0.5±0.1	0.25±0.03	0.30±0.06
$C_8H_{17}O$--C_6H_4-COO--C_6H_4-NO_2	341	1.62	1.5±0.4	0.32±0.03	0.32±0.09
C_5H_{11}-C_6H_4--C(CN)==CH-C_6H_4--OC_6H_{13}	320.4	3.1		0.9±0.1	0.8±0.2

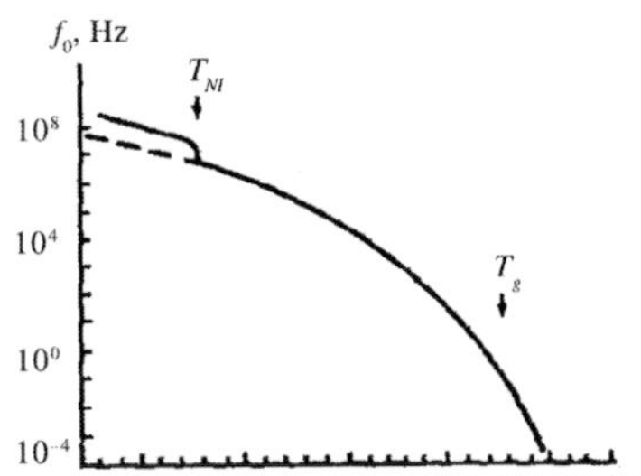

Fig. 4.3.3. Dependence of the frequency of dielectric polarisation f_0 of BF-5 on reciprocal temperature T^{-1}. The dotted line shows processing of results using the Fogel–Fulcher law [167]

the value $(V^a)^{-1}$, included as the co-multiplier in (3.3.1), determines the relative value of the rotational viscosity, and the value ν_0^{-1} determines its temperature dependence. Taking into account (3.1.6) we obtain:

$$\gamma_1 = \frac{4.54 k_B \beta}{\pi^2} \frac{T T_{NI} S^2}{V_N (V^a - V_N)} \frac{1}{\nu_0} = \frac{4.54 k_B \beta}{\pi^2} \frac{G}{\nu_0}. \qquad (4.3.8)$$

Figure 4.3.4 shows the dependences of γ_1 and parameter G on the number

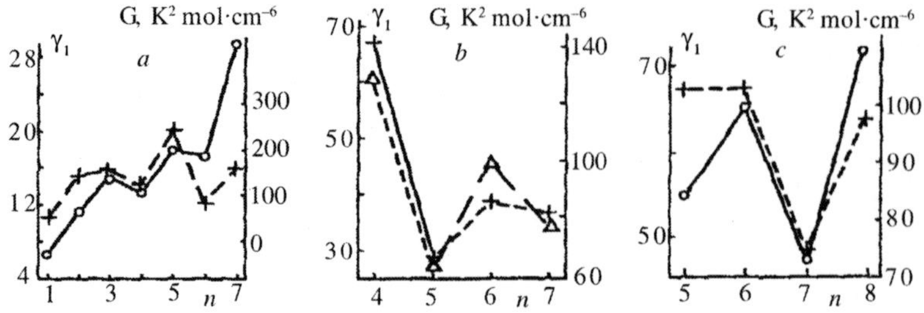

Fig.4.3.4. Dependence of rotational viscosity γ_1 in cP (solid line) and of the parameter of the free volume $G = T\,T_{ni} S^2/[V_N (V^a - V_N)]$ (dotted line) on the number of the homolog for *a*) dialkoxy azoxybenzols ($\Delta T = T_{ni} - T = 10°C$), and *b*) dialkoxy azoxybenzols ($\Delta T = 10°C$ and *c*) alkyl cyano biphenyls ($\Delta T = 5°C$).

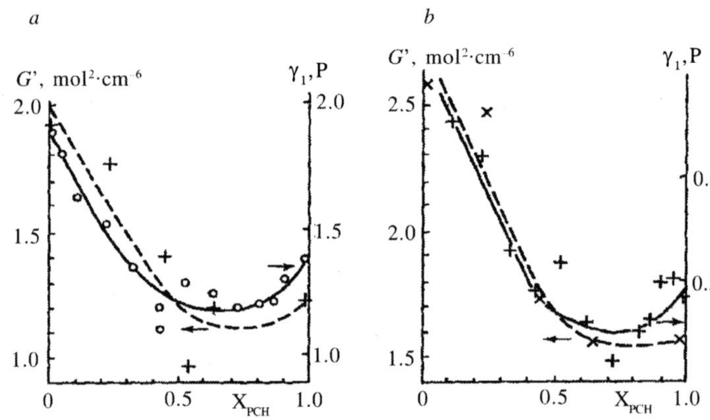

Fig.4.3.5. Dependence of rotational viscosity γ_1 and the free volume parameter ($G = [V_N (V^a - V_N)]^{-1}$ on the concentration of X_{PCH7} in mixtures of PCH-7 with ZhK-440 at *a*) $T = 25°C$ and *b*) $T = 45°C$.

n in the homologous series for the alkylcyano biphenyls BF-*n* (the values of γ_1 were taken from [265], *S* from [266], ρ from [238]), dialkoxy azoxybenzols of the *para*-azoxyanisole series *n*OAB (the values of γ_1 were taken from [168], *S* from [267], ρ from [235] and for dialkylazoxy benzols *n*AB γ_1 from [236], ρ and *S* – 43 [235]. The identical dependences of the measured values of γ_1 and the calculated values of *G* on the number of the homolog indicate the controlling effect of the packing of the molecules of the nematic liquid crystal on the value of γ_1, as also mentioned in the discussion of the free volume model. Using the density data, it is possible to evaluate the relative value of γ_1 for the polar and low-polar substances with alkyl and alkoxy substituents. It appears that the comparison of the values of γ_1 and *G* is useful for explaining not only the temperature but also concentration dependence of the rotational viscosity of the mixtures of the nematic liquid crystals belonging in different chemical classes. Figure 4.3.5

shows the concentration dependences of γ_1 and G for the mixtures of 4-heptyl-4'-cyano phenyl cyclohexane (FCH-7) with azoxy compounds (ZhK-440, see [268]). The values of γ_1 were obtained in [269], $S^* \sim S$ in [270] (for more details see below), the values ρ of in [271, 272]. It appears that these relationships coincide. For more extensive confirmation of the applicability of the Diogo–Martins theory it is necessary to measure the molar volumes of the different nematic liquid crystals and of the mixtures in the isotropic and nematic phases.

Substituting the experimental values of the physical parameters of the nematic liquid crystals into the equations derived in [168] and presented in section 3.3 gives the numerical estimates of some of the values useful for understanding the processes of reorientation of the nematic liquid crystals. For example, the equations (3.3.3) and (3.3.5) can be used to determine the number of molecules N_a, situated under the angle $\theta = \pi/2$ in relation to the director (in the activated state). For the *para*-azoxyanisole (PAA) at $T = 400$ K, $\gamma_1 \approx 0.05$ P, $V^a \approx V_N = 228.2$ cm^3/mole [168], $I_2 = 59 \cdot 10^{-38}$ g·cm^2 [273] and, consequently, $\nu_0 = 2.6 \cdot 10^8$ Hz, $\tau = 2.7 \cdot 10^{-11}$ s, $N_a/N_g \approx 6.4 \cdot 10^{-3}$, i.e. slightly more than 1.5% of the molecules of the nematic liquid crystals are rotated by 90° in relation to the optical axis. This is the consequence of the low value of the angular free volume (see (3.4.36)) which, at $L = 18$ Å, is equal to $d = 4.6$ Å, $k_p = 0.64$, $V_0 = 283$ Å^3, $\Delta\theta = 0.05$ rad $\approx 3°$. These estimates are reliable if it is assumed that only one molecule takes part in the reorientation process. If we discuss the reorientation of a cluster consisting of several molecules, the value N_a/N_g changes; this will be discussed later. Equation (3.3.8) can be used to determine the absolute and relative values $V(0) = V_N(T = T_0)$ of the volume of the nematic liquid crystals, at which the free volume disappears:

$$V(0) = \frac{V_N}{1 + \alpha(T - T_0)}. \tag{4.3.9}$$

For PAA (or 1OAB in the notations used in [168]) $\alpha = 8.9 \cdot 10^{-4}$ K^{-1}, $T_0 =$ 368 K. Consequently, at $T = 400$ K, $V(0) = 0\ .973$, $V_N \approx 223$ cm^3/mol, for 6OAB $T_0 = 313$ K and $V(0) = 0.925\ V_N$. For all members of the PAA series, with the exception of the seventh member, the freezing of the movement of the director ($T = T_0$) takes place at a relatively low value of the molecular packing coefficient, $k_p \approx 0.643 \div 0.651$. For the substances whose molecules contain a fragment with saturated bonds, this value is considerably higher because of the low value of $T_g \sim 200$ K. For example, for PCH-7, the limiting value of k_p is equal to 0.68, and for BCO-5 it is –0.74. It should also be mentioned that the substitution of the physical parameters of the PAA and of the experimental results of the parameter obtained in the fitting of the experimental data, into equation (3.3.10) gives the value of the coefficient

γ which is considerably lower than the theoretical value. This coefficient characterises the overlapping of the free volumes.

If the molecular packing determines the rotational viscosity of the nematic liquid crystals, then its temperature dependence, in particular, activation energy, is determined by the type of intermolecular interaction. It is shown in [229] that for the substances with the identical type of intermolecular interaction $E \sim T_{NI}$. For example, for the cyano derivatives of biphenyl, phenyl pyridine, phenyl pyrimidine $E/T_{NI} \approx 16$, and for the cyan derivatives of phenyl cyclohexane and phenyl bicyclooctane $E/T_{NI} \approx 10$. This is in agreement with (3.1.4), if equation (3.1.6) is taken into account. In [266] on the example of the homologous series of alkyl cyanobiphenyls it was also showed that the variation εS correlates with the variation of T_{NI}. For these substances, with the exception of BP-8, $\varepsilon S/\Delta H_{NI} \approx 6$ in contrast to the ratio $\varepsilon S/\Delta H_{NI} \approx 2$ which was proposed in [162] and corresponds to the potential of the paired interaction. The values of ΔH_{NI} were taken from [274].

Figure 4.3.6 gives the concentration dependence of activation energy E, expressed in eV, and the relative value of the order parameter S^*, measured on the basis of the infrared dichroism of the absorption band of the CN-group of the molecules of PCH-7 in mixtures with ZhK-440. The values of S^* are not equal and are only proportional to the order parameter because the local field is not taken into account [270]. The value of E was calculated from equation (3.1.2) because the substances included in the mixtures differ in the type of intermolecular interaction and their dependence on the order parameter is expressed by different equations (see section 4.1 and Fig. 4.1.3). We can see the qualitative and quantitative agreement of the concentration dependences of E and S^*. The ratio $E/S^* \approx 1.2 \pm 0.2$ is fulfilled for different concentrations of PCH-7. Taking into account the relatively small variation of clarification temperature T_{NI} for the investigated mixtures ($\Delta T_{NI} = 345\ \text{K} \div 330\ \text{K} \approx 4 \div 5\%\ T_{NI}$) it may be assumed that the activation

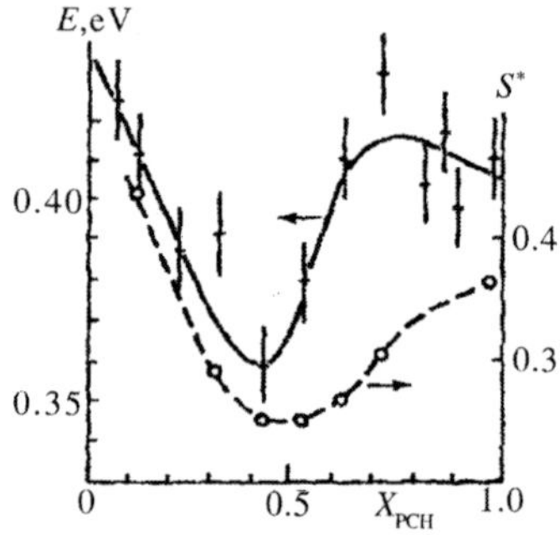

Fig.4.3.6. Dependence of activation energy E of rotational viscosity γ_1 and the relative order parameter S^* on the concentration X_{PCH7} in the mixtures of PCH-7 with ZhK-440.

energy changes mainly as a result of the variation of the order parameter. This is an agreement with (3.1.4) and (3.1.6). The variation of S with the concentration takes place as a result of the formation of associates (solvates) of the molecules of PCH-7 and azoxy compounds, as confirmed by the measurements of the excess volume of the mixtures [271, 272], dichroic spectra of infrared absorption [270] and x-ray diffraction investigations [275].

It was shown in [276] that in the mixtures of components with different temperature dependences of S, the difference in the order parameter, measured by some method for every moment, may reach $\Delta S = S_2 - S_1 \approx 0.1 \div 0.3$. The appropriate model for the distribution function takes into account the presence of the nearest environment of the molecule, including six molecules.

It should be mentioned that the data published in Figures 4.3.4 and 4.3.3 indirectly confirm one of the previously made assumptions, namely: the decrease of the activation energy at approximately identical concentrations of the component of the mixtures correlates with the increase of the free volume, as indicated by the decrease of γ_1 and parameter G in this concentration range.

The proportionality of activation energy in relation to the height of the Maier–Saupe potential is also found in other molecular–statistical descriptions of the temperature dependence of rotational viscosity (in (3.4.43) $J \sim \varepsilon S$, in (3.5.10) activation energy E is described by (3.5.1) which coincides with (3.1.7)). In this case, γ_1 is determined mainly by the value of the molecular friction coefficient ζ using (3.4.43) or (3.5.2). Coefficient ζ is, in turn, the function of the molecular packing coefficient and dimensions of the molecule (see (3.4.42) and (3.5.7)). Using these equations, we estimate the value of ζ for the typical values $k_p = 0.6$, $L = 20$ Å, $d = 5$ Å, $T = 300$ K, $M = 300$ atomic mass units, $V \approx 300$ Å^3, varying activation energy. According to (3.4.42) at the typical value of the energy of intermolecular interaction equal to $G = 5000$ K, $\zeta = 10^{-21}$ g·cm^2 s^{-1}, which coincides with the estimates of the dielectric relaxation time (4.3.1). However, in this case, the activation energy of rotational viscosity should be 15 000 K and this contradicts experiments. Since ζ and γ_1 should have similar values of the activation energy E, substituting into equation (3.4.42) the value $E \approx 3G = 5000$ K, typical of the nematic liquid crystals, we obtain $\zeta = 10^{-33}$ g·cm^2 s^{-1}, which is too low a value.

In [18], the authors presented the values of ζ_γ for a number of substances calculated using equation (3.5.2) and ζ_{mol} and ζ^1_{mol}, calculated from equation (3.5.12). In particular, it is important to mention the selection of the dimensions of the molecules used for the calculations. The equations (3.5.10), (3.5.11–3.5.14) take into account the excluded volume calculated in the contact of two rigid molecules and affecting the different physical–chemical characteristics of substances. The authors of [247] investigated the effect of the excluded volume on the clarification temperature of the nematic liquid crystals. It is shown in [153] that the free volume is taken into account most efficiently if the length of the molecule is determined by the dimensions of the liquid core of the molecule and alkyl substituents, and the width by the dimensions of mainly the alkyl substituents in *trans*-configuration characterised by the highest statistical weight [277, 278]. In this selection the dimensions the nearest environment of the molecule is taken into account together with the rotation of the molecule around the long axis whose rate is three–four orders magnitude higher than the rate of rotation around the short axis. In this case, the mean molecular spacing does not increase and is equal to the value determined by the x-ray diffraction methods and coincides with the width of the core of the molecule d [249] (for aromatic compounds d = 4.5 Å). The values of the length of the interatomic bonds and dimensions of the atoms for the calculations were taken from [252]. The parameters of the nematic liquid crystals used for the calculations and the values of friction coefficient ζ_γ determined on the basis of rotational viscosity (using equation (3.5.2)) and calculated from equation (3.5.12) at different values of d (ζ_{mol} and ζ^1_{mol}), are presented in

Table 4.12. Molecular friction coefficient ζ^γ, calculated from rotational viscosity γ_1 and molecular parameters ζ_{mol}, ζ'_{mol}. The dimension $\zeta = [10^{-22}\,g{\cdot}cm^2{\cdot}s^{-1}]$; $N = (\zeta^\gamma, \zeta'_{mol})^{6/11}$ – estimate of the number of molecules in the reoriented particle (cluster), dialkyl azoxybenzols (nAB); $\Delta T = T_{ni} - T$ = 10 K.

n	γ_1	S	T_{ni}, K	K_p	V_0, Å^3	p
4	0.67	0.496	304.1	0.6314	315	2.57
5	0.28	0.592	339.8	06180	349	2.88
6	0.38	0.611	326.4	0.6228	383	2.39
7	0.36	0.704	343.3	0.6181	417	2.64

n	M	E_{M-3}, K	ζ	ζ_{mol}	ζ'_{mol}	N
4	310	1025	7.7	4.8	0.21	7.1
5	338	1369	5.5	3.5	0.15	7.1
6	366	1357	5.5	7.3	0.12	9.3
7	394	1646	8.7	16.0	0.27	6.7

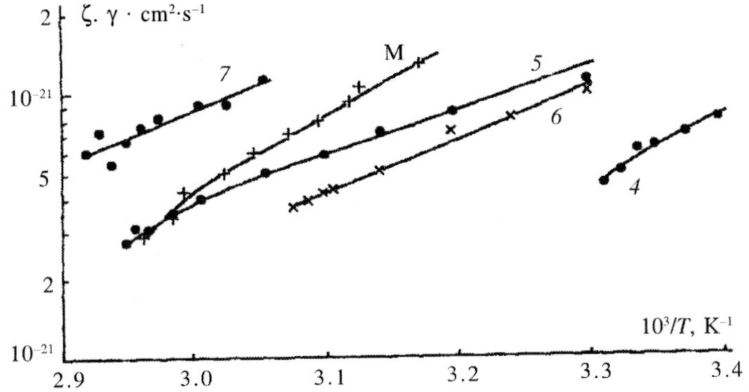

Fig.4.3.7. Temperature dependence of the molecular friction coefficient ζ^γ of dialkylazoxybenzols (nAB, C_nH_{2n+1}–C_6H_4–(N(O)=N–C_6H_4–C_nH_{2n+1}, n = 4÷7) and 4-methoxysalicilidene-4'-butylaniline (CH_3O–C_6H_3(2–OH)–CH = N–C_6H_4–C_4H_9, M). The numbers indicate the number of the homolog nAB.

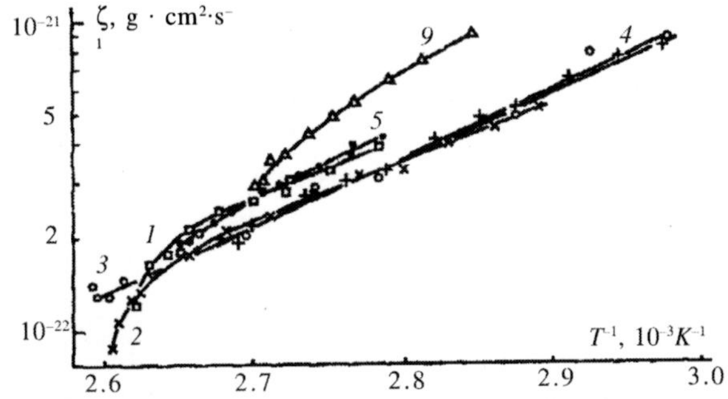

Fig.4.3.8. Temperature dependence of the molecular friction coefficient ζ^γ of series anizilidine-n-aminoxyphenyl acetate (APAPA, or MBnOA CH_3O–C_6H_4–CH=N=C_6H_4–OCO–C_nH_{2n+1}, n = 1÷5, 9, the numbers indicate the number of the homolog).

Table 4.12. Comparison of ζ_{mol} and ζ^1_{mol} indicates that these values are in almost complete agreement, confirming the validity of the selected model. The temperature dependences of ζ^γ for several homologous series are presented in Fig. 4.3.7 and 4.3.8. If the calculations are carried out using $d = d^{mol}$, the values of ζ_{mol} and ζ^1_{mol} differ by approximately a factor of 35. This difference is associated with the fact that a cluster of molecules (not a single molecule) takes part in the reorientation process. The number of the molecules in the cluster can be estimated using equation (3.5.12), assuming that the dimensions of the cluster are proportional to the dimensions of the molecules and the mass is proportional to the cube with the size:

$$\frac{\zeta^{\gamma}}{\zeta^{1}_{\text{mol}}} \approx \frac{\left(m^{1/2}d^{4}\right)^{\text{cl}}}{\left(m^{1/2}d^{4}\right)^{\text{mol}}} \approx \left(\frac{m^{\text{cl}}}{m^{\text{mol}}}\right)^{11/6} \approx 35. \tag{4.3.10}$$

From (4.3.10) $m^{cl}/m^{mol} \approx (\zeta^{\gamma}/\zeta^{1}_{mol})^{6/11} \approx 7$, i.e., the investigated particle consists of seven molecules and this is in satisfactory agreement with the number of the adjacent molecules (~ 6) in 4, 4'-dialkoxy azoxybenzols, determined by x-ray diffraction, or with the number of the molecules in the elementary cell (4–8) of the derivatives of 4-methoxy-4'-alkyltolanes, determined in the measurement of the molecular packing coefficient [278, 279]. The authors of [18] compared the values of the number molecules in the cluster N, determined on the basis of the friction coefficient, with the values $(\xi_2/d_2)^3 = N_{\text{xray}}$, determined in x-ray diffraction analysis of the nematic liquid crystal (ξ_2 is the correlation length, d_2 is the density wavelength equal to the length of the molecule or dimer) [275, 279]. The values, obtained by different methods, can be compared with each other. The supermolecular nature of the reorientation of the nematic liquid crystal has also been observed in the examination of the Kerr effect in [261, 262].

This is probably the reason why the values of the coefficient of rotational diffusion $D_{\perp}$ determined on the basis of the shape of the band of infrared absorption [280] and calculated on the basis of rotational viscosity $D^{\gamma}_{\perp}$ for 4-octyl-4'-ethoxytolane differ by up to 80 times (Fig. 4.3.9).

As regards the order of magnitude of the values ($10^7 \div 10^9$ s^{-1}) the values of $D^{\gamma}_{\perp}$ are close to those presented in [273] for *para*-azoxyanisole ($D_{\varphi} = 5 \cdot 10^7$ s^{-1}, $D_{\theta} = 6 \cdot 10^9$ s^{-1}, $T = 408$ K, φ, θ are Euler angles corresponding to the rotation around the short axis of the molecule).

It should be mentioned that the comparison of the values ζ_{mol} and ζ^{1}_{mol} with ζ^{γ} does not contain any contradiction (one molecule or 6–7 molecules) because the value ζ_{mol} is calculated taking into account the interaction with the nearest molecules.

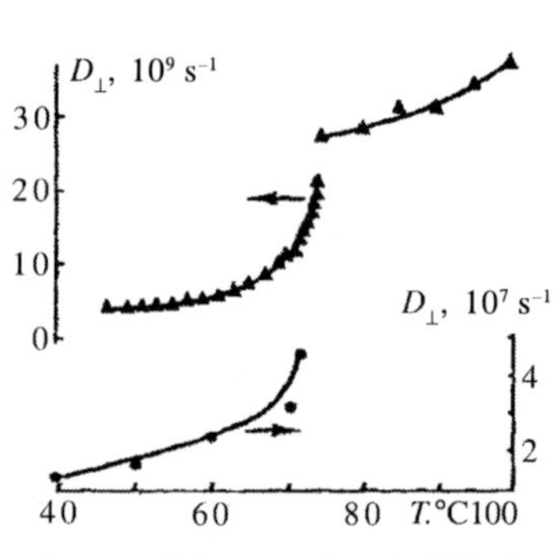

Fig.4.3.9. Temperature dependence of the rotational diffusion coefficient D^{γ} and $D_{\perp}$ for 8TO2.

In [281], investigations were carried out into the effect of the anisotropy of rotational viscosity or rotational friction tensor $\delta\zeta = (\zeta_{\parallel} - \zeta_{\perp})/(\zeta_{\parallel} + 2\zeta_{\perp})$ on the value of the rotational diffusion coefficient $D_{\perp}$ obtained in examination of the infrared spectrum. The time dependence of the function $a(t)$, which is the Fourier transform of the spectrum of infrared absorption $A(\omega)$, in parallel and perpendicular polarisation in relation to the director of polarisation of infrared radiation is described by the following equation:

$$\frac{d}{dt}\ln\frac{a_{\parallel}(t)}{a_{\perp}(t)} = D_{\perp}\frac{9S}{(1-S)(1+2S)}\frac{1+\delta\varsigma(-1+5S+2S^2)/3}{(1-\delta\varsigma S)(1+2\delta\varsigma S)}. \qquad (4.3.11)$$

It was shown in [281] that the anisotropy of molecular friction has only a slight effect on the measured value of $D_{\perp}$.

4.4. Pressure dependence of the viscosity of nematic liquid crystals

Measurements of the viscosity of nematic liquid crystals with the variation of external pressure make it possible to determine the relationship of viscosity with molecular packing and the free volume because the increase of pressure is accompanied by a decrease of the spacing of the molecules. The authors of [67] measured dynamic viscosity by the falling sphere method, and the authors of [94] measured rotational viscosity by the method of the spinning magnetic field (section 2.3).

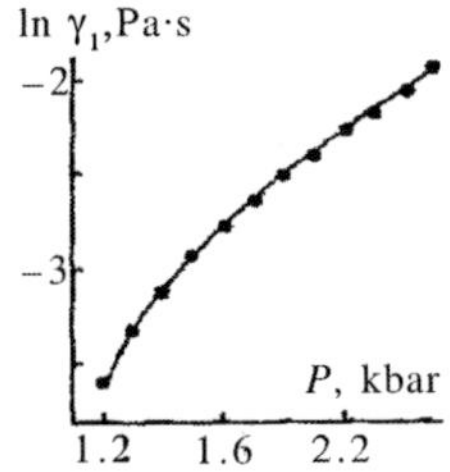

Fig.4.4.1. Dependence of rotational viscosity γ_1 of MBBA on pressure p at T = 85.24°C [94].

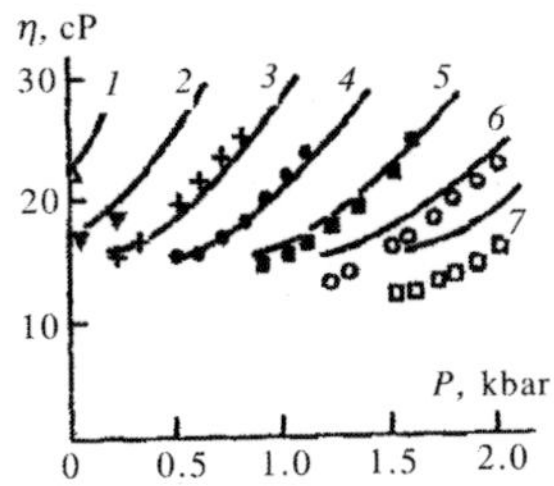

Fig.4.4.2. Viscosity–pressure isotherms for MBBA. Solid lines – calculated in [68], points – experimental data from [67]; T = 30 (*1*), 40 (*2*), 50 (*3*), 60 (*4*), 70 (*5*), 80 (*6*), 90°C (*7*).

The typical form of the dependence $\gamma_1(p)$ is shown in Fig. 4.4.1, $\eta(p)$ in Fig. 4.4.2. The form of the dependence ln $\gamma_1(p)$ is identical to ln γ_1 (T^{-1}) (Fig. 1.1.2), and the relationship is linear at comparatively high pressures ($p > 1.6\cdot10^8$ Pa) with the angle of inclination d (ln γ_1)/$dp = 1.2\cdot10^{-8}$ Pa^{-1} in the SI units and the decrease of ln γ_1 with the decrease of pressure, associated evidently with the decrease of the clarification temperature and the order parameter. For this reason, the pressure dependence of γ_1/χ_a [95–98] is linear because the maximum pressure in these studies reached 10^8 Pa. In [99], the same authors already described the exponential dependence of rotational viscosity on pressure.

The pressure dependence of the viscosity coefficients α_1, α_4, γ_1, γ_2, $\gamma_3 = \alpha_6 + \alpha_5$ is explained in [68] on the basis of the assumptions made in [164, 168]. For example, in [68] viscosity γ_1 is described by the equation similar to (3.3.1):

$$\gamma_1 = \frac{2k_bT}{V}a_1S^2\tau_1, \qquad (4.4.1)$$

where τ_1 is the orientation relaxation time. According to the free volume theory:

$$\tau_i \approx \frac{h}{k_bT}\exp\frac{\Delta V_i}{V_f}, \qquad (4.4.2)$$

where $V_i \sim U_iS^2$ is the volume expansion required for some rotational or translational motion, V_f is the free volume sufficient for rotating the molecule. The expression for V_f differs from (3.3.8) by the presence of the pressure-dependent component which decreases the value of V_f:

$$V_f \approx V_0\left[\alpha(T-T_0)-\beta(p-p_0)\right], \qquad (4.4.3)$$

where α is the thermal expansion coefficient, β is compressibility, p_0 = 1 atm. The dependence of the freezing temperature of motion of the director T_0 on pressure dT_0/dp is determined by the ratio $\beta/\alpha \approx (\beta_N - \beta_K)/(\alpha_N - \alpha_K) \approx 26\cdot10^{-5}$ K/Pa which is in agreement with the pressure dependence of the melting and clarification temperatures (for MBBA and EBBA $35\cdot10^{-5}$ and $32\cdot10^{-5}$ K/Pa respectively [67]). The equations (4.4.1)–(4.4.3) yield the following expression:

$$\tau_i = A_i(p,T)\exp\frac{U_iS^2}{V_0\left[\alpha(T-T_0)-\beta(p-p_0)\right]} \qquad (4.4.4)$$

The equations identical with (4.4.1), with equation (4.4.8) taken into account, were obtained for the viscosity coefficients in [68]:

$$\gamma_1 = g_1 S^2 A_1(p,T)\exp\frac{U_i S^2}{V_0\left[\alpha(T-T_0)-\beta(p-p_0)\right]}. \qquad (4.4.5)$$

When specifying the value of S, it is necessary to take into account the variation of T_{NI} with pressure and also a small decrease of the order parameter at high pressures: 5.6% per 1 kbar = 10^8 Pa. Figure 4.4.2 compares the pressure dependences (calculated using equations (4.4.1) and (4.4.4)) of dynamic viscosity with the experimental values [67]. The values of the pre-exponential multipliers A and energy constants U_i were taken from [168]. These relationships are in good agreement in a wide temperature range. In [94], the equation (4.4.5) describing the dependence $\gamma_1(T, p)$ was replaced by the expressions taking into account the pressure dependence of temperature T^* at which the order parameter is $S = 0$:

$$T^* = T_0^*\left(\frac{p}{a}+1\right)^C. \qquad (4.4.6)$$

for MBBA a = 4339 bar, C = 0.495, T_0^* = 320.0 K. If at the atmospheric pressure $T^* = T_1^*$, the temperature is represented by parameter T' associated with absolute temperature T through the relationship [94]:

$$\frac{1}{T'} = \frac{1}{T} - \frac{1}{T^*} + \frac{1}{T_1^*}. \qquad (4.4.7)$$

Consequently, assuming that $U_1 = B$ and T_0 are independent of pressure:

$$\gamma_1 = A_1\left(1-\frac{T'}{T_1^*}\right)\exp\frac{B}{T'-T_0}. \qquad (4.4.8)$$

With some approximations, this equation may be transformed to the following equation, in a more suitable form:

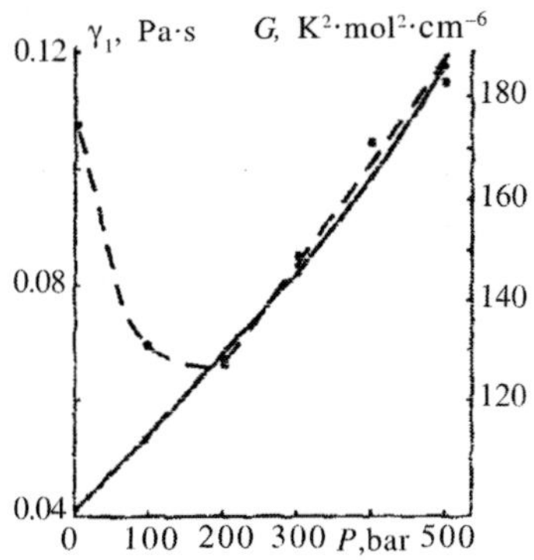

Fig.4.4.3. The dependence of rotational viscosity γ_1 of MBBA on the free volume parameter G on pressure p; T = 50°C.

$$\gamma_1 = A_2\left(1-\frac{T'}{T^*}\right)^{2\beta}\exp\left[c\frac{T^*-T_0}{T-T_0}\right], \tag{4.4.9}$$

The last two equations describe the experimental dependences with sufficient accuracy, with the difference not exceeding 2% [94].

The variation of the free volume with the increase of pressure should be taken into account not only in the index of the exponent (4.4.2) but also, as in the case of (4.3.8), in the pre-exponential multiplier. Figure 4.4.3 shows the dependences on pressure of rotational viscosity γ_1 and parameter G for MBBA at T = 40°C. The values of γ_1 calculated from (4.4.9) at T_0 = 168.3 K, $A_2 = 4.5160\cdot10^{-4}$ Pa·s, $\beta = 0.1520$, $C = 5.4247$; values of G from the relationship $\rho(p)$ [67]. At $p \geqslant 2\cdot10^7$ Pa, the dependences $\gamma_1(p)$ and $G(p)$ satisfactorily correlate with each other. The increase of G at $p < 2\cdot10^7$ Pa is associated with a decrease of the free volume $V^a - V_N$ as the result of the dependence of density on the order parameter in the vicinity of the clarification temperature.

4.5. Temperature dependence of dynamic viscosity and Mesovich η_i and Leslie α_j viscosity coefficients

In the flow of a nematic liquid crystal through a capillary the orientation of the director is approximately parallel to the direction of the flow. This means that $\eta \approx \eta_2$ or, more accurately, according to the Leslie and Ericsson theory:

$$\eta = \eta_2 + (\eta_1 - \eta_2)\sin^2\theta_0 - \eta_2 - \alpha_3 = \frac{1}{2}(\alpha_4 + \alpha_6 - \alpha_3), \tag{4.5.1}$$

where θ_0 is the angle of orientation of the director in the flow (see (1.2.7)). It is rational to link the values of γ_1 and η by a specific the relationship because the measurement of the value of dynamic viscosity is considerably simpler. As indicated by Fig. 1.2.4, the ratio $\eta(\gamma_1/S^2)^{-1}$ depends only slightly on temperature away from the nematic liquid crystal–isotropic liquid

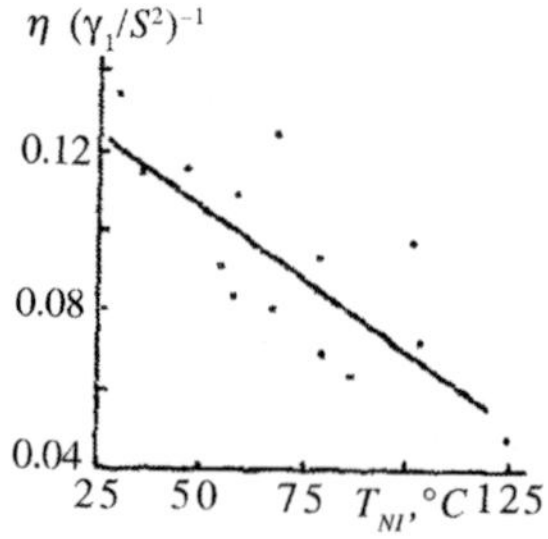

Fig.4.5.1. Relationship of the ratio $\eta(\gamma/S^2)^{-1}$ with clarification temperature.

transition, and the small increase of the the ratio in the vicinity of T_{NI} is explained evidently by the effect of parameter S. At first sight, the ratio $\eta(\gamma_1/S^2)^{-1}$ is independent of the structure of the molecules of the nematic liquid crystal, unless the substances with high T_{NI} have lower viscosity coefficient ratios (Fig. 4.5.1) [26]. The increase of the ratio $\eta(\gamma_1/S^2)^{-1}$ in the vicinity of T_{NI} is explained using the relationship presented in [123]:

$$\frac{\eta_s}{\gamma_1/S^2} = \frac{2}{15}a_6\kappa^2 + \frac{4}{105}\kappa^2 S - \frac{1}{2}S^2, \qquad (4.5.2)$$

where a_6 is a constant which is independent of temperature and is determined by the chemical structure of the nematic liquid crystal, $\kappa = (L^2 - d^2)/(L^2+d^2)$, L and d are the length and the short axis of the ellipsoid of rotation, respectively. This shows that the dependence of the viscosity ratio on the degree of ordering is completely determined by the third term in the right–hand part of (4.5.2), and also that away from temperature T_{NI} $\eta \sim \gamma_1$. The authors of [26] published an empirical relationship linking the values η and γ_1 which holds for different nematic liquid crystals with satisfactory accuracy (the maximum deviation for several substances is of the order of 20%):

$$\gamma_1(T) = S^x\eta_S(T)\gamma_1\frac{T_{NI}-10\text{ K}}{S^x(T_{NI}-10\text{K})}\frac{\eta(T)}{\eta(\text{T}_{NI}-10\text{K})} =$$

$$\left[1-\frac{T}{T_{NI}+0.35}\right]^{0.35}\frac{\eta(T)}{0.1076-5.048\cdot 10^{-4}(T_{NI}[^\circ\text{C}])}. \qquad (4.5.3)$$

The typical temperature dependences of the Mesovich viscosity coefficients are presented in Fig. 1.2.2. The following relationship holds for different substances in the entire nematic range [27]:

$$\sum_i A_i\eta_i = 0, \qquad i = 1, 2, 3, \qquad (4.5.4)$$

Here $A_1 : A_2 : A_3 = 0.08{:}1.1{:}(-1)$ for N_4, MBBA and mixtures of esters and 5BP [23] or 0.114:0.608:(–1) for the MB1OA [27]. Dependence $\eta_i(S)$ may be taken into account by plotting the graph of the dependence of the viscosity coefficients $\eta_{ir} = \eta_i/\eta_a$, reduced to the mean value $\eta_a = (\eta_1 + \eta_2 + \eta_3)/3$, on

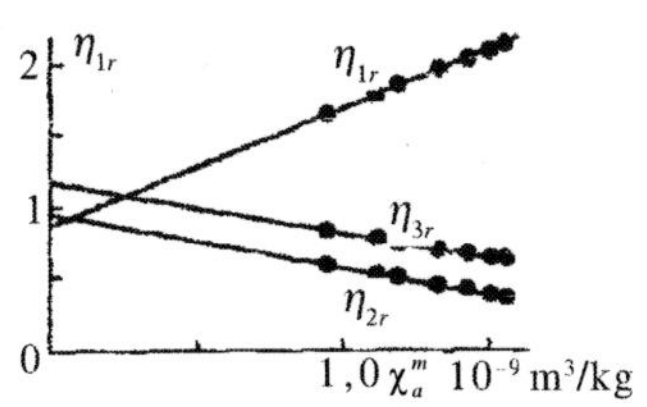

Fig.4.5.2. Relationship of Mesovich viscosity coefficients η_{ir}, reduced to the mean value, with diamagnetic isotropy χ_a^m.

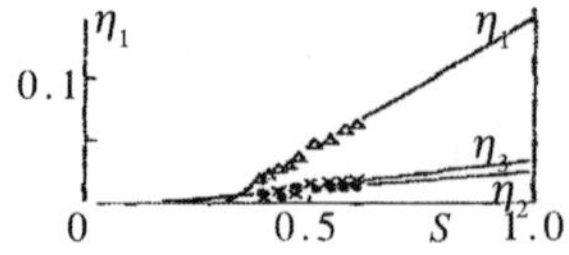

Fig.4.5.3. Relationship of the Mesovich viscosity coefficients η_i of MB1OA with the order parameter S.

the degree of ordering S or a parameter proportional to this degree, for example, diamagnetic anisotropy χ_a (Fig. 4.5.2, [23]). In extrapolation to $\chi_a = 0$ the following relationship is valid with the accuracy to 10%:

$$\eta_{ir} = 1 + \left(\frac{\eta_{i0}}{\eta_a} - 1 \right) S, \tag{4.5.5}$$

where η_{ir} is the viscosity of the absolutely ordered nematic liquid crystal (S = 1). However, the relationship (4.5.5) for the reduced viscosity coefficients is not valid for MB1OA [27]. For this substance, it is more rational to use the absolute values of viscosity (Fig. 4.5.3, [27]), instead of the reduced values:

$$\eta_i = \eta_i^0 + \beta_i S, \tag{4.5.6}$$

Here $\eta_1^0 = -0.068$, $\beta_1 = 0.214$; $\eta_2^0 = -0.007$, $\beta_2 = 0.029$, $\eta_3^0 = -0.012$, $\beta_3 =$ 0.042 Pa·s.

This method of processing the experimental data gives the equal values of η_i at $S = 0.33$ (at $T = T_{NI}$ $S \sim 0.4$). A simpler approach to describing the temperature dependence of viscosity in the flow of a nematic liquid crystal through a right-angled capillary was published in [282]. For each of the coefficients η_i for BP-5 the authors of [282] determined the thermal energy of activation which is equal to $E_1 = 0.49 \pm 0.06$, $E_2 = 0.21 \pm 0.04$, $E_3 = 0.2$ $\pm$ 0.04 eV. Thus, the height of the potential barrier appears to be linked with the flow geometry and also correlates with the viscosity values. The height is maximum when the viscosity is maximum, and vice versa. Possibly, the ordering effect of the magnetic field is important in this case.

In [170] it is shown for the Mesovich coefficients of a number of substances (biphenyls, azoxycompounds, phenylbenzoates, etc.) that in the entire nematic range:

$$0.5 \left(\frac{\eta_3}{\eta_2} + \frac{\eta_3}{\eta_1} \right) \approx 1. \tag{4.5.7}$$

The temperature dependences of the viscosity coefficient η_S and η_B, corresponding to the deformation by transverse and longitudinal bending, can be described, like γ_1, in accordance with the activation mechanism, for example, using the Diogo–Martins equation (3.1.7) (Fig. 4.5.4). As expected,

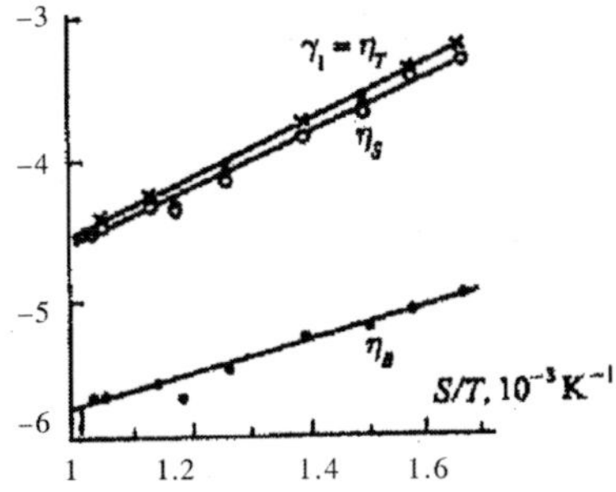

Fig.4.5.4. Dependence of splay viscosity η_S, bend viscosity η_B and twist viscosity (rotational) γ_i of MB1OA on the S/T [27].

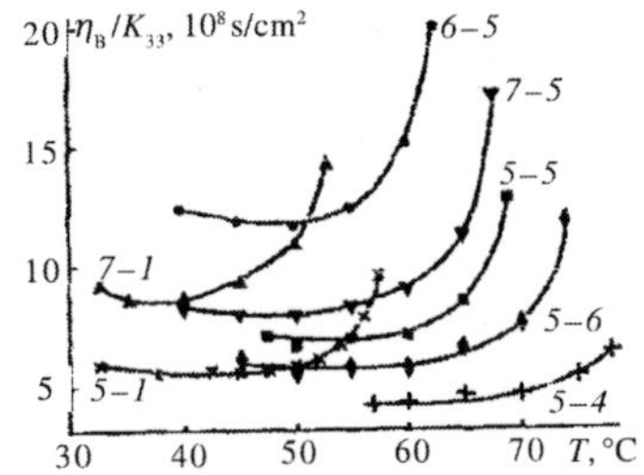

Fig.4.5.5. Temperature dependence of the ratio η_B/K_{33} of toluene derivatives nTOm.

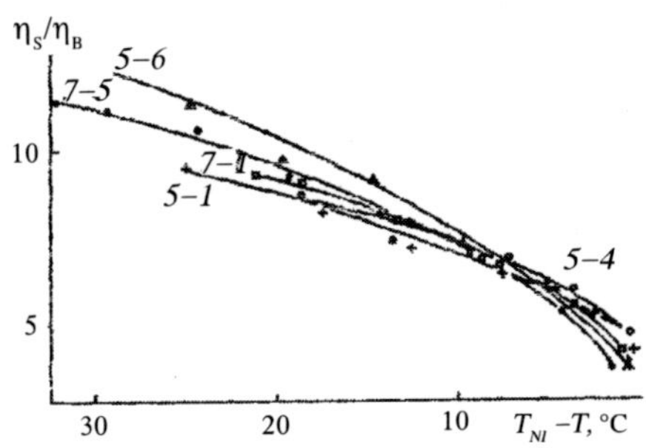

Fig.4.5.6. Dependence of ratio $\eta_S/\eta_B = \eta_1/\eta_2$ on reduced temperature $\Delta T = T_{NI} - T$ for toluene derivatives nTOm.

$\eta_S \approx \gamma_1$. The value of η_B is considerably lower than γ_1. In addition to this, η_B is also characterised by lower activation energy.

Figure 4.5.5 shows the temperature dependence of ratio η_B/K_{33} for the tolane derivatives nTOm [153]. It is determined by the simultaneous increase of η_B and the decrease of K_{33} in the vicinity of T_{NI} as a result of the variation of the order parameter S. If the ratios η_S/K_{11} (chapter 5) and K_{33}/K_{11} are available, it is possible to determine the ratio η_S/η_B equal to η_1/η_2 (1.2.13)). The dependence of the ratio on the temperature reduced to T_{NI} is shown in Fig. 4.5.6. Plotting the ratio η_B/η_S as a function of the parameter of the theory [123] gives:

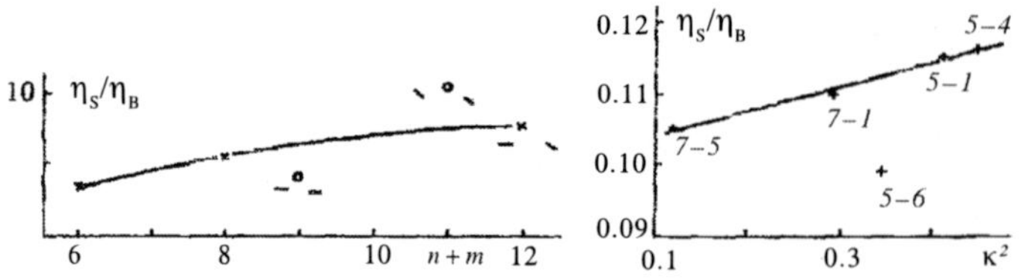

Fig.4.5.7. Dependence of ratio η_B/η_S for toluene derivatives nTOm on the sum n+m (left) and parameter κ^2; κ = $[(L/d)^2-1]$ / $[(L/d)^2+1]$ (on the right). The strokes indicatw the orientation of the final C–C groups in alkyl- and alkoxy substituents.

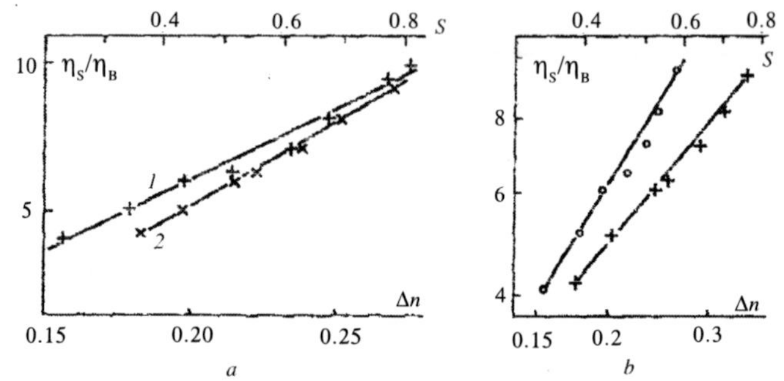

Fig.4.5.8. Dependence of ratio η_S/η_B of 5TO1 on birefringence Δn (1) and order parameter S (2) plotted in a) linear and b) logarithmic scales.

$$\eta_B / \eta_S = 1 + 341\kappa^2 + 0.038, \tag{4.5.8}$$

if the dimensions of the molecule are determined by the model (Fig. 4.5.7) described in chapter 5. Thus, (4.5.8) is identical with (4.5.2) at a_6 = 10.06 S. The values of the ratios $\eta_B/\eta_S \approx 0.1 \div 0.12$ for different substances [153] correspond to the value of the ratio $\eta(\gamma_1/S^2)^{-1} \sim 0.065$ [26].

The temperature dependence of ratio η_S/η_B is identical with the appropriate dependence of order parameter S (Fig. 4.5.6). Taking from [254] the variation of S and Δn with temperature for the substance 5TO1, we determine the appropriate dependence for η_S/η_B (Fig. 4.5.8). For the dependence on Δn plotting on the linear scale gives $\eta_S/\eta_B = -3.6 + 47.7\,\Delta n$, plotting on the logarithmic scale $\eta_S/\eta_B \sim (\Delta n^2)$. The dependence on S is the same for both processing methods and is presented in the following form:

$$\frac{\eta_s}{\eta_B} = 0.4 + 10.5S. \tag{4.5.9}$$

This equation is identical to the linear dependences of the ratios of the Mesovich coefficients on the order parameter, published in [27]. The equation (4.5.9) can be substantiated by dividing the equations in (3.4.49) for η_1, η_2 (see [193, 194]):

$$\frac{\eta_1}{\eta_2} = \left(1 + \frac{103}{35}\beta S\right)\left(1 - \frac{4}{7}\beta S\right)^{-1} = 1 + \frac{123}{35}\beta S + O(\beta^2 S^2) = 1 + 3.51\beta S + O(\beta^2 S^2). \quad (4.5.10)$$

Assuming that $\beta \sim 3$, it can be seen that (4.5.10) describes the experimental relationships quite satisfactorily.

The results of measurements of the Leslie viscosity coefficients α_i were presented in Fig. 1.2.3 for the MBBA and BBBA [23, 22] and in Fig. 4.5.9 for MB1OA (APAPA) [27]. In the case of MBBA, the temperature dependence of all coefficients, with the exception of α_3, is of the activation type with similar values of activation energy. In the vicinity of T_{NI} the dependence of all the coefficients, with the exception of $\alpha_4 = 2\eta_3$, which changes to the isotropic phase without any interruption, is affected by the variation of the order parameter. This is not contradicted by the dependences α_5/α_4 and $-\alpha_2/\alpha_4$ for APAPA, and also $|\alpha_3/\alpha_2|$ and α_6/α_5 (Fig. 4.5.10). The absolute value of the last two relationships is small and this confirms the conclusions drawn in [283] on the relationship of the dimensions of the

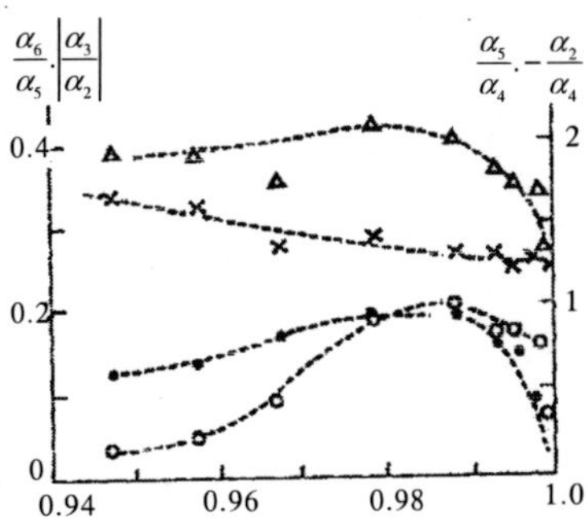

Fig.4.5.9. Dependence of the ratios of the Leslie viscosity coefficients α_i/α_j of MB1OA on reduced temperature T/T_{NI} [27]; $|\alpha_3/\alpha_2|$ – •, α_6/α_5 – o, α_2/α_4 – ×, α_5/α_4 – Δ.

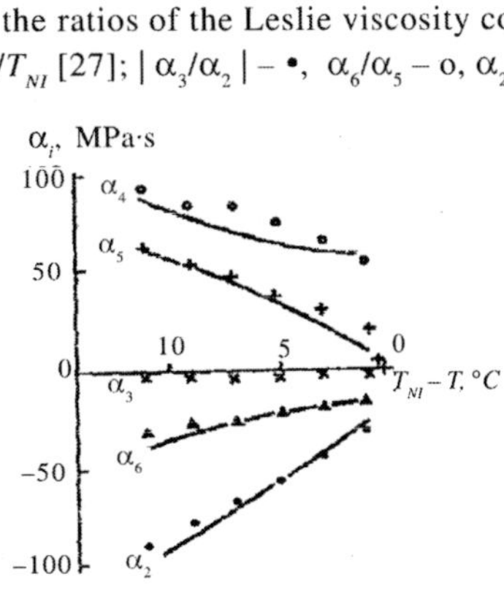

Fig.4.5.10. Dependence of Leslie viscosity coefficients α_i of DP-5. Points are data from [284], lines – data from [285].

molecules with the viscosity of the nematic liquid crystal, for example $\alpha_3/\alpha_2 = (L/d)^2$.

The dependences $\alpha_i(T)$ for BBBA appear to be, in the majority of cases, a continuation of the appropriate relationships of the MBBA. This indicates the small effect of the value of the alkyl chain on the value of α_i for the given homologous series. For the cyano biphenyls, on the other hand, $|\alpha_6/\alpha_5| \sim 1$, and in contrast to the substances belonging to other chemical classes, $|\alpha_2/\alpha_4| < 1$ [284, 285] (Fig. 4.5.10).

Table 3.2 compares the experimental values of η_i and γ_1, γ_2 for different substances with the values calculated by the molecular–dynamic methods [223, 224]. The application of these methods gives the exact order of the reduced quantity for $\eta_1^*, \eta_3^*, \gamma_1^*$ and η_{12}^* and the values of η_2^* reduced by a factor of 2–5. The estimates of α_i made at $Q = 7/3$ using the equations (3.6.13), give the following relationships: $\alpha_1 = -0.95\alpha_4$, $\alpha_2 = -\alpha_5 = -2.22\alpha_4$, $\alpha_3 = \alpha_6 = -0.41\alpha_4$; the quantities $|\alpha_i/\alpha_4|$, with the exception of $|\alpha_6/\alpha_4|$, are too high in comparison with the experimental values. Values closer to the given ratios are obtained at $Q = \sqrt{3}: \alpha_1 = 0.57\alpha_4$, $\alpha_2 = -\alpha_5 = -\alpha_4$, $\alpha_3 = \alpha_6 = -0.44\alpha_4$.

The ratios $(\eta_1 - \eta_2)/\eta_3$, calculated from equation (3.4.50), increase with the increase of the length of the molecule. This is in agreement with the experimental values. For example, for the PAA, BP-5 and MBBA ($L \sim 18$ Å) this ratio is equal to 2, and for hexyloxy benzilidene amino benzonitrile 2.67, and for *para*-azoxyphenetol ($L \sim 21$ Å) $(\eta_1 - \eta_2)/\eta_3 = 2.8$.

Comparison of the experimental data and the data calculated in accordance with [27] for the temperature dependences $\alpha_j(T)$ shows that if the viscosity coefficients are normalised using the value α_2, the accurate ratio is obtained for $|\alpha_5/\alpha_2|$ at any J/k_bT, $|\alpha_1/\alpha_2|$ at $J/K_bT \sim 6$ (as reported deviously, $\varepsilon S/k_bT \sim 6$ for alkylcyanobiphenyls). For $|\alpha_3|$ the theoretical estimate is 10–30 times higher, and for $|\alpha_6|$ it is 4–5 times lower.

The coefficients η_i, α_j, describing the flow of the nematic liquid crystal, can be used to describe the processes of translational self-diffusion. For the components parallel and perpendicular to the director, the viscosity coefficients α_j are included in the expression for the self-diffusion coefficient in the form of additions to molecular friction ζ [286]:

$$\begin{aligned} D^{\parallel} &= k_bT\Big[(\mu\zeta)^{-1} + 6\pi\mu^2\varphi(2+S)\times \\ &\qquad (4\alpha_1 + 1.5\alpha_2 + 6.5\alpha_3 + 6\alpha_4 + 10\alpha_5\Big], \\ D^{\perp} &= k_bT\Big[(\mu\zeta)^{-1} + 12\pi\mu^2\varphi(5-S)\times \\ &\qquad (-3.75\alpha_2 + 2.25\alpha_3 + 6\alpha_4 - 1.5\alpha_5)\Big], \end{aligned} \tag{4.5.11}$$

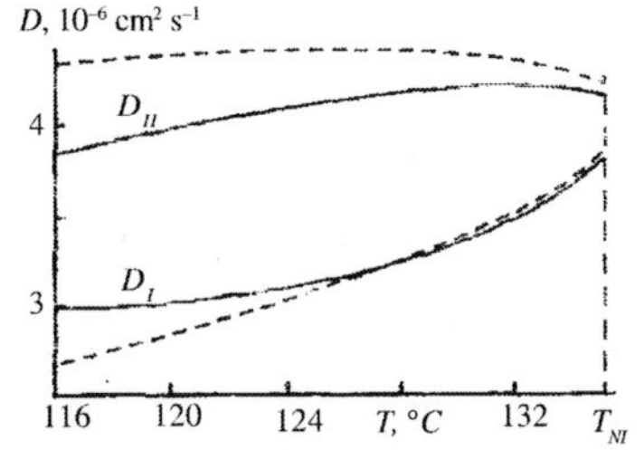

Fig.4.5.11. Temperature dependence of the translational diffusion coefficients, $D_\parallel$, $D_\perp$ of PAA. The solid lines – experimental data, dotted lines – calculated from (4.5.7)

Table 4.13. Values of Leslie viscosity coefficients ($\alpha_1 \div \alpha_6$) of MBBA (T_{NI} = 45.1°C), expressed in Pa·s [24]

T, °C	α_1	α_2	α_3	α_4	α_5	α_6
20.0	–0.0215	– 0.1534	– 0.000773	0.1095	0.1071	– 0.0471
25.0	– 0.0181	– 0.1104	– 0.001104	0.0826	0.0779	– 0.0336
30.0	– 0.0141	– 0.0800	– 0.001513	0.0644	0.0572	– 0.0244
35.0	– 0.0095	– 0.0573	– 0.001939	0.0515	0.0417	– 0.01763
40.0	– 0.0054	– 0.0387	– 0.00223	0.0422	0.0285	– 0.01244
42.0	– 0.0036	– 0.0310	– 0.00218	0.0394	0.0244	– 0.01086
44.0	– 0.0012	– 0.0210	– 0.001790	0.0374	0.01360	– 0.00917

where μ and φ are the parameters characterising the geometry of the molecule (for PAA μ = 2.95, φ = 1.47·10^{-9} cm). The values of $D^\parallel$, $D^\perp$, calculated from equation (4.5.11), are in good agreement with the experimental data (Fig. 4.5.11, [286]).

To conclude the section, the tables taken from [23, 24] for the coefficients α_j for MBBA, and η_i for MBBA, BP-5, BP-8, N4 (BMAOB) and EM (mixtures of phenyl benzoates C_5H_{11}–C_6H_4–COO–C_6H_4–OC_mH_{2m+1}, m = 1 and 3), are also included, Tables 4.13 and 4.14.

Table 4.14 Values of the Mesovich viscosity coefficients η_1, η_2, η_3, η_{12}, and rotational viscosity γ_1 of different NLC expressed in Pa·s [23, 24]

Substance	T, °C	η_1	η_2	η_3	η_{12}	γ_1
MBBA 45.1	20	0.1891	0.0308	0.0548	-0.0297	0.1526
	25	0.1361	0.0240	0.0413	-0.0181	0.1093
	30	0.1007	0.01927	0.0322	-0.0141	0.0785
	35	0.0741	0.01596	0.0257	-0.0095	0.0554
	40	0.0534	0.01377	0.0211	-0.0054	0.0365
	42	0.0456	0.01320	0.01972	-0.0036	0.0289
	44	0.0367	0.01323	0.01871	-0.0012	0.01920
	46			0.01852		
	50			0.01585		
	55			0.01323		
N4	20	0.277	0.0275	0.0515		0.209
	25	0.1708	0.0215	0.0396		0.1575
	30	0.1383	0.01715	0.0311		0.1204
	40	0.0862	0.01152	0.0204		0.0735
	50	0.0565	0.00832	0.01431		0.0461
	60	0.0376	0.00640	0.01058		0.0287
	68	0.0260	0.00550	0.00866		0.01805
	71	0.0220	0.00529	0.00815		0.01407
	73	0.01900	0.00526	0.00784		0.01090
	74	0.01711	0.00532	0.00777		0.00876
	75.5			0.00806		
	80			0.00719		
	85			0.00635		
	90			0.00566		
EM (5O1 and 5O3) 49.0	20	0.385	0.0458	0.0780		0.312
	25	0.260	0.0347	0.0576		0.221
	30	0.1880	0.0272	0.0444		0.1592
	35	0.1351	0.0220	0.0349		0.1146
	40	0.1031	0.01829	0.0282		0.0807
	45	0.0747	0.01588	0.0235		0.0527
	47	0.0634	0.01534	0.0221		0.0408
	48	0.0543	0.01530	0.0217		0.0334
	49.5			0.0234		
	55			0.01887		
	60			0.01572		
	65			0.01326		
BP-5 35.4	23	0.1296	0.0229	0.0374	-0.0111	0.0968
	26	0.1052	0.0204	0.0326	-0.0060	0.0777
	29	0.0869	0.01855	0.0287	-0.0055	0.0607
	32	0.0685	0.01720	0.0256	-0.0021	0.0450
	34	0.0581	0.01667	0.0240	-0.0020	0.0334
	34.9	0.0529	0.01689	0.0235	-0.0024	0.0262
	36			0.0253		
	41			0.0205		
	46			0.01684		
	51			0.01402		
BP-8 40.9	34	0.0967	0.0664	0.0291	0.6510	0.1112
	35	0.0892	0.0335	0.0280	0.1342	0.0836
	37	0.0790	0.0233	0.0260	0.0382	0.0618
	39	0.0667	0.01971	0.0244	0.0138	0.0447
	40	0.0594	0.01882	0.0239	0.0078	0.0337
	40.5	0.0547	0.0161	0.0237	0.0060	0.0250
	42			0.0253		
	47			0.0208		
	52			0.01731		
	57			0.01455		

Chapter 5

Effect of the molecular structure on the viscosity of nematic liquid crystals

5.1 Effect of the dimensions and form of the molecule and of its fragments on viscosity and its temperature dependence

The data on the viscosity of the substances, belonging in 320 chemical classes, have been published in [240]. Special attention has been given to the effect of the shape of the molecule and its dimensions on the viscosity coefficients η_S and η_B, on the ratios η_S/K_{11}, η_B/K_{33}, η_S/η_B and on the temperature dependence of these parameters (Fig. 5.1.1, 5.1.2, 5.1.3, and also Fig. 4.2.7, 4.5.6) for derivatives of tolane *n*TO*m* [153, 231].

In this section, attention is given to the variation of the investigated physical parameters in relation to the dimensions and configurations of the molecules. Plotting the dependence of the parameters on the sum $n+m$ at a fixed temperature shows that the viscosity coefficient η_S changes in the range 34–64 MPa·s, η_B in the range 4.0–6.8 MPa·s.

In the homologous series *n*TO*m*, the viscoelastic ratios η_S/K_{11} and η_B/K_{33} provide an even–odd alternative (Fig. 5.1.2, 5.1.3). The value of this ratio is minimum when the direction of the last intercarbon bond C–C in the alkyl and alkoxy substituents is parallel to the molecule axis; there is an intermediate value when one of the last bonds is situated under an angle in relation to this axis; the value is maximum when both bonds are directed under an angle to the longer axis of the molecule.

For 4, 4'-dialkyl and 4, 4'-dialkoxyazoxybenzols (*n*AB and *n*OAB), 4-methoxy benzilidene-4-alkaniloxyanilines (MB*n*OA) and for the derivatives of phenylpyridines the ratio $\eta_S/K_{11} \sim (W/L)^2$ is satisfied with sufficient accuracy (Fig. 5.1.4) [153, 231]. This approach has made it possible to explain the decrease of viscosity with the increase of the length of the molecule for the 2,3-dicyano-4-alkoxy phenyl esters of 4'-alkyl cyclo-hexane carbonic acids (Fig. 5.1.4) [287]. It is shown that the viscosity $\gamma_1 \cong \eta_S$ in the homologous series *n*TO*m* is proportional to the parameters F_1, F_2 which include the geometrical parameters of the molecules, according to the theory of rotational diffusion proposed by Osipov and Terent'ev (see (3.4.40), (3.4.43)) (Fig. 5.1.5).

The dependence of ratio η_S/η_B, which is identical to the ratio of the Mesovich viscosity coefficients η_1/η_2, on the shape of the molecules and temperature (see Fig. 4.5.6, 4.5.7) is described by the relationships:

$$\frac{\eta_B}{\eta_S} = 1.341\kappa^2 + 0.038;$$
$$\eta_S/\eta_B = 0.4 + 10.5S, \tag{5.1.1}$$

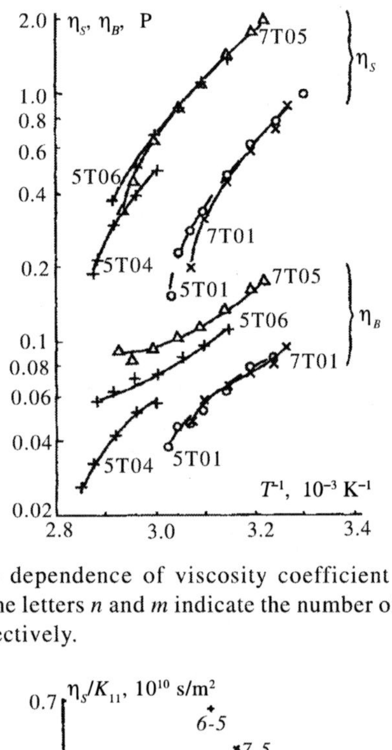

Fig 5.1.1. Temperature dependence of viscosity coefficient η_S and η_B of 4-alkyl-4'-alkoxytolanes *n*TO*m*. The letters *n* and *m* indicate the number of carbon atoms in the alkyl and alkoxy chains, respectively.

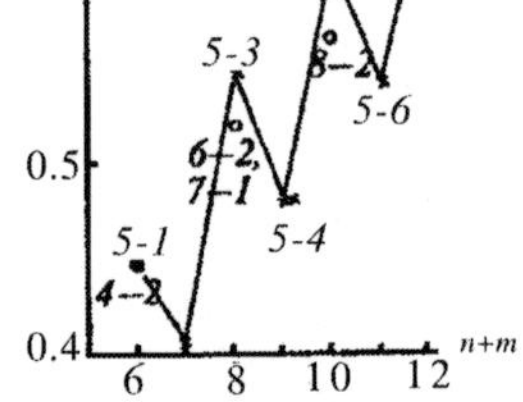

Fig 5.1.2. Dependence of the viscoelastic ratio η_S/K_{11} on the total number of carbon atoms in the alkyl and alkoxy chains *n+m* on the tolane derivative *n*TO*m* (*n–m*).

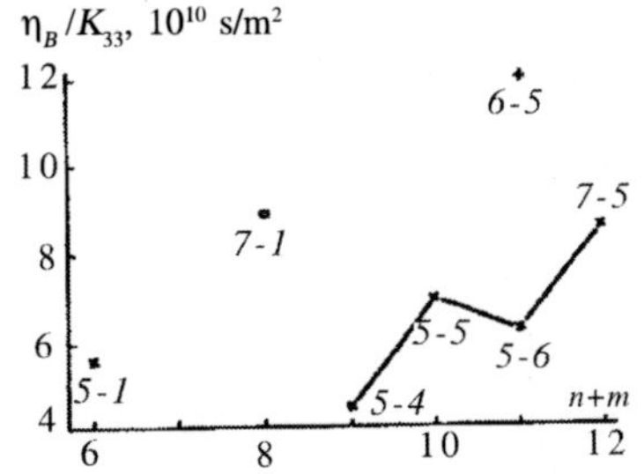

Fig 5.1.3. Dependence of the viscoelastic ratio η_B/K_{33} on the total number of carbon atoms in the alkyl and alkoxy chains $n+m$ on the tolane derivative nTOm (n–m).

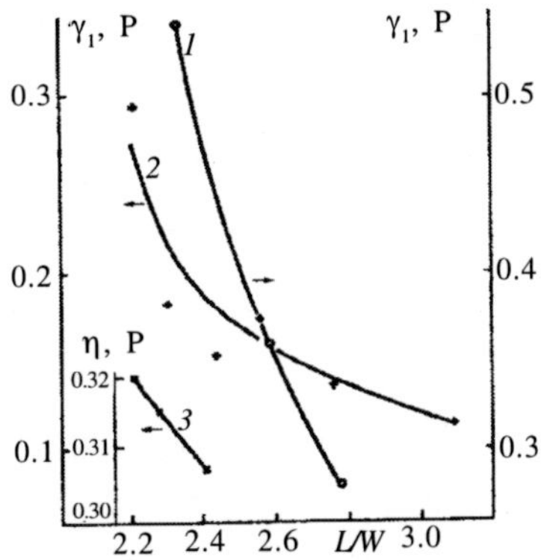

Fig 5.1.4. Dependence of rotational viscosity γ_1 of 4,4'-dialkylazoxybenzols (*1*) 4,4'-dialkoxyazoxybenzols (*2*) and 4-(2,3-dicyano)-phenyl esters of 4'-pentyl cyclohexane carbonic acid (*3*) on the form factor $p = L/W$ of the volume occupied by the molecule.

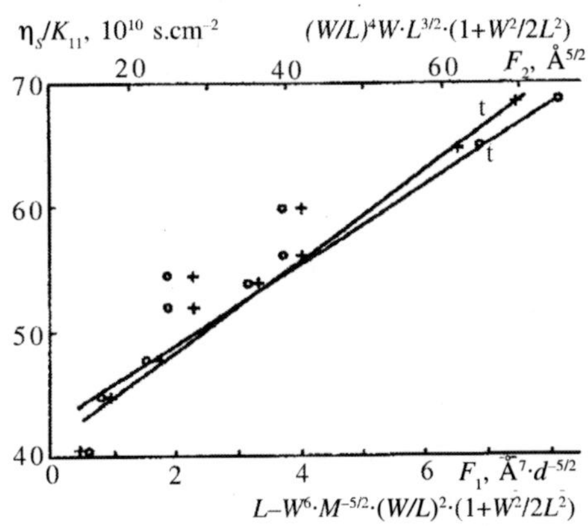

Fig 5.1.5. Relationship of viscosity ratio η_S/K_{11} with the geometrical parameters, included in the expression (3.3.4) from Osipov and Terent'ev theory [186].

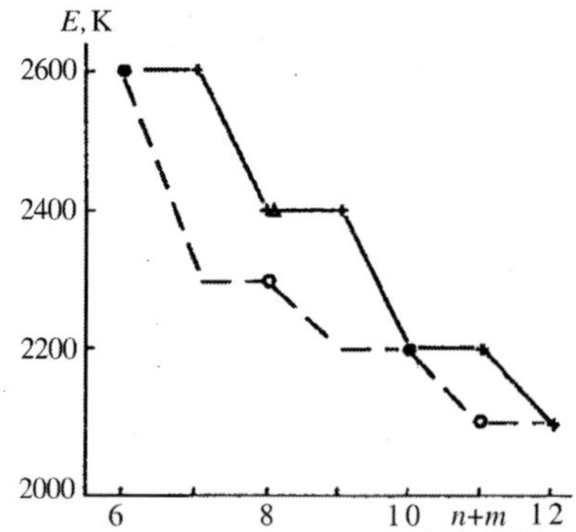

Fig 5.1.6. Dependence of the activation energy E and ratio η_S/K_{11} on the total number of the carbon atoms in the alkyl and alkoxy chains $n+m$ of tolane derivatives nTOm. Solid line indicates substances with odd n, dotted line substances with even m.

where $\kappa=[(L/W)^2-1]/[(L/W)^2+1]$, S is the order parameter. This is in satisfactory agreement with the theoretical results obtained by Hess [124] and Tskhai [194].

The increase of the alkyl substituents is accompanied by a decrease of the exponent of the dependence of viscosity η_S on the order parameter ($\eta_S \sim S^x$) from $x = 2$ for the homologs with the sum $n+m \leqslant 8$ to $x \sim 1$ for the homologs with very long chains ($n+m \geqslant 11$).

The effect of the configuration of the alkyl chains on the variation of activation energy E_1 of the ratio η_S/K_{11} was detected for the first time in the homologous series nTOm (Fig. 5.1.6) [153]. If in comparison of the characteristics of two substances with the values of $n+m$ differing by unity in the substance with the higher value of $n+m$ the last inter-carbon bond C–C is parallel with the long axis of the molecule, there is no formation of new gosh-configurations, i.e. the variation of conformation energy $\Delta E_{t-t} = 0$ which also corresponds to the detected variation of activation energy. However, if the last bond C–C is oriented under a valence angle in relation to the long axis of the molecule, a new *gosh*-configuration may form with the variation of conformation energy $\Delta E_{t-g} = 412$ kcal/mol = 207 K. The detected variation of activation energy $\Delta E_1 = 200$ K corresponds to this value.

5.2. Effect of the polarity of the molecule and its fragments on the viscosity of liquid crystal substances

The distribution of the electronic density in a molecule is described by the components of the polarisability tensor and by the value of dipole moment μ. For low-polarity compounds, the controlling contribution to the dielectric

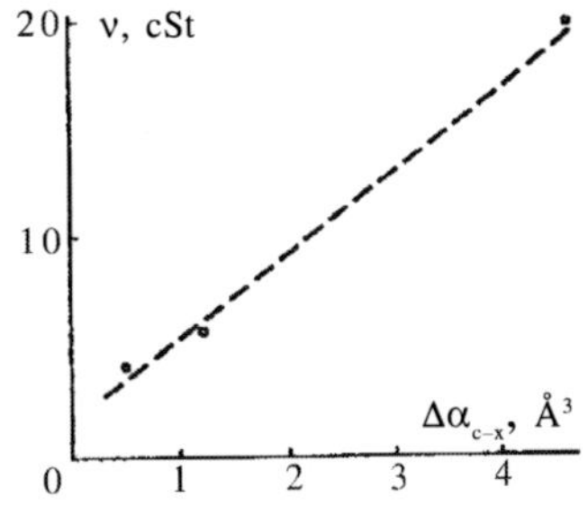

Fig 5.2.1. Dependence of the kinetic viscosity ν of the derivatives of phenylcyclehexane 3PCH –X (X=F, CH_3, CN) on the anisotropy of polarisability of the substituent X$\Delta\alpha_X$.

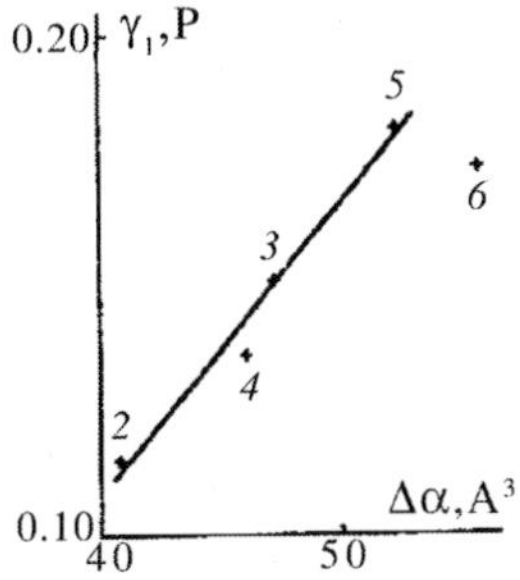

Fig 5.2.2. Dependence of the rotational viscosity γ_1 of 4,4'-dialkoxyazoxybenzols on the isotropy of polarisability of the molecule Δα. These numbers indicate the number of carbon atoms in alkoxy chains.

permittivity comes from electronic polarisability and, consequently, it is convenient to compare the viscosity with the anisotropy of polarisability of the molecule $\Delta\alpha=\alpha_{\parallel}-\alpha_{\perp}$ or with the anisotropy of polarisability of the bond $\Delta\alpha_{C-X}$, where X is the substituent (Fig. 5.2.1).

In the homologous series in which the variation of the anisotropy of polarisability is determined by the configuration of the alkyl substituent, the approximate ratio is γ_1~Δα (Fig. 5.2.2). The values of Δα, determined by the refractometric method, were taken from [111].

The dynamic or kinematic viscosity changes in the same fashion with increasing polarisability of the bond $\Delta\alpha_{C-X}$ or its dipole moment. On the example of derivatives of tolane, azomethines, azoxycompounds, phenyl benzoates and salicyliden anilines it has been shown that the increase of the dipole moment of a bridge fragment results in an increase of rotational viscosity γ_l and its activation energy E (Table 5.1). In heterocyclic compounds (phenyl dioxanes, phenyl pyridines, phenyl pyrimidines), the variation of the dipole moment of the heterocyclic fragment is very small but the increase of the coupling strength of π-electrons, especially in cyano derivatives, results in a large increase of the anisotropy of dielectric permittivity for

Table 5.1. The effect of the dipole moment μ of the core of the molecule with the bridge group on rotational viscosity γ_1 of the substance and its activation energy E

Substance	γ_1, P(25°C)	E, K	μ, D
	1.2	3600	0.3
	1.09	5820	1.57
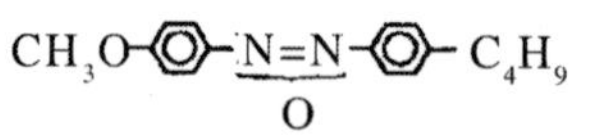	1.57	4550	1.70
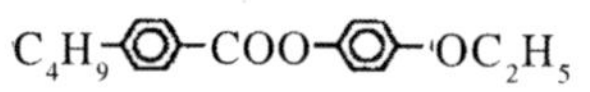	2.20	5110	1.90
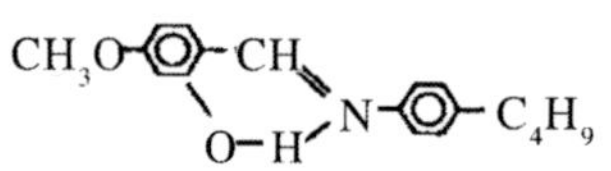	(6)	7350	2.39

Table 5.2. The effect of the structure of the heterocyclic fragment with the dipole moment μ and dielectric anisotropy Δε and rotational viscosity γ_1

Substance	γ_1, P	μ, D	Δε(ΔT = 10K)
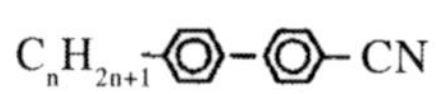	1.02	~0	11.73
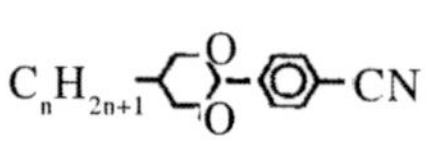	1.34	2.14–2.55	13.30
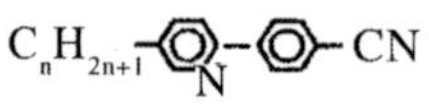	1.84	2.11–2.20	16.60
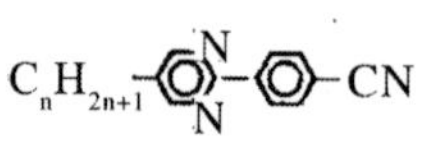	2.01	2.00–2.42	19.72

these substances in comparison with the derivatives of diphenyl and in an approximately proportional increase of their rotational viscosity (Table 5.2). In [240], on the example of liquid crystal materials produced by Hoffmann -La Roche, it is shown that at Δε=3÷20 and viscosity η=15÷50 cP the following linear relationship may be written:

$$\eta=A+0.02\Delta\varepsilon, \qquad A=0.105\pm0.05. \qquad (5.2.1)$$

Table 5.3. Physical properties of 4-alkylbenzoic (A; butyl/hexyl = 50/50%), 4-alkyl cyclohexane carbonic acids (B; butyl/hexyl = 50/50%) and their mixture (C; A/B=50/50%). The activation energy determined from the linear part of the dependence ln γ_1 $(1/T)$. All data at T=25°C

Mixture	T_m, °C	T_{NI}, °C	$\varepsilon_{\parallel}$	$\varepsilon_{\perp}$	n_e	n_o	10^7 K_{11}, dyne	γ_1, P	E, eV
A	72.5	109.5	3.10	2.72	1.609	1.471	8.75	0.30	0.36
B	–27	93.0	2.35	2.25	1.482	1.43	7.10	0.35	0.34
C		102.3	3.08	2.63	1.545	1.457	11.55	0.40	0.325

On comparative examples of 4-alkoxyphenyl esters of 4'-alkylbenzoic and 4'-alkylcyclohexane carbonic and also 4-alkylbenzoic acids, 4-alkylcyclohexane carbonic acids and their equimolar mixtures, it has been shown that the simultaneous presence in the mesogenic molecule of alkyl cyclohexane and alkyl phenyl fragments results in the appearance of a non-compensated dipole moment [172, 290]. Consequently, the type of the potential of intermolecular interaction changes (members, proportional to the Legendre polynomials of not only the even orders but also of the first order appear) and the height of the potential barrier for the reorientation of the molecule decreases. The physical parameters of the mesogenic acids are presented in Table 5.3. The experiments show that the value of $\Delta\varepsilon$ for the mixture of the acids in which the dimers of the type

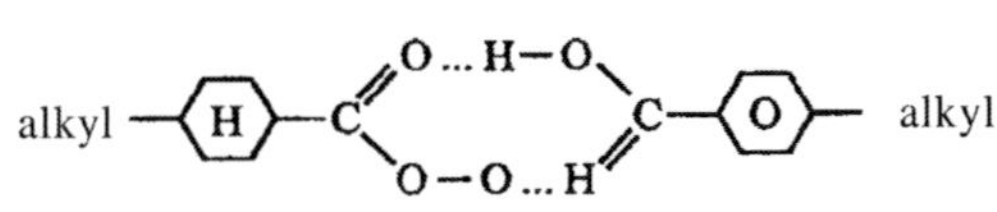

form is higher in comparison with the alkyl benzoic or alkyl cyclohexane carbonic acids, each taken separately. At the same time, the activation energy of rotational viscosity for the mixture is lower and it is in agreement with the theoretical considerations [172]. A similar decrease of the activation energy (from 0.52 to 0.41eV) and also of the rotational viscosity c (from 2.20 to 1.34 poise) is also found in phenyl benzoates and cyclohexyl carboxylates [172, 240].

5.3. Effect of the structure of molecule fragments on the viscosity and other properties of nematic liquid crystals

The substitution of oxygen atoms in dioxane derivatives by one or two sulphur atoms results in a consecutive increase of the viscosity of the derivatives

Table 5.4. Effect of substitution of oxygen atoms in derivatives of dioxane, oxathine and dithine on viscosity η at 20°C; C_nH_{2n+1}–X–Ph–N

n	X	η, P	T_{NI}, K
3(30%), 5(40%), 7(30%)	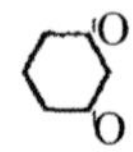	0.41	318.2
4(40%), 7(60%)	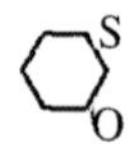	[0.54]	(288)
4(40%), 7(60%)	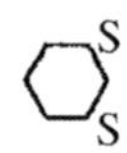	[0.74]	

Table 5.5. Effect of phenylcyclohexane R_1–Cy–Ph–R_2 with alkyl substituents

R_1	R_2	η, mPa·s	T_{NI}, K	Reference
C_3H_7	C_3H_7	[7]	[217]	296
C_3H_7	OC_2H_5	[13,5]	358	296
C_3H_7	CH=CH-CH_3	[17]	329.7	269
C_3H_7-CH=CH-	OC_2H_5	11.8		297

of oxathiane and dithiane (Table 5.4) [291–293]. This is also observed in the substitution of oxygen in the –COO-fragment [294, 295]:

$$\nu(20^\circ\mathrm{C})\{C_4H_9-Cy-COO-Ph-OC_2H_5\} = 0.17\,\mathrm{St} <$$
$$< \eta(25^\circ\mathrm{C}\{C_5H_{11}-Cy-COS-Ph-OC_2H_5\} = (0.20\,\mathrm{P})$$

Tables 5.5 and 5.6 show the effect of the position of the –C=C-bond in alkenyl substituents on the physical properties of nematic liquid crystals [294, 296–298]. If the alkenyl group is combined with a benzol ring, the viscosity of the liquid crystal increases in comparison with the case of the alkyl substituent. On the other hand, the viscosity decreases if the alkenyl chain is bonded with the cyclohexane fragment. In the case of the PCH derivatives, the maximum viscosity is found if the vinyl group is separated from the rigid core by the ethylene group, and the minimum value is obtained when the vinyl group is bonded directly with the cyclohexane fragment. However, the mixture with the first component has the lowest values of the

Table 5.6. Physical properties of 4-alkyl-(4'-cyanphenyl) cyclohexane T=22°C; [294, 297, 298] C_nH_{2n+1}-CH=CH-C_mH_{2m}-Cy-Ph-CN. Approximation $\ln \gamma_1 = \ln \gamma_0 + E/T$.

n, m	γ_1, P	η, P	γ_1/K_t, 10^{-10} s·cm^{-2}	T_{NI}, K	$-\ln \gamma_0$	E, K	Δn (ΔT=10 K)	t_{off}, ms
n = 3, m = 0	1.34	0.221	1.11	331.7	16.62	4990	0.116	40
n = 3, m = 0 (60 %) n = 2, m = 0 (40 %)	1.21	0.228	1.23	325.4	16.32	4870	0.117	42
n = 3, m = 0 (60 %) n = 0, m = 2 (40 %)	1.19	0.209	1.15	324.1	16.33	4870	0.115	37
n = 1, m = 2 (60 %) n = 0, m = 2 (40 %)	1.50	0.190	0.83	338.4	15.73	4760	0.127	32
n = 3, m = 0 (60 %) n = 0, m = 3 (40 %)	0.97	0.278	1.59	308.3	20.10	5920	0.105	67
n = 0, m = 5	0.99	0.215	2.02	305.4	14.65	4320	0.086	43
PCH-5	1.28	0.263	1.14	328.1	16.67	4990	0.099	55
BP-5	1.01		1.18	308.1	21.24	6270	0.178	

Table 5.7. Rotational viscosity, activation energy E and relaxation time of the twist effect t_{off} of bicyclic polar compounds.

Compound	γ_1, P	E, K	t_{off}, ms
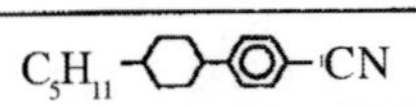	1.28	4870	43
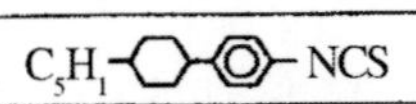	0.83	4300	41
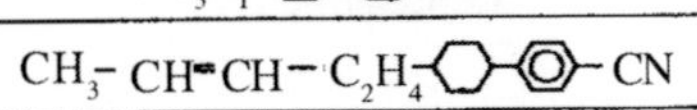	1.50	4760	32
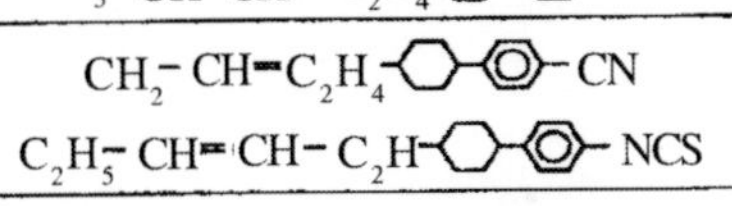	1.00	4820	32
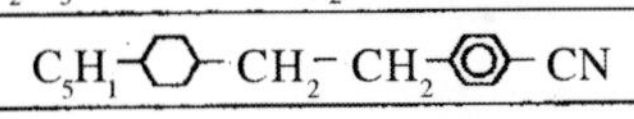	1.80	4870	43
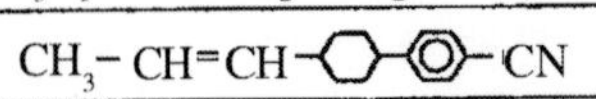	2.15	4990	29
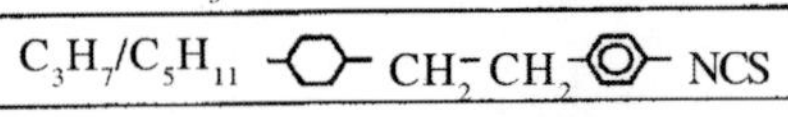	1.05	4820	21
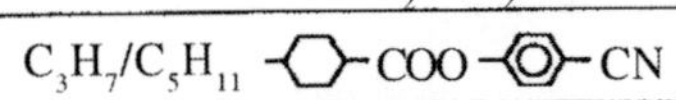	2.69	4430	29
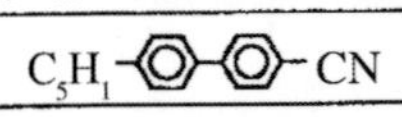	1.01	6270	55
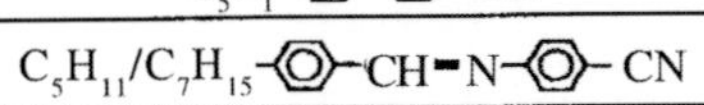	3.60	5360	39
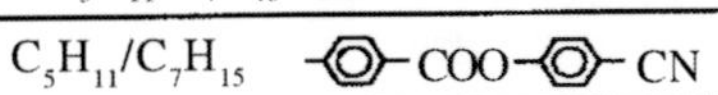	4.10	6480	70

viscoelastic ratio γ_1/K_t (K_t is the effective coefficient of elasticity in the twist cell) and, correspondingly, the lowest value of deactivation of the twist effect t_{de}.

The effect of the consecutive substitution of the alkyl substituent, the polar CN group and of the bridge group on the decrease of the relaxation

Table 5.8. Viscosity of cyclohexane and bicyclohexane derivatives R_1–Cy–X–R_2 (trans), R_3 (cis)

X	R_1	R_2	R_3	Viscosity, P or St (20 °C)	T_{NI}, K	Reference
*	C_3H_7	C_5H_{11}	-	η=[0.079]	300	300
**	C_3H_7	C_5H_{11}	-	η=[0.079]	295	300
Cy	C_3H_7	C_5H_{11}	-	η=[0.065]	370	300
Cy	C_3H_7	OCH_3	-	ν=[0.07]	290	301
Cy	C_3H_7	CH_2OCH_3	-	η=[0.05]	324.2	302
Cy	C_5H_{11}	$COOCH_3$ (50%)	-	η=[0.14]	351	303
		$COOC_3H_7$ (50%)				
Cy	C_5H_{11}	CN	-	ν=[0.65]	358	304
Cy	C_nH_{2n+1}	C_5H_{11}	CN	η=0.70	338	305
Cy	C_5H_{11}	C_2H_4-CH=CH_2	CN	η(22°C)=[0.50] γ_1(22°C)=[1.58] E=[6620K]	[325.8]	297
CH_3-Cy-COO-C_6H_2(2.3-CN)-OC_5H_{11}				η=[1.5]	(329.4)	287

*cyclohex-1-en (double bond at the alkyl fragment);
**cyclohex-3-en (double bond at the adjacent cyclohexane fragment)

Table 5.9. Viscosity of derivatives of cyclobutane, spiro [3.3] heptane and cyclopentane: 20°C; C_nH_{2n+1}–X–$(COO)_m$–Ph–$(COO)_p$–Ph–CN.

n	X	Position of alkyl group	*m*	*p*	η, P	T_{NI}, K
3		cis	1	0	[1.33]	336.2
3		trans	1	0	[0.74]	414.7
3	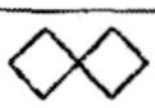	-‖-	1	0	[1.70]	428.0
5		-‖-	0	0	ν = [0.62] Ct	334
5		-‖-	0	0	ν = [0.62] Ct	495
3		-‖-	0	1	ν = [0.80] Ct	363
3		-‖-	0	1	ν = [1.00] Ct	500

time of the twist effect t_{de} is shown in Table 5.7. The rotational viscosity and relaxation time decrease if the isothiocyanate group (NCS), or an ethylene bridge or an alkenyl substituent is added to the molecule. The combined effect of all the previously mentioned fragments enables all the control parameters to be minimised. The decrease of the kinematic viscosity from 50 to 18 cSt and the increase of the clarification temperature from 43.5 to 98.5°C in the substitution of the CN group by the NCS the group in 5-pentyl-2-(4'-cyano or 4'-isothiocyanatephenyl, was described in [299].

In the derivatives of cyclohexane, the viscosity is independent of the position of the –C = C– double bond inside the cyclohexane fragment (Table 5.8). If the molecular core consists only of cyclohexane fragments, the viscosity of the cyan derivatives increases as a result of denser packing. However, the viscosity in bicyclohexane and derivatives with alkyl and alkoxy substituents and ester chains with the oxygen atom between two carbon atoms is very low (Table 5.8). The substitution of the hydrogen atom in the equatorial position by a cyano-group enables substances to be produced with high values of the negative dielectric anisotropy and with the viscosity lower in comparison with a 2, 3-dicyanphenyl esters of alkylcyclohexane carbonic acids.

The data presented in Table 5.9 illustrate the effect of the structure of other cyclic fragments on the viscosity. The axial substitution of the alkyl group in the cyclobutane fragment reduces the viscosity of the nematic liquid crystal in comparison with the same substitution in the equatorial position [306]. The addition of a spiropyran fragment instead of the cyclobutane one increases the viscosity of the substance [306]. In comparison with identical cyclohexane derivatives, the substances with the cyclopentane fragments are characterised by lower viscosity and clarification temperature [307].

Table 5.10 compares the data on the viscosity of complex esters with the direct and reversed sequence of the atoms in the –COO– group. The substances with the fragment C_nH_{2n+1}–Ph–O– are characterised by lower viscosity in comparison with the substances with the fragment C_nH_{2n+1}–Cy–O–. For the alkoxy compounds with the –COO– bridge, the viscosity in the core of the molecule decreases if the benzol ring is situated between two oxygen atoms. 4-methylphenyl ester 4'-propylbicyclohexane of carbonic acid is characterised by the lowest value of viscosity η amongst the three-ring slightly polar compounds, and also by a high clarification temperature. In the group of the low-viscosity polar compounds it is important to note 1-(4'-butylcyclohexyl)-2-(4''-(3''',4''', 5'''-trifluorophenyl)phenyl)ethane (η = [16.1 mPa·s], $\Delta\varepsilon$ = [11.8] and T_{NI} = 51.8°C [50]).

The fluorine derivatives have become the main component of the materials for active matrix liquid crystal displays. The effect of the position

Table 5.10. Viscosity of several complex esters R_1–A–$(COO)_n$–B–$(COO)_m$–D–R_2

R_1	A	n	B	m	D	R_2	η, P (20°C)	T_{NI}, K	Literature
C_5H_{11}	Cy	1	-	0	*	F	[0.194]	(258)	308
C_5H_{11}	Cy	1	-	0	**	F	[0.17]	(282)	309
F	*	1	-	0	Cy	C_5H_{11}	[0.70]	(267.2)	310
F	**	1	-	0	Cy	C_5H_{11}	[0.62]	(289.9)	310
C_5H_{11}	Cy	1	Ph	0	Cy	C_3H_7	(0.40)	466	301
C_3H_7	Cy	0	Ph	1	Cy	C_3H_7	[1.30]	433	311
C_3H_7	Cy	0	Cy	1	Ph	CH_3	[0.16]	477	295
C_3H_7	Cy	0	Cy	1	Cy	CH_3	[0.19]	406	295

*–Ph(3,F) and ** – Ph(2,F) – the benzol ring with fluorine in the *ortho-* and *meta*-position, respectively.

of the fluorine atoms on the physical properties of the three-ring nematogens is shown in Table 5.11. Basically, viscosity γ_1 increases with the increase of polarisability and the size of the core. The minimum value the viscosity have been recorded for 3,4-difluorophenyl and 3-fluorine-4-perfluoromethoxy phenyl derivatives. The additional substitution by the fluorine atom in the orthoposition reduces the viscosity of the substances with the fluorine atom in the terminal position and increases it for the compounds with the OCF_3 group.

Table 5.12 shows the effect of the position of the nitrogen atom in a pyridine fragment on the viscosity and dielectric anisotropy of different substances with phenyl and cyclohexane rings [313]. The nitrogen atom in the orthoposition increases the dipole moment of the molecule and, correspondingly, rotational viscosity, whereas, being in the *meta*-position, its dipole is directed in the side opposite to that of the dipole generated by the CN-group.

Table 5.11. Physical properties of tricyclic compounds with fluorine atoms at 20°C

[312] C_3H_7-Cy-A-B-R; Ph(F) = 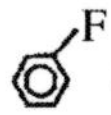 ; Ph(F,F)= 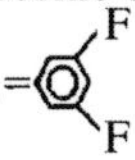

A	B	R	$[\gamma_1]$, MPa·s	T_{NI}, °C	[Δn]	U_p, V
Cy	Ph(F)	F	138	121	0.079	1.00
Ph	Ph(F)	F	138	81	0.142	0.95
Ph(F)	Ph(F)	F	165	50	0.139	0.88
Ph(F,F)	Ph(F)	F	220	48	0.135	0.83
Ph(F)	Ph(F,F)	F	153	50	0.135	0.81
Ph(F)	Ph(F,F)	$OCHF_2$	225		0.147	0.89
Ph(F)	Ph(F,F)	OCF_3	198		0.123	0.85
Ph(F)	Ph(F,F)	$OCHFCF_3$	232		0.138	0.82
Ph(F)	Ph(F,F)	OC_2F_5	290		0.135	0.88
Ph(F)	Ph(F,F)	$OCH_2C_2F_5$	220		0.138	0.85
Ph(F)	Ph(F,F)	$OCH_2CF_2CHFCF_3$	252		0.121	0.80
Ph(F,F)	Ph(F,F)	F	143			0.93

Table 5.12. Physical properties of polar derivatives of pyridine C_nH_{2n+1}–X–Pyr–CN at T = 25°C

n	X	Pyr	γ_1, P	E, K	T_{NI}, K	Δε
3(30%), 5(40%), 6(30%)	Ph	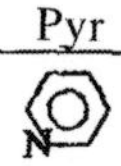	3.50	5860	335.9	7.3
5(40%), 7(60%)	Ph		2.20	5740	322.4	11.3
3(30%), 5(40%), 7(30%)	Cy	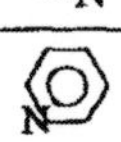	1.40	5630	330.5	3.3
5(40%), 7(60%)	Cy		1.78	5570	331.2	12.4

Chapter 6

Viscosity of mixtures of nematic liquid crystals

In the development of liquid crystal materials for various devices it is necessary to know not only the physical parameters of the individual components but also the variation of these components in mixing of the nematic liquid crystals with other substances. If the 70s of the previous century were characterised by the tendency for formation of the liquid crystal materials from components with similar chemical structures, recently to produce various useful properties of the liquid crystal materials extensive use has been made of mixtures of polar and non-polar substances belonging in different chemical classes, and also of the introduction of non-mesogenic and optically active conditions. The present chapter is concerned with the examination of the viscosity properties of the four main types of the mixtures of liquid crystals: from substances, having the identical molecular structure; from polar and non-polar substances; from systems of mesogen–non-mesogen mixtures; from the mixtures of nematic liquid crystals with optically active additions.

6.1. Viscosity of mixtures of mesogens

Examples of the mixtures of nematic liquid crystals consisting of chemically active substances have been examined in [314, 315]. Figure 6.1.1 shows the concentration dependence of the logarithm of rotational viscosity of mixtures of two azometins: ethoxy- and methoxybenziliden butyl aniline (EBBA and MBBA). These dependences are almost linear. The difference from linearity is 9%, which is 2.5 times smaller than the deviation from the linearity of the concentration dependence $\gamma_1(x)$. The identical situation is found for the concentration dependence of the quantity $\gamma_1(x)$ of the mixtures of ethyl hydroquinonadi-4-hexyl benzoate and alcoxy phenyl alcoxy benzoates [316] (in this system of the mixtures, the clarification temperature T_{NI} is almost constant) and also the values of η of mixtures. The

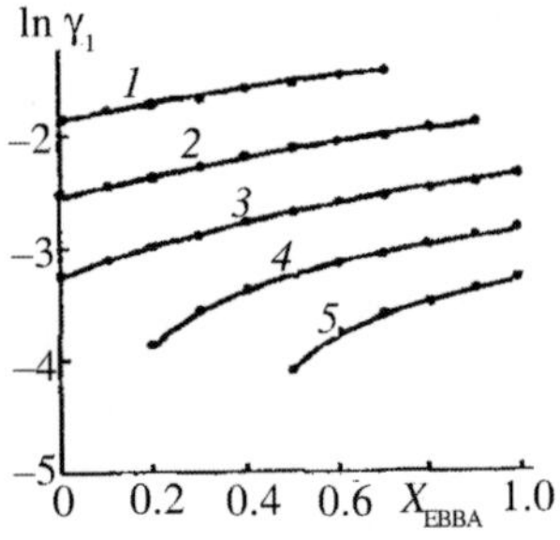

Fig. 6.1.1. Dependence of the logarithm of rotational viscosity γ_1[Pa·s] of mixtures of MBBA and EB-BA on the concentration of EBBA, X_{EBBA}: $T = 10°C$ (*1*), 25°C (*2*), 30°C (*3*), 20°C (*4*), 42°C (*5*) [314].

concentration dependences of this type of mixtures are strongly affected by the free volume.

A more complicated dependence is found in the case of the mixtures of polar and non-polar substances. This problem has been studied extensively in [269, 317]. Figures 6.1.2 and 6.1.3 show the concentration dependences of rotational viscosity γ_1, clarification temperature T_{NI}, the activation energy E and the coefficient of elasticity of deformation in transverse bending K_{11} of two systems of mixtures-non-polar ZhK-440 (mixtures of

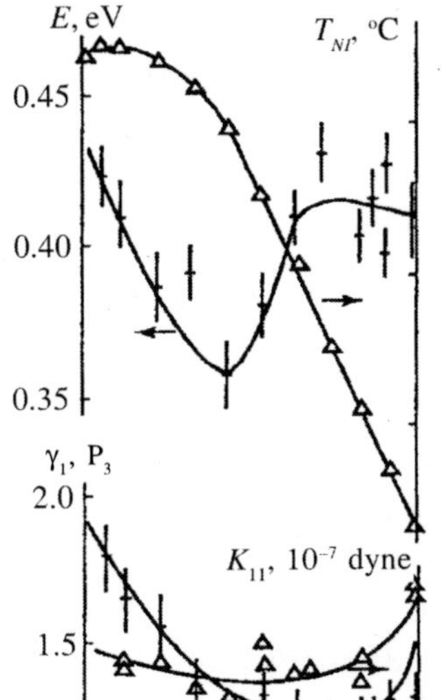

Fig. 6.1.2. Dependence of the rotational viscosity γ_1, its activation energy T, elasticity coefficient K_{11}, and clarification temperature T_{NI} of mixtures of ZhK-443 (azoxy compounds BMAOB and BGAOB [268]) with PCH-7 on the concentration of the polar component.

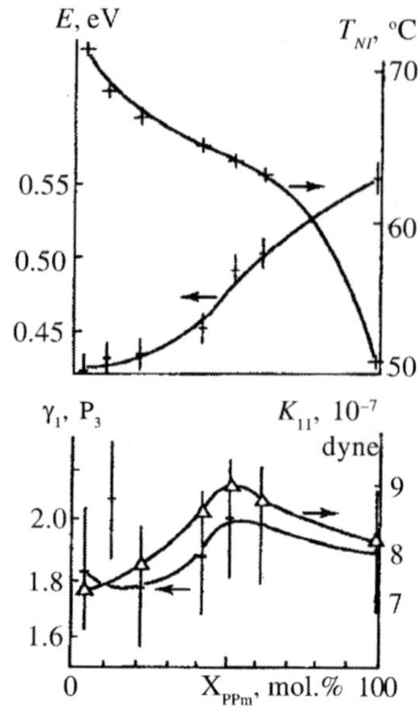

Fig. 6.1.3. Dependence of rotational viscosity γ_1, its activation energy T, elasticity coefficient K_{14}, and clarification temperature T_{NI} of mixtures of ZhK-940 (azoxy compounds BMAOB and BGAOB [268]) with FPm-5 (pentyl (cyanophenyl) pyrimidine) on the concentration of the polar component.

azoxycompounds $\varepsilon_{\perp} = 5.5$, $\Delta\varepsilon = -0.5$), and the substances, containing the polar CN-group – 4-heptile-(4'-cyanophenyl) of cyclohexane (PCH-7) and 4-pentyl-(4'-cyanophenyl) of pyrimidine. It should be mentioned that the last two substances differ in the type of intermolecular interaction. Both mixture systems are characterised by the considerably non-additive behaviour of the concentration dependences of all parameters. The dependence of γ_1 on the concentration of 4-cyanophenyl ester of 4'-heptyl benzoic acid in the mixtures with ZhK-440 was investigated in [318]. At low concentrations of the polar component, the viscosity of the mixtures changes only slightly. However, this is not associated with the slight effect of the additions, but with the mechanism discussed below.

The dependences $\gamma_1(x)$ and $E(x)$ will be examined in greater detail. At low concentrations of the slightly polar component, the dimers of cyanoderivative molecules still exist [279]. Consequently, the energy of the potential barrier at the higher level does not change. The increase of the concentration of the azoxycompounds is accompanied by two processes: failure of the dimers of the cyanoderivative and formation of associates from polar and non-polar molecules, as indicated, for example, by the x-ray diffraction measurements of the wavelength of the density of fluctuations of the shock range order [279]. The stoichiometry of the associates is equal to 1:1 and, consequently, the maximum deviation of all physical quantities is detected at the concentration of 40÷50 mol.%. The formation of associates results in the appearance of a non-compensated dipole moment reducing the height of the potential barrier required for rotation of the molecule, i.e., the activation energy decreases together with the degree of ordering S [271] (see also 4.1.3 and Fig. 4.3.4). In addition, depending on the presence or absence of the fragment with saturated bonds, the structure and concentration of the associates also change. This leads to changes and, correspondingly, to a decrease or retention of the value of γ_1 on the previous level.

Figure 6.1.4 shows the concentration dependence of the excess volume V^E of mixtures of five homologues of cyanoderivatives of diphenyl (BP-5) with 4-butyl-4'-methoxyazoxobenzol (BMAOB) which is the main component of ZhK-440 [268]. It may be seen that in mixing of azoxycompounds with the substance characterised by higher values of the energy of intermolecular interaction(BP-5, E = 0.55 eV, γ_1 = 0.82 P), the formation of associates is accompanied by 'densening' of the mixture, and viscosity changes slightly or extensively in comparison with calculations using the rule of additivity (at $X_{BP\text{-}5}$ = 0.4, γ_1 = 1.25 P [271]; in the mixtures of FPm-7 with ZhK-440 the viscosity is almost constant). With a lower energy of intermolecular interaction (PCH-5, E = 0.41eV) and with approximately the same size of the molecules, the concentration of the polar compound of 40÷50 mol.% is characterised by the 'disintegration' of the mixture, and the viscosity of the mixture decreases (Fig. 4.3.3). In the

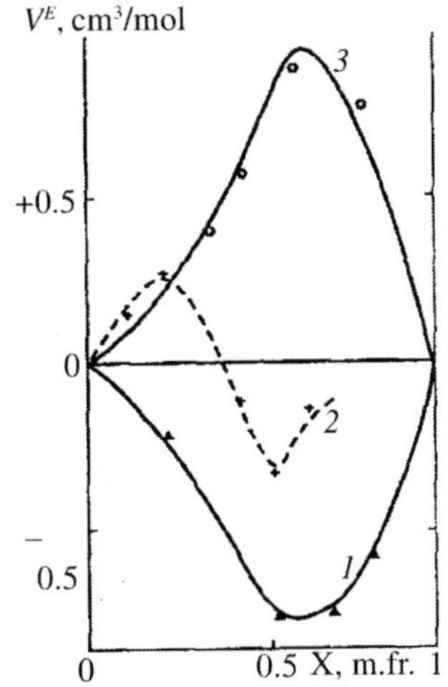

Fig. 6.1.4. Dependence of the excess volume V^E of mixtures of BMA of BMAOB (azoxy, analogue N7) with BP-5 (*1*), PCH-5 (*2*) and BCO-5 (*3*) of the concentration of the polar component.

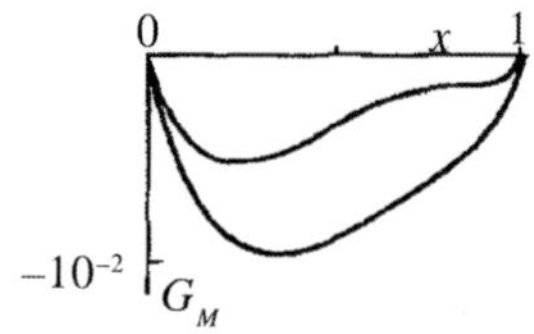

Fig. 6.1.5. Dependence of free energy G_M, calculated in [319] for the liquid mixture of charged solid particles at different ratios of particle sizes.

mixtures with BCO-5 it is natural to expect a decrease of viscosity in comparison with the viscosity calculated on the basis of the additivity rule, although the general viscosity of the mixture may slightly increase because of the higher viscosity of the BCO-5 which, as mentioned previously, is a consequence of the effect of the form of the bicyclooctane fragment on molecular packing.

In the mixtures of polar and non-polar substances forming an induced smectic phase, for example, BP-5 and MBBA, the variation of the concentration dependence of the parameters of the nematic liquid crystal is difficult to detect because of the rapid increase of the viscosity in the vicinity of the phase transition between the nematic and smectic crystals A [278]. At low concentrations of BP-5 in the mixtures with the EBBA, the activation energy slightly increases, as indicated by equation (3.1.2).

Thus, it may be seen that the variation of the concentration dependences of the physical parameters of the mixtures of the nematic liquid crystals with different polarity is determined greatly by the appearance of the non-compensated dipole moment leading to a decrease of the height of the potential barrier. This is agreement with [172] and may be formally described using the theory proposed by Diogo and Martins [168] (section 4.3 and 5.2). It is also important to note that the form of the dependence $E(x)$ is similar to the dependence of the free energy, calculated in [319] for a liquid mixture of charged solid particles (Fig. 6.15). The non-additive dependence of E on the concentration in the mixtures of polar and slightly polar substances is also in agreement with the non-additive variation of the coefficient of surface tension in the mixtures of BP-5 with MBBA [320].

6.2. The viscosity of liquid crystal solutions of non-mesogenic compounds

In the developed liquid crystal materials not all the components are mesogenic compounds. This applies to, for example, alkyl and alkoxy derivatives of phenylcyclohexanes, added for reducing the viscosity of the mixture, or different dyes for the visitor–master effect whose molecules have a

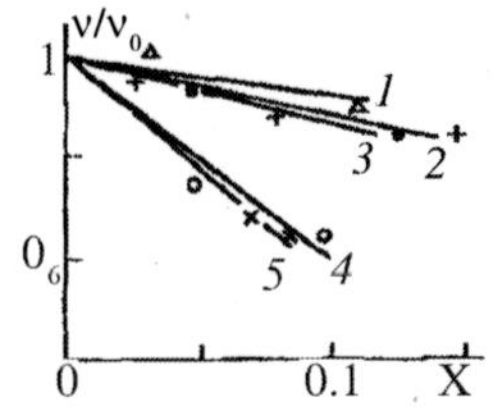

Fig. 6.2.1. Concentration dependencies of relative viscosity ν/ν_0 of the mixtures of *n*-xylol with five different liquid matrices.

complicated electronic structure. In [321–325], using binary mixtures of 'mesogen–non-mesogen' type, investigations were carried out into the effect of different additions and matrices on the viscosity of the mixtures, and physical mechanisms have been proposed for explaining the concentration dependence of kinematic viscosity ν, or molar fluidity $\varphi_m = (V\nu)^{-1}$, where V is the molar volume.

Figure 6.2.1 shows the concentration dependences of the relative viscosity ν/ν_0 of the mixtures of *n*-xylol with five different liquid crystal matrices [323]. The authors of [323] also calculating the concentration dependences of the molar fluidity φ_m of the mixtures H-23 (chemical formula $C_4H_9OCOO–C_6H_4–COO–C_6H_4–OC_2H_5$) with the hydrocarbons with the general formula C_8H_{10} and *n*-nitroaniline $H_2N–C_6H_4–NO_2$. The concentration dependences of the viscosity of the mixtures are described by the relationship:

$$\nu = a_1 + a_2 X, \qquad (6.2.1)$$

where X is the concentration of the non-mesogen. Table 6.1 [323] shows the values of the coefficients a_1 and a_2 for the mixtures of *n*-xylol with different matrices [323]. The decrease of a_1 is accompanied by a decrease of a_2, i.e. the viscosity in a more viscous matrix decreases to a greater

Table 6.1. Values of coefficients a_1 and a_2 in equation (6.2.1), describing the concentration dependence of the viscosity of solutions of *n*-xylol in different NLC

LC matrix	a_1, 10^{-5} m²/s	$-a_2$, 10^{-5} m²/s	T, °C
Cholesterine myristate	8.67 ± 0.54	32.9 ± 8.7	80
IVb	3.70 ± 0.08	15.50 ± 1.30	30
IIIb	3.15 ± 0.02	4.02 ± 0.29	22
IVg	1.42 ± 0.03	1.66 ± 0.51	66
IIIe	1.33 ± 0.02	1.40 ± 0.22	40

extent with the addition of the non-mesogen.

The concentration dependence of fluidity can also be described by the relationship identical with (6.2.1). In [324], the concentration dependences of viscosity and fluidity were investigated for the correspondence to the equation derived in [248] from the theory of the free volume for isotropic liquids:

$$\frac{\varphi_M - \varphi_0}{\varphi_0} = b_1 + b_2 \frac{V - V_0}{V_0}, \tag{6.2.2}$$

where V_0 and φ_0 is the molar volume and molar fluidity of the liquid at the melting point, and also for the correspondence with the equation:

$$\frac{\varphi_M - \varphi_0}{\varphi_0} = c_1 + c_2 V_f, \tag{6.2.3}$$

where V_f is the free volume of the individual solvent (mesogen) or liquid crystal solution (mesogen–non-mesogen mixture). The value of V_f is calculated from the density of the components, if the mechanism of formation of the solution is available (implantation, substitution mechanisms, or a combined mechanism), or from the density of mixes, if the solution formation mechanism is not available.

For for all mixtures, the value of coefficient b_1 from equation (6.2.2) is similar to zero, or more accurately it is equal to $1.40 \cdot 10^{-5}$ (Table 6.2) [324]. This value of coefficient b_1 is obtained if the temperature in the equation is assumed to be equal to the melting point. The value b_1 (as a_2 in 6.4.2) is determined mainly by the nature of both the solvent and the addition. The values of the coefficient b_1, presented in Table 6.2, showed that as the polarity of the non-mesogenic addition decreases, the fluidity of the mixture increases (the viscosity of the mixture decreases).

The correlation coefficients, determined in the approximation of the experimental data using (6.2.2), are close to unity, indicating that the approach, described in [248] for isotropic liquids, can also be used for anisotropic systems.

The equation (6.2.3) is not as universal as (6.2.1) or (6.2.2) since the increase of the concentration of the non-mesogen for a number of systems of mixtures (mainly with H-23) is accompanied by a variation of the free volume who sign is opposite to the sign of fluidity variation.

The latter fact is associated with the result which shows that in [323, 324] the specific intermolecular interaction was not taken into account in deriving (6.2.3) based on the geometrical factors associated with the implantation or substitution of the molecules.

The typical dependences of kinetic viscosity ν and activation energy of the viscous flow ΔH_p^m on the concentration of the non-mesogenic addition

Table 6.2. Values of the coefficients of equation (6.2.2) describing the relationship of fluidity φ_m and the molar volume V of solutions of different non-mesogens in H-23 ($C_4H_9OCOO{-}C_6H_4{-}COO{-}C_6H_4{-}OC_2H_5$) at T = 66°C

Non-mesogen	b_1, 10^{-5}	b_2, 10^{-5}	ρ, g · cm^{-3}
Ethylbenzyol	1.40	3.37	0.979
Octane	1.39	2.84	0.996
Chlorobenzol	1.38	2.59	0.904
Decane	1.41	2.31	0.995
Hexane	1.41	2.27	0.986
Toluene	1.40	2.20	0.996
m-xylol	1.42	2.19	0.992
Ethyl acetate	1.40	2.01	0.983
Benzol	1.38	1.78	0.960
Acetonitrile	1.40	1.76	0.980
n-xylol	1.43	1.66	0.956
o-xylol	1.38	1.48	0.996
Diphenyl	1.42	1.40	0.948
3-methylbutanol-1	1.38	1.17	
Nitrobenzol	1.41	1.15	0.991
Chloroform	1.40	0.95	0.967
Cyclohexane	1.43	0.39	0.997
Aniline	1.39	0.31	0.753
Butanol-1	1.41	–0.20	0.664
n-nitroaniline	1.43	–0.92	0.898

are shown in Fig. 6.2.2, and 6.2.3 [326–329]. Both these quantities increase at low concentration of the additions. Evidently, this is associated with a decrease of the free volume in the formation of the solution by occupying the vacancies by the molecules of the addition. With the increase of the

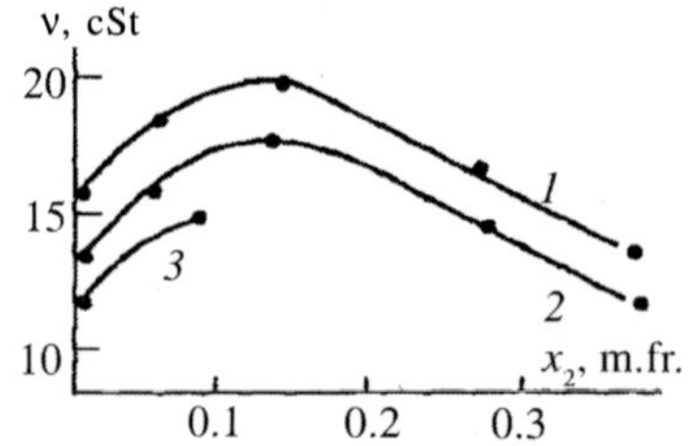

Fig. 6.2.2. Dependence of the kinematic viscosity of the flow ν on the concentration of the non-mesogenic addition N-23. 1) nitroaniline, 2) octane, 3) linear hydrocarbons.

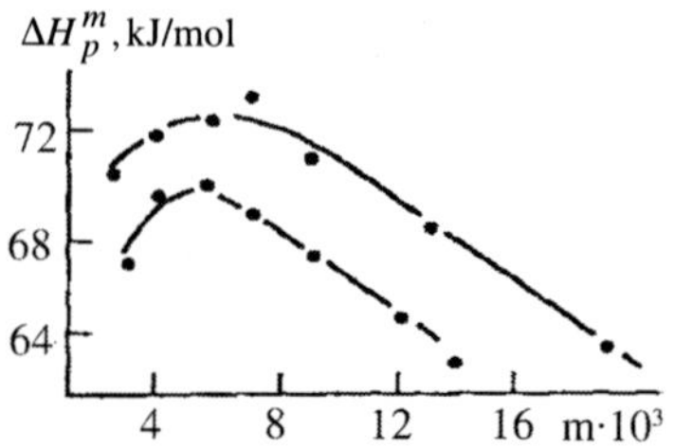

Fig. 6.2.3. Dependence of activation energy ΔH_p^m of the viscous flow on the concentration of the non-mesogenic addition to N-63. 1) nitroaniline (upper) 2) octane (lower).

concentration of the non-mesogen, the values of ν and also E decrease. In turn, this is caused by the increase of the free volume.

The authors of [330] presented the concentration dependences of rotational viscosity γ_1 with n-eikezan ($C_{20}H_{42}$). The variation of γ_1 with the increase of the concentration of the mesogen is several times greater than the appropriate variation of ν. Similar mixtures with hydrocarbons, with the length of the non-branched chain of 10–35 carbon atoms, have been proposed for decreasing the switching time of the twist-effect.

The data presented in Fig. 6.2.2 show that the octane with linear molecules and no dipole moment greatly increases the fluidity and reduces the viscosity of the solution, and nitroaniline, whose molecules have a high dipole moment $\mu = 6.8\ D$, slightly increases the viscosity of the mixture. The addition of 4 mol.% of the linear hydrocarbons C_nH_{2n+2} ($n = 6, 8, 10$) has almost no effect on the viscosity of N-23. The activation energy of the viscous flow E rapidly increases with the increase of the length of the hydrocarbon chain (Fig. 6.2.4, [322]). Evidently, this is associated with the increase of the flexibility of the molecules, the formation of non-linear confirmations and the decrease of the size of the free volume. As reported in section 4.3, this may increase the effective activation energy. The addition of 6 mol.% of *ortho*- and *meta*-xylol in EBBA does not change the viscosity of the solution, but the addition of para-xylol in the amount of 8 mol.%

reduces the viscosity of the solution from 13.7 to 11.7 cSt. The latter is the consequence of the more elongated form of the molecule of *n*-xylol and of the absence of the dipole moment in the molecule. The value of E increases with the increase of the ratio of the length of the molecule to its width: in 6% solution of *o*-xylol, $E = 3176$ K, in *m*-xylol $E = 3415$ K, in 8% solution of *n*-xylol $E = 3458$ K, in the case of EBBA $E = 3255$ K. These ratios may change for the mixtures of the isomers of xylol with other matrices.

The activation energy of the viscous flow is sensitive to the very small additions of benzol derivatives characterised by methyl or ethyl substituents, and the value of E increases more appreciably with the addition of the compounds characterised by intrinsic dipoles. For example, in the mixtures with N-23 ($E = 3501$ K), the addition of 3 mol.% of benzol increases the value of E up to 3591 K, in the case of *n*-xylol up to 3782 K, toluol ($\mu = 0.37$ D) up to 4120 K, and ethyl benzol ($\mu = 0.39$ D) up to 3990 K. The viscosity changes only slightly. The presence of the dipole moment in the molecules of the non-mesogen has different effects on activation energy. For example, if the molecules of the addition result in the failure of the complexes (associates), the value of E decreases (3.6 mol.% of chlorobenzol (μ = 1.60 D) or nitrobenzol ($\mu = 3.9$ D) reduces the activation energy of the viscosity of the solution of the addition to N-23 to 3270 K). If the molecules of the addition are capable of forming a hydrogen bond with the mesogen molecules, the value of E increases (3.6 mol.% solution of aniline (μ = 1.5 D) in N-23 gives the value $E = 3592$ K, and for 2.8% of nitroaniline ($\mu = 6.8$ D) in the same matrix $E = 4343$ K). Until now we have discussed the aromatic non-mesogens. The compounds whose molecules are tetrahedral slightly increase the activation energy of the viscosity of the solution and this increase increases with the increase of the dipole moment of the molecule: 5 mol.% solution of chloroform ($\mu = 1.24$ D) in N-23 has the value $E = 3533$ K, and in acetonitrile ($\mu = 3.51$ D) the value of E = 3854 K.

In [316], investigations were carried out into the effect on rotational

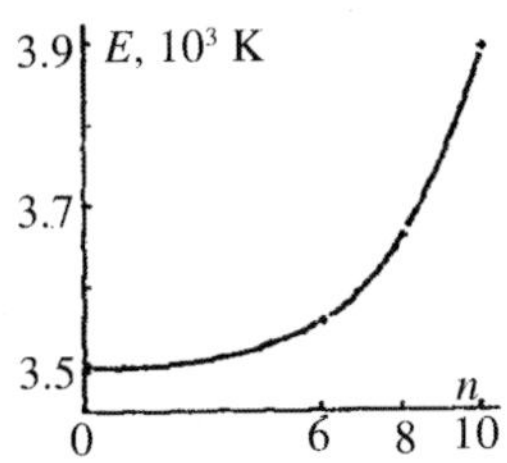

Fig. 6.2.4. Dependence of activation energy E of the flow of liquid crystal mixtures with *n*-alkanes C_nH_{2n+2} on the length of the hydrocarbon chain.

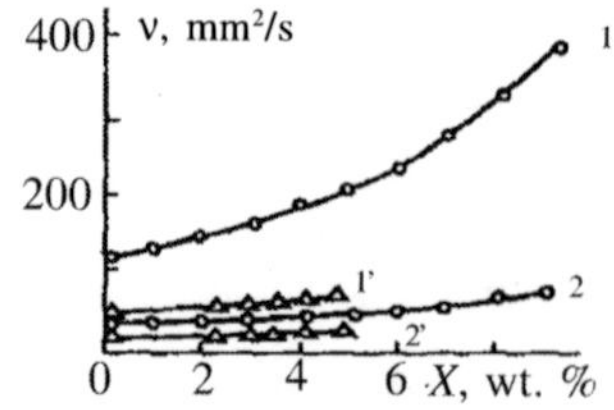

Fig. 6.2.5. Dependence of viscosity ν of wide temperature LCM for external applications on the concentration of dyes X at lower (1,1') and higher (2,2') temperature.

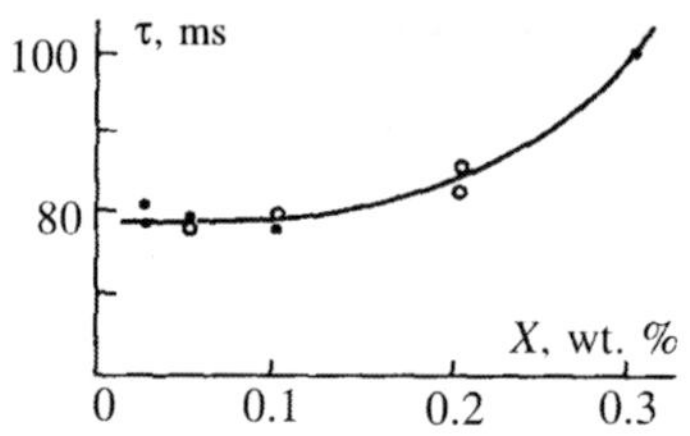

Fig. 6.2.6. Variation of the relaxation time of the twist effect τ with the increase of the concentration of the dye (methylred in MBBA).

viscosity γ_1 of more complex additions forming complexes with the molecules of the liquid crystal (a mixture of the Schiff bases and the free radical of 2,2,6,6-tetramethyl-4-oxypiperidine-1-oxyl (T)). In the room temperature range, the addition increases the viscosity of the solution, but with the increase of temperature the free volume of the liquid crystal phase increases as a result of complex formation, and the viscosity decreases 2.3 times at the concentration $T = 1\%$.

Figure 6.2.5 [331] gives the dependence of the viscosity of the wide-temperature liquid crystal materials for external application on the concentration of dyes. The form of the relationships is identical with the temperature dependence of viscosity at low temperatures and, consequently, it is natural to assume that the increase of viscosity with the increase of concentration is associated with a decrease of the free volume. In [332] in the measurement of the characteristic relaxation time values it has been shown that the increase of the concentration of the dye (methyl red in MBBA) reduces the diffusion coefficient, evidently as a result of dimerisation of the dye molecule (Fig. 6.2.6 [332]). The higher values of the viscosity of material ZLI-2327 indicate the higher polarity of the molecules which form the material. Consequently, the solubility of the dye increases but, on the other hand, the viscosity of the mixture with the dye increases at a higher rate.

Thus, by selecting the non-mesogenic additions it is possible to control the viscosity and activation energy of the liquid crystal mixture, and the viscosity value can be calculated if we know the molar volume (density). In producing liquid crystal materials, it is important to take into account the effect of the addition on the thermal stability of the mesophase.

6.3. Viscosity of mixtures of nematic liquid crystals with optically active additions

To remove regions with different directions of the optical twisting in the liquid crystal indicators, and also to reduce the switching time in the liquid crystal materials for the twist effect, it is necessary to add a small amount of optically active substances resulting in preferential twisting of the specimen of the liquid crystal in a specific direction. These additions are the most important component of the material for the supertwist effect, characterised by extremely high multiplex characteristics.

In this section, in contrast to the previous section in which attention was given to the viscosity of the mixtures of nematic liquid crystals with non-mesogens, we also take into account the dependence of the viscosity on a parameter describing the twisting of the structure, i.e. the pitch of the cholesteric helix P_0. Figure 6.3.1 [333] shows the dependence of the viscosity of the solutions ZhK-440 with the optically active additions (OAA), characterised by the following values of the twisting capacity $A = (PC)^{-1}$ (C is the molar concentration of the optically active addition): $A = 0.037$ μm^{-1} (L-menthyl chloroacetate), $A = 0.095$ μm^{-1} (L – menthyl benzoate), $A = 0.036$μm^{-1} (L – menthyl oxalate). For comparison, data are also given for the viscosity for the mixtures with cyclohexyl oxalate, i.e. the substance identical with the latter but not optically active. At the concentration of the optically active addition of $C < 3$ mol.%, the viscosity of the mixtures increasing together with their twisting capacity and at higher concentrations

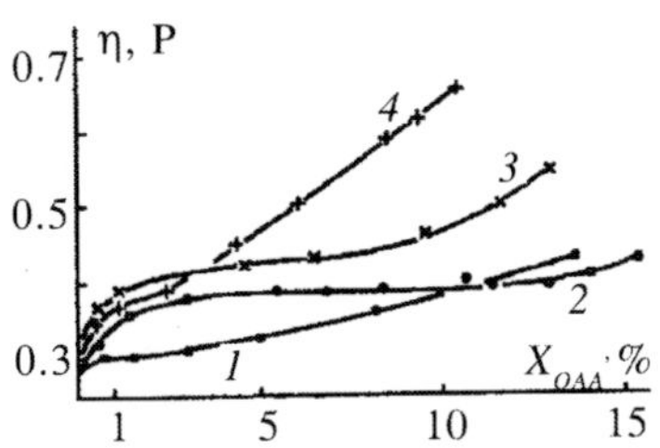

Fig. 6.3.1. Dependence of viscosity η of solutions of ZhK-440 with optically active additions (OAA): L – menthylchloroacetate (2), L – menthylbenzoate (3), L – menthyloxalate (4) and mixtures with cyclohexyloxalate (1) on the concentration of the addition X_{OAA}.

a significant effect is exerted by the larger dimensions of *L*-menthyl oxalate, containing two menthol rings. The increase of the viscosity of the mixtures with the optically active additions has a strong effect on the switching time of the electrical optical effect (in [333] – dynamic scattering (Fig. 6.3.2)).

Taking into account the fact that the flow of the cholesteric liquid crystal through a capillary is the consequence of the helical structure, the increase of apparent viscosity has been described in detail by Chandrasekhar [2] who examined a very important case of the reorientation of the planar structure of the cholesteric liquid crystal, more accurately, mixtures of the nematic liquid crystals with the derivatives of cholesterine or with other non-steroid cholesterics. In this case, the effective values of rotational viscosity γ_1^* and of elasticity coefficient K_{11}^* which determine the switching time depending a complicated manner on thickness, in particular, on the number of the Granjon zone m and on the position inside the zone (the ratio $2L/P_0$) (Fig. 6.3.3). The latter is determined by the difference between the induced pitch of the helix P and the equilibrium value P_0, described by the relationships:

$$P = \frac{P_0}{m} \qquad m - \frac{1}{2} < \frac{L}{P_0/2} < m + \frac{1}{2}, \quad m = 1,2,3..., \tag{6.3.1}$$

$$P = \frac{P}{m + 1/2} \qquad m < L\frac{P_0}{2} < m + 1, \quad m = 0,1,2... \tag{6.3.2}$$

for a planar texture with the parallel (6.3.2) or perpendicular (6.3.2) directions of orientation on the substrate. For Granjon zones with low numbers ($m \leqslant 3$) the pitch of the helix differs from the equilibrium value and, consequently, the value of deformation in this case is calculated numerically, and the effective values of coefficients K_{11}^* and γ_1^* calculated from the threshold stress of deformation and the relaxation time of the

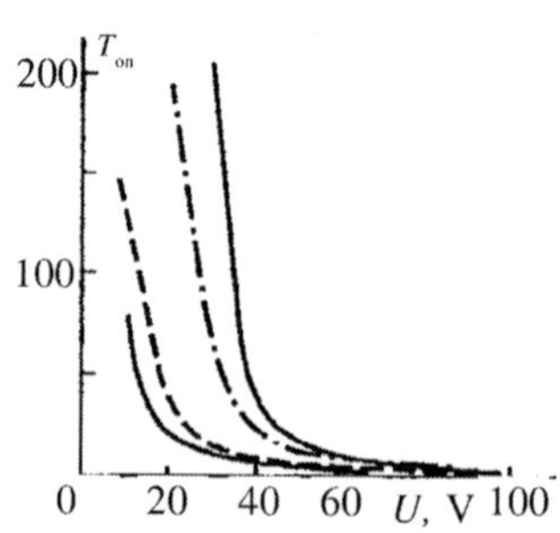

Fig. 6.3.2. Switching time of the electrical effect of dynamic scattering of light τ_{on} in relation to high voltage. Upper solid line – cyclohexyloxalate, dot-and-dash line – *L*-menthylchloroacetate, dotted line – *L*-menthylbenzoate, lower solid line – *L*-menthyloxalate.

phase delay, in accordance with (2.4.13). For zones with high numbers ($L \gg P_0$) the values of K_{11}^* and γ_1^* can be determined analytically. The reorientation of the planar texture in the stationary electrical field was examined in [334–336], the dynamics of reorientation in [335, 336].

Figure 6.3.4 and 6.3.5 shows the dependences of the deformation of the planar texture of the cholesteric liquid crystal on relaxation time τ_R for different Granjon zones, and also of the relative effective value of rotational viscosity γ_1^*/γ_1^N, calculated for the zones with the number $m \leqslant 3$ from the relationships:

$$\tau_R = \gamma_1^* \frac{L^2}{\pi^2 K_{11}^*}, \qquad (6.3.3)$$

$$\gamma_1^* = \tau_R U_t^2 \frac{\Delta\varepsilon}{2\pi L^2}, \qquad (6.3.4)$$

Here U_t is the threshold stress of deformation, and for the zones with high numbers it was calculated analytically [335, 336] (see below). In the zones 0 and 1, relaxation time τ_R coincides with the relaxation time of deformation of the planar-oriented layer of the name at the liquid crystal without the

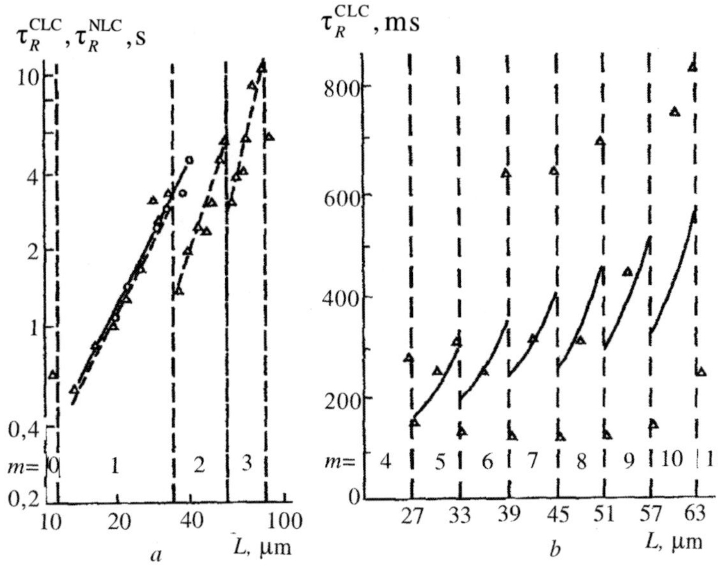

Fig. 6.3.3. Dependence of relaxation time of reorientation of CLC in wedge-shaped cells with a thickness corresponding to the first 11 Granjon zones on thickness : a) τ_R^{NLC} – solid line, τ_R^{CLC} – dotted line, mixture of MBBA and 3% 4-cyanophenyl ester or 4'-heptylbenzoic acid (nematic mixture *A*), dotted lines – mixture *A*+0.2% cholesteril caprinate (AC) – *b*) mixture *A*+1% CC, solid line as indicated, results of calculations using equation (Fig.6.3.6.).

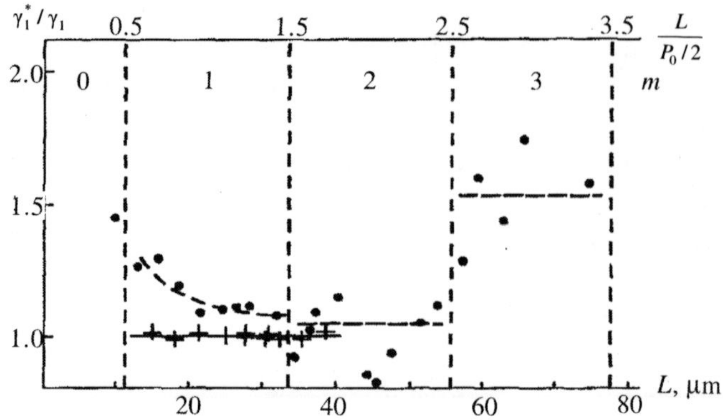

Fig. 6.3.4. Dependence of effective rotational viscosity γ_1^* of the CLC (mixture A+0.2% cholesteril caprinate; dotted line) normalized for rotational viscosity γ_1 of NLC (mixture A; solid line), in the first four Granjon zones.

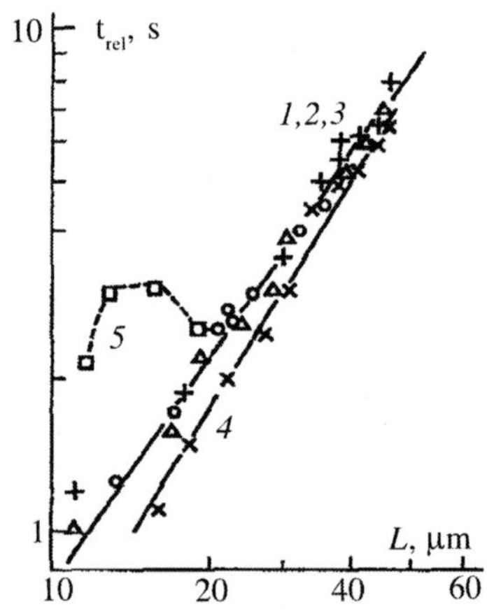

Fig. 6.3.5. Dependence of relaxation time t_{rel} of the twist effect on thickness: $U = 6U_n$; *1, 2, 3* ZhK-614 (NIOPiK) – multicomponent mixture of phenyl benzoate with $\Delta\varepsilon = +23$ and viscosity of approximately 4 P (+ – +); concentration of CC $C = 0.01\%$ (o–o); $C = 0.028\%$ (Δ–Δ); *4* – $C = 0.055\%$ (×–×); *5* – $C = 0.2\%$ (□–□).

chiral (optically active) addition, although the value of γ_1^* is almost twice the rotational viscosity of the pure nematic crystal γ_1^N. Starting with the second zone, τ_R suddenly decreases at the boundary of the zone and increases in proportion to the square of thickness within the limits of the zone. In zone 3, the dependence $\tau_R(L)$ differs from the quadratic dependence because of the larger decrease of the threshold stress U_t in the zone. The value of γ_1^* differs only slightly from zone to zone. The ratios γ_1^*/γ_1^N may be used for estimating the effective viscosity of the materials for the supertwist effect, since the twisting of the director in the SBE-, STN- and OMI effects corresponds to the Granjon zones 0 and 1.

For the Granjon zones with high numbers (m>4), the values of τ_R measured in the central points of the zones ($P = P_0$) are proportional to the thickness of the layer of the liquid crystal L and not to the square of the thickness, as in the nematic liquid crystals. This is indicated by the proportionality of the square of the threshold stress of deformation of the planar texture to the thickness [334]:

$$U_t^2 = \frac{4\pi^3 K_{11}^*}{\Delta\varepsilon} = \frac{8\pi^3}{\Delta\varepsilon}\left[(6K_{22}K_{33})^{1/2} + 4K_{22}\left(\frac{L}{P} - \frac{L}{P_0}\right)\right]\frac{L}{P}, \quad (6.3.5)$$

from which

$$\tau_R = \frac{1}{2\pi^2}\frac{\gamma_1^* LP}{(6K_{22}K_{33})^{1/2} + 4K_{22}(L/P - L/P_0)}. \quad (6.3.6)$$

The effective value of γ_1^* at $L \gg P_0$ was obtained by solving the Navier–Stokes equation for a cholesteric liquid crystal taking the inverse flow into account [335, 336]:

$$\gamma_1^* = \gamma_1 - \frac{2\alpha_2^2}{2\alpha_4 + \alpha_5 - \alpha_2}. \quad (6.3.7)$$

Because of the relative low value of γ_1^* (for MBBA γ_1/γ_1^* =2.6), the relatively high value of K_{11}^* (four MBBA at $L/P = K_{11}^*/K_{11}$= $2(6K_{22}K_{33})^{1/2}/K_{11}$= 6.7) and also for the linear dependence of τ_R on thickness $L \gg P_0$ the relaxation time of low deformation of the planar texture is considerably smaller than the values for the nematic liquid crystals without chiral additions. For example, for the centre of the zone 5 calculations give τ_R^X/τ_R^n =14.7, which corresponds to the experimental value of 14.

Taking into account the inverse proportionality of the pitch of the helix P_0 to the concentration of the optically active addition $C(C = A/P_0)$, the form of the dependence of γ_1^* on ($2L/P$), i.e. on the number of the Granjon zone (the increase of γ_1^* in the zones with small numbers and decrease at $L \gg P_0$) is identical with the dependence of viscosity ν on the concentration of the non-mesogenic addition (Fig. 6.2.3).

The dynamics of reorientation of the planar texture of the cholesteric liquid crystal was also investigated in [337], but no attention was given to the variation of effective viscosity and elasticity with the increase of the ratio L/P_0, resulting in the simplified relationship of the relaxation time with the physical properties of the liquid crystal. However, this relationship is not in agreement with the experiments.

An important factor for application in the liquid crystal indicators is the availability of planar structures with twisting of 90° and 270° (Granjon zones

0 and 1), which corresponds to the twist, SBE- and STN-effects (supertwist), and with twisting of ~180° (OMI-effect). We examine the effect of the concentration of the optically active addition C on the thickness dependence of the relaxation time of the twist effect t_{rel} in the zones 0 and 1 for different liquid crystal matrices [338] (relaxation time depends to a lesser extent on the control voltage in comparison with the switching time). Figures 6.3.5 and 6.3.6 show the thickness dependences of t_{rel} of the mixtures of ZhK-614 (mixture of phenyl benzoates with a higher value of $\Delta\varepsilon$) and ZhK-654 (mixture of phenyl benzoates and azoxycompounds) with cholesterol caprinate (CC). The twisting capacity of CC in ZhK-614 is 0.25 $(\mu m\%)^{-1}$, and in ZkH-654 it is 0.125 $(\mu m\%)^{-1}$. For low concentrations of the optically active addition in ZhK-614 the relationship $t_{rel} \sim L^{1.5}$ is efficiently satisfied (in [250] $t_{rel} \sim L^{1.6}$), and the value of t_{rel} is independent of C. At the concentration of the CC of C = 0.2% in the mixture with ZhK-614 at the boundary of the zone 0 the value t_{rel} slightly decreases. In the mixtures of the CC with ZhK-654 at a low concentration of the addition, the value of t_{rel} in the centre and at the edge of the zone 0 slightly decreases. At the concentration C = 0.1% (P_0 = 80 μm) the largest decrease (by a factor of 6) of the value of t_{rel} is detected in the thickness range 20÷–40 μm with the minimum at t_{rel} at L =40 μm (the boundary of the zones 0 and 1). With the increase of the concentration of the CC in the mixture with ZhK-654 the value of t_{rel} in contrast to the mixture of CC with ZhK-614 decreases in comparison with the value t_{rel} for the pure nematic liquid crystal. At the edges of the zones 0 and 1, the value of t_{rel} decreases. On the whole, the variation of the relaxation time of the optical response with thickness corresponds to the identical variation of the deformation relaxation time associated with the viscoelastic properties of the liquid crystal. Evidently, the differences are due to the interaction of the optically active addition

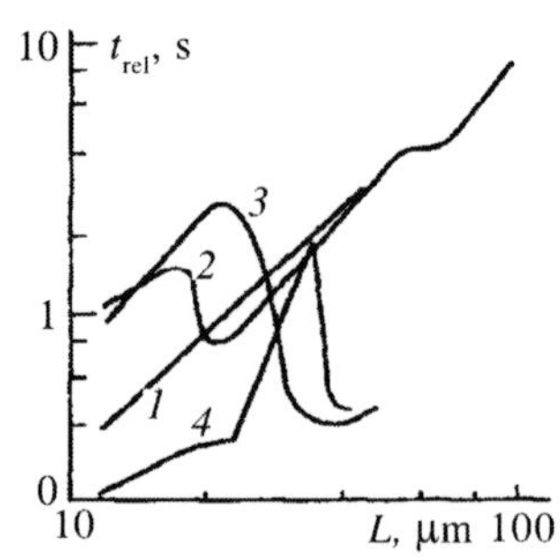

Fig. 6.3.6. Dependence of relaxation time t_{rel} on the electrooptical effect of cells with twisting to 90° and 270° on thickness; ZhK-654 (NIOPiK) – mixture of azoxycompounds (ZhK-440) and 40% of 4-cyanophenyl ester of 4'-heptyl benzoic acid (*1*); concentration of CC C = 0.05% (*2*); C = 0.1% (*3*); C = 0.2% (*4*); U = 30V.

and the nematic matrix and also to the very high value of deformation of the layer of the liquid crystal.

Figure 6.3.7 shows the oscillograms of the relaxation of the twist effect in the mixture with ZhK-654 with 0.1% CC, obtained at different thicknesses of the wedge-shaped cell and sweeps of the screen of the oscilloscope. It appears that the decrease of t_{rel} in the vicinity of the boundaries of the zones 0 and 1 is associated with the disappearance of the characteristic 'surge' of intensity on the oscillograms of the twist effect, formed as a result of the interaction of the flow of the liquid with the boundary layers

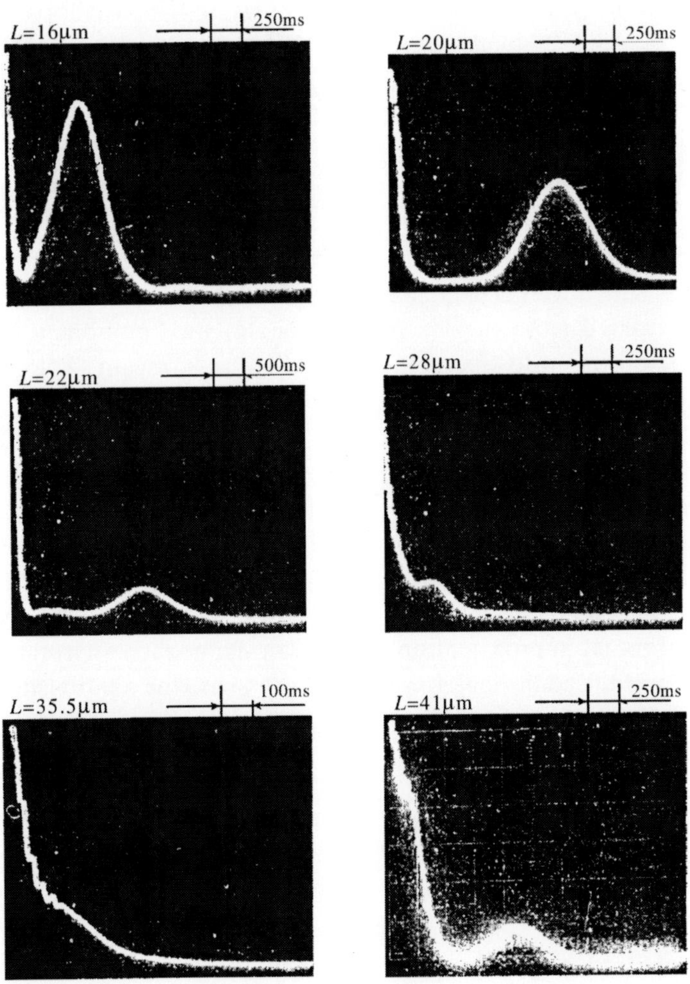

Fig. 6.3.7. Oscillograms of relaxation of the electrooptical effect of the cells with the twisting to 90° and 270° at different thicknesses of the LC cell L. Mixture ZhK-654 and 0.1% CC. Time sweep of the screen of the oscilloscope is shown.

and leading at the initial moment of relaxation to a situation in which the reorientation angle of the director exceeds 90° [339]. Taking into account the fact that the pitch of the cholesteric helix at the edge of the zone 0 is twice the equilibrium value, it may be concluded that the increase of the elastic energy associated with twisting of the structure prevents the initial reorientation of the director in the direction opposite to the directional relaxation of deformation.

However, it should be mentioned that although at the thicknesses $P_0/2>L>P_0/4$ the time t_{rel} decreases but at the same time in this thickness range the curvature of the volt–contrast characteristics of the twist effect decreases [338, 339]. Evidently, this effect determines the shorter duration of switching of the intensity also in the cells with a twisting of 180°.

6.4. Viscosity of liquid crystal materials

6.4.1. Calculation of the viscosity of the developed materials

In the development of liquid crystal materials consisting of several (*N*) components with different chemical composition, the viscosity of the material can be calculated from the viscosity values of the component shown in Chapter 5, taking into account their molar (or weight) fraction x_i in accordance with the expression:

$$\eta_{LCM} = \sum_{i=1}^{N} \eta_i x_i. \tag{6.4.1}$$

In Chapter 5, η_i values for many substances were calculated using the linear relationship, i.e. by extrapolation of the viscosity of the solution. In some casesin which the addition greatly decreased the viscosity of the solvent and linear approximation resulted in the negative virtual value of the viscosity of the additional, it was necessary to use the relationship linking the viscosity logarithms:

$$\ln \eta_{LCM} = \sum_{i=1}^{N} x_i \ln \eta_i. \tag{6.4.2}$$

For small differences between the viscosity of the solution and the viscosity of the solvent, the equations (6.4.1) and (6.4.2) yield almost identical values of the viscosity of the additions. Although the equation (6.4.2) results in smaller deviations from the experimental values in a larger number of cases (section 6.1), the simpler equation (6.4.1) is used more extensively.

It should be mentioned that both equations should be used with extreme care because the same addition can decrease or increase the values of the viscosity of the solution depending on the matrix used. For example, in the mixtures of ethyl-(pentylcyclohexyl)diphenyl ($C_5H_{11}-C_6H_{10}-C_6H_4-C_6H_4-C_2H_5$, BCD-52) with phenylcyclohexanes the viscosity of the solution does not change, and indeed mixtures with cyclohexyl carboxylates the viscosity slightly increases, but in the mixtures with the materials whose components do not contain the cyclohexane fragment with E7 (diphenyls) and SP-21 (diphenyls and phenyl benzoates), viscosity increases appreciably (Fig. 6.4.1) [340]. The approximation of the graphs to the concentration of the addition equal to zero would give four greatly differing values of the viscosity of BCD-52. In chapter 5, there are different values of η for several substances, obtained in the measurement of the viscosity of the substances in solutions with different matrices. In the given example, these concentration relationships of the viscosity of the solution are associated to some degree with the specific intermolecular interaction and, unfortunately, to a higher degree with the special features of the geometrical construction of the molecules.

Figures 4.3.3 and 6.1.2 also illustrate the possibility of the inaccurate determination of the viscosity of the substance in extrapolation of the concentration dependence of the viscosity of mixtures.

Nevertheless, as shown later, equation (6.4.1) can be used efficiently for the evaluation of the viscosity of the liquid crystal material, including the material consisting of the components whose molecules belong to different chemical classes, to polar and non-polar compounds, etc. Evidently, linear approximation can also be carried out in this case because of the fact that the concentration of the individual substance in the currently available multicomponent liquid crystal materials is not high and these materials can be processed more efficiently by the conclusions of the theory

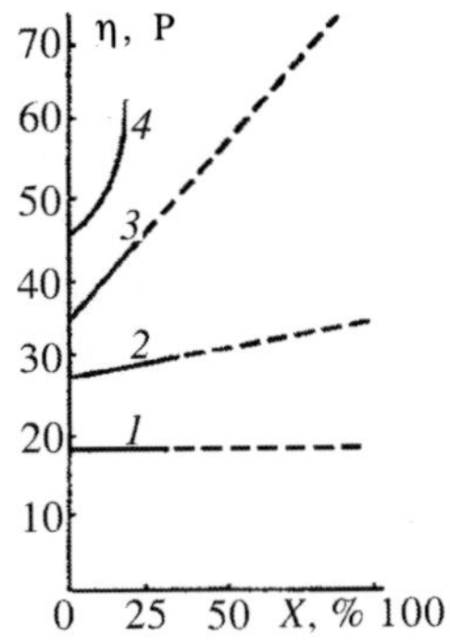

Fig. 6.4.1. Concentration dependences of the measured viscosity η of the mesogen solid solution for different LC matrices. Mixture of ethyl (pentyl cyclohexyl) diphenyl (BCG-52) with phenyl cyclohexyl (1), cyclohexyl carboxylates (2), E7 (diphenyl, 3) and SP-21 (material containing diphenyls and phenyl benzoates, 4).

of ideal solutions in comparison with two-component mixtures.

6.4.2. Approaches to the development of liquid crystal materials for various applications from the viewpoint of their viscosity properties

The highest viscosity properties (values of γ_1, η or ν and the temperature dependence of these quantities) are shown by different derivatives of cyclohexane. However, the developed liquid crystal materials with minimum viscosity are not suitable for a wide range of applications because of one of the following reasons: comparatively high working stresses; the insufficiently high curvature of the volt–contrast characteristic of the twist effect; low value of birefringence; low solubility of dichroic dyes. In a small number of cases, the high value of the viscosity of the liquid crystal material is essential. Therefore, there is no universal liquid crystal material which would satisfy the requirements, and to ensure specific properties it is necessary to sacrifice a viscosity. In this section, we use the data on sections presented previously in patent literature as an example to show the variation of the properties of the mixtures in relation to composition. The current principles of construction of liquid crystal materials have been described in sufficient detail in [341].

The main requirements for liquid crystal indicators for external application is a wide range of working temperatures (from –50÷–40°C to +90÷100°C) and short switching time (no more than 1 s at T = –30÷–40 °C). These requirements are satisfied by the liquid crystal materials developed on the basis of phenylcyclohexanes: alkyl- and alcoxyderivatives ensure low values of the melting point because of the low value of the enthalpy of melting and low values of the relaxation time at low temperatures due to the low value of the activation energy or glass transition temperature; tri- and tetracyclic – high values of clarification temperature without increasing viscosity; cyan derivatives – low working stresses.

Table 6.3 shows the compositions of the mixtures on the basis of derivatives of phenylcyclohexane and their properties [342–344]. It may be seen that with the decrease of the polarity of the molecules in the mixture the viscosity of the mixture decreases. The highest viscosity is shown by mixture 1 prepared only from cyan derivatives. Substitution of the cyan-group in the tricyclic compound by an ethyl group reduces the viscosity immediately by a factor of 1.5 (mixture 2). The mixtures containing cyclohexyl phenylbenzoate GP-33 have viscosity in the range 56÷72 MPa s (centi Poise), and the lowest value of η (mixture 5) is recorded for the largest amount of PCH-32, the non-mesogenic non-polar diluent

Table 6.3. Composition viscosity of wide temperature LCM

Substance	Notation of component	Viscosity of component	1	2	3*	4	5	6	7	8	9	10	11	12	13**	ZLI 1957/5
C_3H_7-CH-B-CN	PCH-3	0.18	24	24							22	19	23	18	10	
C_5H_{11}-CH-B-CN	PCH-5	0.21	36	36	33	29	29	35	34	29	28	24	30	22		
C_7H_{15}-CH-B-CN	PCH-7	0.26	25	25												
C_3H_7-CH-B-C_2H_5	PCH-32	0.04			32	36	36	30.5	32	31	20	25	20	16	22	
C_3H_7-CH-B-OC_2H_5	PCH-3O2	0.05												15	10	
C_3H_7-CH-B-B-C_2H_5	BCH-32	0.22													8	
C_5H_{11}-CH-B-B-C_2H_5	BCH-52				16	14.5	17	18	18	20	24	25	18	23	15	
C_3H_7-CH-B-COO-C_3H_7	HP-33	0.60			6	5	5	3.5	4	6						
C_5H_{11}-CH-B-COO-C_3H_7	HP-53														8	
C_5H_{11}-CH-B-B-CN	BCH-5	0.9	15		6	4.5	5.5	6.5	7	6						
C_5H_{11}-CH-B-B-CH-C_3H_7	BCH-53	0.42			7	8	7.5	6.5	5	8	6	7	9	6	9	
C_5H_{11}-CH-B-B-F	BCH-5Φ	0.2		15												
C_3H_7-CH-B-B-(F)B-C_2H_5	BCH-32(Φ)***	0.27													18	
	T_{NP} °C		70	65	86.5	81.4	83	86	82.1	95.6	83.1	83.5	84	83.8	86	93
	T_m, °C		-6	-10	-30	-30	-40	-30	-30		-30	-30			-17	-40
	ν(20°C), mm²/s		28	24											18	18
	ν(0°C)		97	68	59.5	57	56	60	59	72	51	49	52	48.5	47	46
	ν(-20°C)				274	264	259	295	280	415	237	232	242	229	225	200
	ν(-30°C)															580

*At T=–30°C and thickness L=9.3μm the mixture 3 has the following values of the switching time of the twist effect: t_{on}=0.54s, t_{off}=0.54s, in the presence of the chiral addition the times decreased to 0.33 and 0.50s, respectively.

**Threshold voltage of the twist effect U_t=2.8V,

***–(F)B– benzolrine with the fluorine atom in the meta position.

(ε = 2.8) and the lowest amount of bi- and tricyclic polar compounds PCH-5 and BCH-5. The highest value of η (mixture 8) is recorded with the increase of the concentration of tri- and tetracyclic compounds. In the group of the mixtures 9–12 which do not contain GP-33, the viscosity is considerably lower in comparison with the previous group, η = 48.5÷52 cP at 0°C, η = 229÷242 cP at –20°C. The amount of the diluent PCH-32 is not very large (16÷25% in comparison with 30.5÷36% for the previous group of the mixtures 3–8). The main reason for the decrease of viscosity is evidently the replacement of the tricyclic polar compound BCH-5 by bicyclic PCH-3. Although the amount of the DCH-32 in this group of the mixtures is minimum, the total amount of the diluent increased as a result of PCH-302. In mixture 13, viscosity is low because of the following reasons: the small amount of the polar compounds, large amount of bi- and tricyclic diluents, and the compound BCH-32(F) as a result of fluorine in the *meta*-position results in less intense packing of the molecules and in the failure of the dimers of the cyan derivative PCH-3. Consequently, the dielectric anisotropy of the liquid crystal material increases and the threshold voltage of the twist effect of the material is not very high (2.8 V). The viscosity characteristics of mixture 13 are already close to those of the material of Merck company (Germany) ZLI-1957/5, designed for external application, and also as the solvent for external application, and also as the solvent for the 'visitor–master' effect.

Table 6.4 gives the data for the liquid crystal materials developed by Russian and foreign organisations in different periods. These data, presented for better information also in Fig. 6.4.2, make it possible to compare the level of development of the liquid crystal material in different countries and also reflects the dynamics of development. For example, over a period of 6 years, the Merck company reduced viscosity by 3 cP at room temperature and almost halved the viscosity at 30°C. Evidently, this has been achieved as a result of the development of the chemistry of fluorine derivatives. In Hoffmann-La Roche company, the viscosity was also reduced by 3 cP a period of less than 1.5 years. In the 80s and 90s of the previous century, this company and subsequently the ROLIK company which separated from the former, carried out investigations to develop compounds with the ethanol bridge, the NCS group and alkenyl substituents characterised by slightly higher viscosity but also by shorter switching time as a result of reducing the viscoelastic ratio (Table 5.4). The new products by NIOPiK (LC-1390 and LC-1391) were made without fluorine derivatives and characterised by the minimum viscosity of all the mixtures, developed from the derivatives of phenylcyclohexanes without any complication in the lateral chain or polar substituents. Although the materials of BDH company (United Kingdom) have the viscosity η (20°C) values higher than the values of the materials

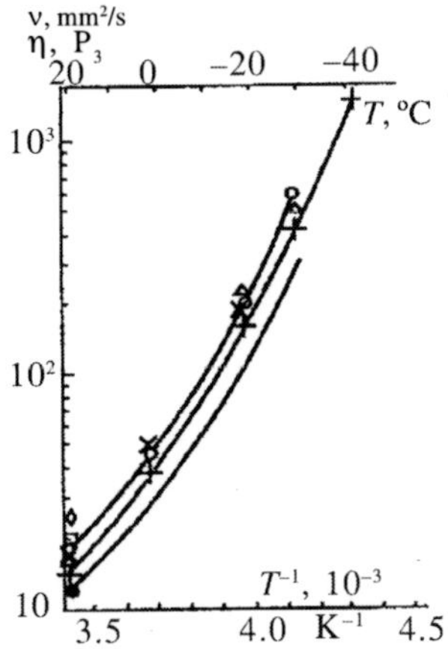

Fig. 6.4.2. Temperature dependences of kinematic viscosity ν of low-temperature LC materials developed by companies Merck (Germany; x, Δ, •) and Hoffmann-LaRoche (Switzerland; □ ,+) in 1984, 1985 and 1987, ∇ – materials NIOPiK LC–1391 [345–349].

produced by other companies, these smectic crystals are characterised by extremely low smectic–nematic temperatures (less than –50°C) and the short switching time at low temperatures. In the twist cell with the thickness of 7 μm at $U = 6$ V and $T = -20$°C, $t_{on} = 0.35$ s, $t_{off} = 0.50$ s, at $T = -30$°C, $t_{on} = 10.1$ s, $t_{off} = 1.50$ s for the liquid crystal material E200. The materials E200, E203, E210 and E220 have been developed on the basis of a mixture of the compounds with the general formula C_nH_{2n+1}–C_6H_{10}–CH_2–CO–C_6H_4–C_6H_3F–C_mH_{2m+1} (26%, $n = m = 3$, 35% $n = 3$, $m = 5$, 39% $n = 5$, $m = 2$; fluorine is substituted in the meta-position, i.e., closer to the adjacent benzol ring) [350]. The German claim [343] for a liquid crystal material corresponding to mixture 3 in Table 6.3, has the following values of the switching time: $t_{on} = 0.54$ s, $t_{off} = 0.88$ s at $T = -30$°C and the thickness of the layer of 9.3 μm and in the presence of the chiral addition these values decrease to 0.33 s and 0.50 s, respectively.

Figure 6.4.3 shows the ranges of the variation of the values of the rotational (γ_1) and kinematic (ν) viscosity in relation to polarity. Polar liquid crystal materials are those with the high value $\Delta\varepsilon = 10.2 \div 14.7$ (ZLI-1982, 1840, 1691), and slightly polar ZLI-1565, 1694, 1957/5, with the values in the range 4.5÷6.5.

Considerable changes took place in the market of the liquid crystal materials in the 90s. The main developer and supplier of the liquid crystal materials for information imaging became the company Merck (Germany with branches in Great Britain, Japan, South Korea, USA) with the range including approximately 450 commercial mixtures [351]. The viscosity characteristics did not change greatly. The extremely low kinematic viscosity

Table 6.4. Viscosity of LCM with a wide temeprature range

	E. Merck (Germany)					BDH (England)			Hoffman-LaRoche (Switzerland)		DI-Nippon (Japan)		Niopik (former USSR)	
	ZLI-1957/5	ZLI-2584	ZLI-3119	ZLI-222-000	ZLI-222-100	E-200	E-202	E-220	RO-2875	RO-3652-1	EXP-002	RDX-5078	LC-1351	LC-1391
T_{NI}, °C	93	85	86	66	66	88.5	85.9	88.6	71	87	67.6	91.6		
T_m, °C	-40	-40	-40	-40	-40	-50	-50	-50	-20	-30	-55	-32		
ν, mm²/s														
+20 °C	18	15	14	14	12	29.7	24.7	25.5	19**	16**	16.6	20.4	19	21
0 °C	46	41	38	35	30	103*			(22 °C)		47.1	65.9		
-10 °C						202*								
-20 °C	200	180	159	144	110	500*		530*	316*	217**	189.4	291.7		
-30 °C	580	470	420	350	280	1420*				522**				
-40 °C		1700	1470	1200	1650									
data shown in the firm literature	1982	10.04.1986		3.08.1989					1.04.1985	1.08.1986	1988			

*Values of T_{off}, MS are given,
** Values of η are given.
*** The composition of EXP-002 includes bi-and tetracyclic compounds with the fragments - CH–CH_2–CH_2–B–, CH–CH_2–CH_2–CH; compounds alkyl – CH–B–B–X, X=CN, alkyl, alkyl–CH–CH–B–O–alkylnil; compounds alkynil –CH–CH–X, X=CN, O–alkyl; bi-and tricyclic compounds with the phenylcyclohexene fragment . The composition of RDX-5078 includes bi-and tri-cyclic compounds with the fragment – B–C=C–(2XB)–, X=CH_3, F in the meta position.

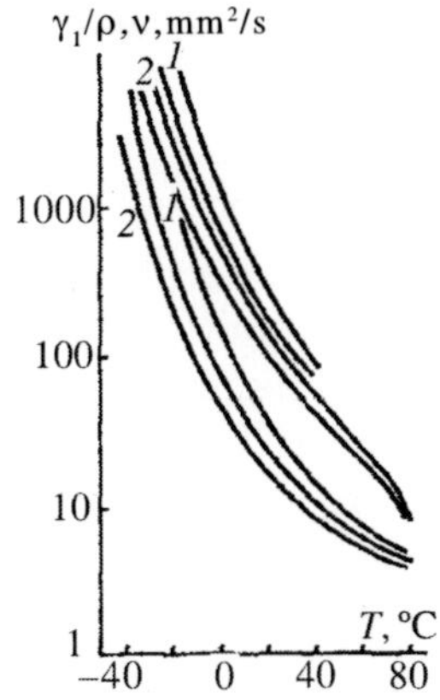

Fig. 6.4.3. Ranges of variation of the values of rotational (γ_1/ρ) kinematic (ν) in relation to their polarity; *1*) polar materials, *2*) slightly polar materials.

is typical of the mixture ZLI-2222-100 with ν (20°C) = 12 mm^2/s and ν (−30°C) = 280 mm^2/s, but with the threshold voltage of 3.64 V. The directions of development of promising liquid crystal materials for various applications are indicated in [352]. The ROLIK company was formed from the Swiss company Hoffman-la Roche and specialises only in new types of effects in liquid crystals, mainly photo-orientation [353]. In the British company BDH the liquid materials are not being developed at the present time. The position of NIOPiK has weakened considerably. A wide range of the liquid crystal materials is also produced by the Japanese company Chrisso.

The effect of the molecular structure of the components of the developed liquid crystal materials on their viscosity and other properties will be examined in greater detail.

It has already been mentioned that the presence of a halogen atom in the terminal or lateral position reduces the viscosity of the liquid crystal material. This is especially evident in Table 6.5 which gives the data on the mixtures containing chlorophenylcyclohexanes and derivatives of 2-fluorodiphenyl [354]. These compounds do not form dimers and prevent their formation in cyanoderivatives present in the mixture. Consequently, the dimensions of the particle sliding in the environment of the neighbours decrease to the size of the single molecule, the free volume increases and, consequently, the fluidity $\varphi = \eta^{-1}$. Attention should be given to the extremely low value of the viscosity of mixture 7 associated, in addition to the previously mentioned reasons, also with the absence of cyanoderivatives (see also mixture 1).

It was shown in [355] that the addition of compounds with the general formula C_nH_{2n+1}–C_6H_{10}–COO–C_6H_4–F and C_mH_{2m+1}–COO–C_6H_4–COO–C_6H_4–C_qH_{2q+1} to a concentration of 30% does not change the viscosity of the initial mixture. This is accompanied by a decrease of the melting point from –11 to –40°C (of course, the clarification temperature also decreases by the same extent, from 91 to 63°C), the relaxation time decreases from 107 to 69 in the second and the multiplexicity improves. The parameter $\gamma = V_{15°,\ 50\%}/V_{45°,\ 10\%}$, describing this property, changes from 1.38 to 1.26. The diluent is usually represented by PCH-32 (C_3H_7–C_6H_{10}–C_6H_4–C_2H_5) whose molecules are characterised by the same form as the molecules of the mesogens. However, this addition may also be represented by the compounds in which to form of the molecules at first sight does not support their orientation in the nematic liquid crystal. For example, the solution of bis-pentylcyclooctane in the nematic matrix, as the same temperature dependence of viscosity, as the solution in the same matrix with the same amount of PCH-32. The clarification temperature does not change, and dielectric anisotropy even slightly increases (from 3.5 to 4.0).

The number of the addressed lines of the liquid crystal display (the degree of multiplexicity) is determined by the curvature of the volt–contrast characteristic of the electrooptical effect [356] which depends mainly on the ratio of the elasticity coefficients K_{33}/K_{11} and, to a lesser degree, on the birefringence Δn and the ratio $\Delta\varepsilon/\varepsilon_\perp$. For the liquid crystal material ZLI-1957/5, characterised by a satisfactory degree of multiplexicity, the ratio K_{33}/K_{11}= 0 .99. To increase the number of the addressed lines, compounds with a low value of K_{33}/K_{11} added to the liquid crystal material, in particular, the derivatives of phenyl pyrimidines [357]. Table 6 .6 gives the data on the composition and physical properties of the liquid crystal materials with a low value of the K_{33}/K_{11} of the ratio, i.e. with high multiplexicity, containing from 26 to 37% of alkyl (alkoxy phenyl) pyrimidines C_nH_{2n+1}–$C_4H_2N_2$–C_6H_4–OC_mH_{2m+1} [358]. The diluent is represented by Demus esters C_nH_{2n+1}–C_6H_{10}–COO–C_6H_4–OC_kH_{2k+1}. The highest viscosity is recorded for the material containing cyano phenyldioxanes as the polar component. Of course, because of the low value of the enthalpy of melting of these liquid crystal materials (mixture 9), the melting point is minimum. In mixture 1, containing cyan phenylpyrimidines, the viscosity is lower. The lowest value of the viscosity η and the ratio K_{33}/K_{11} is recorded in mixture 2 characterised by the maximum concentration of the derivatives of phenylcyclohexanes. The viscosity of the alkyl (alkoxy phenyl) pyrimidines is 50 cP and, evidently, the viscosity of the commercial liquid crystal materials for matrices is high as a result of this [345–348, 359].

As indicated by Gooch and Tarry [360], the transmission of the twist cell in the parallel polaroids is described by the following equation:

Table 6.5. Viscosity and the properties of liquid crystal materials containing 4-chlorophenyl fragment

substance	notation	mixture							
		1	2	3	4	5	6	7	8
C_2H_5-Ch-B-Cl	FCH-2X							25	
C_3H_7-Ch-B-Cl	FCH-3X	30		27					27
C_4H_9-Ch-B-Cl	FCH-4X				25				
C_5H_{11}-Ch-B-Cl	FCH-5X		30						
C_7H_{15}-Ch-B-Cl	FCH-7X					27			
C_3H_7-Ch-B-C_2H_5	FCH-32		15					15	
C_5H_{11}-Ch-B-C_2H_5	FCH-52		15						
C_3H_7-Ch-B-OC_2H_5	FCH-302							20	16
C_3H_7-Ch-B-CN	FCH-3				18	18	15		18
C_4H_9-Ch-B-CN	FCH-4						11		
C_5H_{11}-Ch-B-CN	FCH-5				9		21		
C_5H_{11}-B-(F)B-C_5H_{11}	BP-55(P)						16		
C_5H_{11}-B-B-CN	BP-5		40		30				
C_5H_{11}-DO-B-CN	PDO-3			10					
C_4H_9-DO-B-CN	PDO-4			10					
C_3H_7-Ch-COO-B-OC_2H_5	PECHK-302	20		18					
C_4H_9-Ch-COO-B-OC_2H_5	PECHK-402	20							
C_5H_{11}-Ch-COO-B-OC_2H_5	PECHK-501	20							
C_5H_{11}-Ch-B-B-C_2H_5	BCH-52					22	21	15	22
C_5H_{11}-Ch-B-(F)B-C_2H_5	BCH-52(P)*					23			
C_5H_{11}-Ch-B-B-CN	BCH-5				8		4		7
C_5H_{11}-Ch-B-B-Ch-C_3H_7	CBC-53				10	10	12	10	10
C_4H_9-Ch-B-COO-B-C_3H_7	HP-43	10		17					
T_{HP}, °C		55	54	57	75	95	90	62	86
$-T_m$, °C		3	7	6	9	9	20	13	15
$\Delta\varepsilon$		0.9	6.7	7	9.4	3.8	10.5		
Δn			0.18		0.18				
η(20 °C), mPa·s		12	14	16	21	19	20	9	15
η(0 °C), mPa·s						55			48

* (F)-B-benzol ring with fluorine atoms in the metaposition.

Table 6.6 K_{33}/K_{11}

substance	notation	mixture		
		LCM 1	LCM 2	LCM 3
C_5H_{11}-Pn-B-CN	PPm-5	7		
C_7H_{15}-Pn-B-CN	PPm-7	8		
C_2H_5-DO-B-CN	PDO-2			5
C_3H_7-DO-B-CN	PDO-3		3	8
C_4H_9-DO-B-CN	PDO-4			8
C_5H_{11}-DO-B-CN	PDO-5			7
C_5H_{11}-B-B-CN	BP-5		6	
C_5H_{11}-Ch-B-B-CN	BCH-5	8		
C_3H_7-Ch-B-B-C_2H_5	BCH-32	4	7	
C_5H_{11}-Ch-B-B-C_2H_5	BCH-52	6	8	
C_3H_7-Ch-B-B-Ch-C_3H_7	CBC-33			5
C_5H_{11}-Ch-B-B-Ch--C_3H_7	CBC-53			7
C_6H_{13}-Pn-B-OCH_3	PPm-61	7		
C_6H_{13}-Pn-B-OC_5H_{11}	PPm-65		5	6
C_6H_{13}-Pn-B-OC_6H_{13}	PPm-66		5	6
C_6H_{13}-Pn-B-OC_7H_{15}	PPm-67		5	6
C_6H_{13}-Pn-B-OC_9H_{19}	PPm-69		4	5
C_7H_{15}-Pn-B-OC_7H_{15}	PPm-77		5	5
C_7H_{15}-Pn-B-OC_9H_{19}	PPm-79	6	4	5
C_8H_{17}-Pn-B-OCH_3	PPm-81	6		
C_9H_{19}-Pn-B-OC_9H_{19}	PPm-99	7	5	
C_3H_7-Ch-COO-B-OCH_3	PECHK-31	7	5	4
C_3H_7-Ch-COO-B-C_2H_5	PECHK-32	7		
C_4H_9-Ch-COO-B-OCH_3	PECHK-41	7	7	6
C_5H_{11}-Ch-COO-B-OCH_3	PECHK-51	7	8	7
C_3H_7-Ch-COO-B-C_4H_9	PECHK-34	7		
C_5H_4-Ch-B-C_3H_7	FCH-53		17	10
C_3H_7-Ch-B-C_2H_5	FCH-32	6		
C_3H_7-Ch-B-$OOCC_3H_7$	FCH-3-m3		6	
T_m, °C		-15	-15	-30
T_{NP}, °C		68.1	62.5	62.8
η(20 °C), mPa·s		40.6	33.7	
η(-30 °C), mPa·s		2986	1580	3900
K_{33}/K_{11}		0.8	0.71	0.79

$$T = \sin^2\left[\frac{1}{2}\pi\frac{\sqrt{1-u^2}}{1+u^2}\right], \tag{6.4.3}$$

where $u = 2L\Delta n/\lambda$. If birefringence Δn of the liquid crystal material is selected to ensure that $u = 3^{1/2}$ or $15^{1/2}$, the transmission of the cell becomes minimum, i.e., contrast increases and, consequently, the degree of multiplexicity and the viewing angle also increase. The authors of [361] presented the compositions of liquid crystal materials for operation in the Gooch–Tarry regime, i.e., for these materials, the optical path difference is $\Delta nL \approx 0.5$ μm which corresponds to the maximum contrast at $\lambda = 0.55$ μm. At the thickness of the layer of the liquid crystal $L = 5$ μm, $\Delta n \approx 0.1$. Table 6.7 gives the composition and properties of the liquid crystal materials characterised by the minimum relaxation time at room temperature and by a relatively high value of T_{NI}. The required value of birefringence is achieved by adding cyclohexyl carboxylates.

When powering liquid crystal displays by autonomous sources, in particular, solar batteries, it is necessary to use liquid crystal materials with the minimum working voltage. This may be achieved using substances with high dielectric anisotropy (in particular, phenyl pyrimidines) or with a low value of the elasticity coefficient in the deformation of transverse bending K_{11} (for example, phenyl dioxanes). This approach is also used in [362]. Almost all previously mentioned liquid crystal materials contain derivatives cyclohexyl (pyrimidinyl) of benzonitrile TsPmB with the general formula C_nH_{2n+1}–C_6H_{10}–$C_4H_2N_2$–C_6H_4–CN.

In the group of the tricyclic polar compounds TsPMbB, viscosity and activation energy are minimum and clarification temperature is maximum. The mixture containing mainly phenyl dioxanes (>59%) has the combination of viscosity and threshold voltage close to optimum. The minimum viscosity is recorded for a mixture with the increased concentration of diluents (C_nH_{2n+1}–C_6H_{10}–COO–C_6H_4–OC_mH_{2m+1}). A very low value of threshold voltage $U_t = 0.76$ V is the characteristic feature of a mixture without slightly polar compounds and with a high concentration of phenyl dioxanes and azomethines. Of course, viscosity greatly increases in this case. In the mixture with the relatively low concentration of phenyl dioxanes and without slightly polar compounds, dielectric anisotropy is very high and viscosity is 4, but the minimum threshold voltage is not reached. A decrease of the concentration of phenyl dioxanes results in a further increase of U_t but this may be accompanied by a slight decrease of viscosity because of the introduction of the diluent and by a large increase of the clarification temperature because of the increase of the concentration of the cyclohexyl (pyrimidinyl) benzonitriles.

For liquid crystal indicators, utilising the visitor–master effect, the solvents for dyes include materials with both positive dielectric anisotropy (ZLI-1957/5, 2444 [345, 351] and other materials based of phenyl cyclohexanes [363]) and with negative dielectric anisotropy. In the latter case, additions of the compounds of molecules characterised by a higher value of the transverse dipole moment are added to the mixture. Table 6.8 gives the data on the composition and properties of the mixtures containing derivatives of hydroxyquinone (–C_6H_2(2,3–CN)-)-2,3-cyan-4-alcoxyphenyl esters of alkylcyclohexacarbonic and alkylcyclohexylbenzoic acids [TsFETsGK 2,3TsFETsBK [364]. All these mixtures have a wide nematic range (T_m = –10÷–20°C, T_{NI} = 62÷67°C). With the increase of the concentration of hydroxyquinone additions the molecules grow linearly, irrespective of length, and viscosity and dielectric anisotropy also increase.

The addition of stilbene derivatives instead of hydroquinone derivatives increases the viscosity of the mixture and reduces dielectric anisotropy. These materials may also be used for the effect of dynamic scattering of light in doping with conducting additions. In [342], the composition of the liquid crystal materials for the effect of the dynamic scattering of light is given, with the nematic range from –11 to 78°C, η(20°C) = 65 cP, $\Delta\varepsilon$= –0.6; the liquid crystal material contains: 12% $CH_3O–C_6H_4–COO–C_6H_4–C_3H_7$, 23% $CH_3O–C_6H_4–COO–C_6H_4–OC_5H_{11}$, 9% $C_5H_{11}–C_6H_{10}–C_6H_4–CN$, 21% $C_3H_7–C_6H_{10}–C_6H_4–COO–(2–CN)C_6H_3–C_4H_9$, 35% $C_5H_{11}–C_6H_{10}–C_6H_4–(2–F)C_6H_3–C_2H_5$.

Materials have been developed for similar applications containing the derivatives of cyclohexane with the cyan-group in the *cis*-position. Consequently, the negative electrical anisotropy of the compound was higher. Table 6.9 gives the compositions and properties of these mixtures [365]. The lowest viscosity is typical of mixture 1 which includes the derivatives of *cis*-cyan-bicyclohexane, the phenylcyclohexane diluents, and also a large number of the derivatives of tolan which have low viscosity and are characterised by high clarification temperatures and low enthalpy of melting. This greatly widens the nematic range and makes it possible to use devices with these liquid crystal materials for external applications. The presence of tolan derivatives also increases birefringence Δn and this influences the level of light scattering by the liquid crystal element in the switched-on condition, i.e., its contrast. The viscosity of mixture 2, based on the tricyclic compounds with an ethanol bridge and fluorine in the *meta*-position, is comparatively low. An increase of the length of the molecule with the cis-cyan-group increases the viscosity of the mixtures 3 and 4.

This type of liquid crystal material with negative electrical anisotropy is the subject of special attention because the displays with the vertical (homeotropic) orientation of the liquid crystals have a number of advantages

Table 6.7. Properties and composition of LCM for operation in Gooch–Tarry regime (T_{NI} = 79.5°C, Δn = 0.1029, U_t = 2.3V, t_{on} = 98 ms, t_{off} = 14 ms)

substance and notation	amount, %
C_3H_7-Cy-Ph-CN (PCH-3)	20
C_3H_7-Cy-Ph-OCH_3 (PCH-3O1)	20
C_3H_7-Cy-Ph-OC_2H_5 (PCH-3O2)	15
C_3H_7-Cy-Ph-OC_4H_9 (PCH-3O4)	14
C_3H_7-Cy-Ph-COO-Ph-C_3H_7 (BCHP-33)	10
C_3H_7-Cy-Cy-COO-Ph-C_5H_{11} (BCHP-35)	10
C_3H_7-Cy-Ph-Ph-Cy-C_3H_7 (CBC-33)	4
C_5H_{11}-Cy-Ph-Ph-Cy-C_5H_{11} (CBC-55)	2
C_5H_{11}-Cy-Ph-Ph-Cy-C_3H_7 (CBC-53)	6

in comparison with conventional liquid crystal displays based on the twist effect, in particular, the high operating speed and contrast. The fraction of these liquid crystal devices on the world market in the year 2003 was 18% [366]. The authors of [367] developed a number of mixtures based on 2,3-difluorophenyl derivatives {C_nH_{2n+1}–$(C_6H_{10})_k$–(X)–C_6H_2(2.3–F)–OC_2H_5, where k = 1 or 2, X is the bridge group –COO–, –CH_2O– or a simple bond}. These mixtures are characterised by increased negative dielectric anisotropy and reduced viscosity. Table 6.10 gives the physical and electro-optical characteristics of new mixtures in comparison with the properties of Merck mixtures. In mixture D optical anisotropy is increased by the addition of 10% 4-propyl-(4'-propoxytolane)-cyclohexane CH_3–C_6H_{10}–C_6H_4–C≡C–C_6H_4–OC_3H_7. Attention should be given to the low freezing point and low values of the viscoelastic ratio of the new mixtures.

The hydroquinone derivatives are used actively in the development of materials with the inversion of the dielectric anisotropy sign at a low frequency of f_0~10 kHz. These liquid crystal materials operate in the control regime with forced relaxation of the liquid crystal layer by supplying to the cell a radiopulse filled with the voltage with the frequency of $f > f_0$ [368]. This gives the switching times of the twist effect up to 0.25 ms at the pulse amplitude of 70 V [368], the S-effect up to 20÷25 μs at an amplitude of 100 V [369, 370]. The components with a high transverse dipole moment are included in the composition of these liquid crystal materials for increasing the value of $\varepsilon_{\perp}$. To ensure low-frequency relaxation $\varepsilon_{\parallel}$, tri- or tetracyclic compounds with a high longitudinal dipole moment are added to the liquid crystal material. Because of these reasons, these liquid crystal materials are characterised by high viscosity, because the low frequency of relaxation of dielectric permittivity denotes the high value of the reorientation time (the equation (3.3.1)).

The compositions and properties of the mixtures with the low-frequency inversion of the sign were published in [371]. Doubts are cast of the $\Delta\varepsilon$ data obtained for mixture 3, i.e., in the absence of components with the transverse dipole and also at a low concentration of the compounds with the longitudinal dipole low-frequency relaxation should be detected at a higher frequency. In the group of other mixtures, the lowest values of η and f_0 and also the highest amplitude of variation $\Delta\varepsilon$ (f) are exhibited by mixture 5. The composition and properties of the liquid crystal materials for identical applications, containing 3-chlorophenyl fragment ($-C_6H_3$ (3-Cl)–) are given in Table 6.11 [372]. Attention should be given to the mixtures 7–9 with the naphthalene derivatives characterised at a relatively low viscosity by low frequency f_0 of inversion of the sign of $\Delta\varepsilon$ and a high clarification temperature. Of other materials of the same type it is important to mention the Swiss material RO-ZF-2857 with η (22°C) = 51 cP [347], and Russian materials ZhK-999, ZhK-1000 with η (25°C) = 36.4 and 62.1 cP, respectively [373]. The difference in the viscosity values is determined by the selection of the diluent, i.e. azoxycompound in ZhK-999 and phenyl benzoate in ZhK-1000.

In [374] a liquid crystal device addressed with an electron beam is described. The device contains eutectic mixtures BP-5, BP-7, PCH-3, PCH-7 (75%) and CBC-53 (25%), with the characteristics $T_m < -3$°C, $\varepsilon_{\parallel} = 12.4$, $\Delta n = 0.171$ and $\eta = 1.15$ cP. Evidently, this high viscosity is essential to store the image in the recorded element whilst other elements are still being written to. The switching time of the cell is 25 ms, at a thickness of 10 μm, the writing speed 164 cm/millisecond. The image contrast reached 59:1, the resolution capacity 225 lines with the aperture of 25mm.

At present, the principle of construction of binary liquid crystal materials is used in the development and application of liquid crystal materials when the developer produces two mixtures with different values of some characteristic and the user, mixing these two liquid crystal materials, obtains the required value of the characteristic. This principle is shown in Fig. 6.4.4 [375]. Table 6.12 gives the characteristics of the binary liquid crystal materials, i.e. the ranges of variation of a specific property.

The advanced liquid crystals are produced by the so-called spacer technology. In most cases, spherical particles of a specific size are selected and these particles define with high accuracy (< 0.1μm) the required

Table 6.8(a) Viscosity of LCM with negative dielectric polarisation *

Substance	1	2	3	4	5	6	7	8	9	10	11	12	13	14
C_3H_7-Ch-B-OC_2H_5	20	20	10	20	20	45	35	33	20	28	31	20	30	25
C_6H_{13}-Ch-B-OC_2H_5			10	10		45								
C_3H_7-Ch-B-OC_4H_9	20	20	10	10	20		22	21	20	20	20	20	10	10
C_3H_7-Ch-B-OC_5H_{11}	55	50	40	40	55				13			10	10	10
C_5H_{11}-Ch-B-OC_6H_{13}				5										
C_3H_7-Ch-COO-B(2,3-CN)-OC_5H_{11}	2.5	5	10	5	5	10			2		3	5		3
C_5H_{11}-Ch-COO-B(2,3-CN)-OC_5H_{11}	2.5	5	10	5			5	3	3	5	3	5		5
C_4H_9-Ch-COO-B(2,3-CN)-OC_4H_9			10	5				3	3	5	3	5	5	3
C_5H_{11}-Ch-COO-B(2,3-CN)-OC_4H_9										3	2		5	3
C_5H_{11}-Ch-B-COO-B(2,3-CN)-OC_4H_9								1.5	2	3	2		5	3
C_5H_{11}-Ch-B-COO-B(2,3-CN)-OC_5H_{11}								1.5	2	3	2			3
C_3H_7-Ch-B-C_2H_5								2	3	3			5	2
C_5H_7-Ch-B-C_3H_7											4	5		3
C_5H_{11}-Ch-B-B-C_2H_5							28	27	25	25	25	25	20	20
C_5H_{11}-Ch-B-B-Ch-C_3H_7							10	8	7	5	5	5	10	10
η(20 °C), cP	20	24	38	29	19	26	21	24	29	35	28	25	27	34
$\Delta\varepsilon$(20 °C)	-1.5	-2.7	-6.3	-4.0	-1.6	-2.6	-1.5	-2.6	-3.2	-4.9	-4.0	-3.9	-4.1	-5.1

*-B(2.3-CN) - the benzol ring with cyano groups in *meta*- and *ortho* -position.

Table 6.8(b) Comparison of the properties of the LCM with negative dielectric anisotropy containing derivatives of hydroquinone and α-cyanostilbene

Substance	1	2	3	4
C_5H_{11}-Ch-COO-B-OCH_3	50		30	45
C_3H_7-Ch-COO-B-OC_5H_{11}	35		20	35
C_5H_{11}-Ch-COO-B-OC_2H_5			20	
CH_3O-B-N(O)=N-B-C_4H_9		40		
C_2H_5O-B-N(O)=N-B-C_4H_9		40		
C_5H_{11}-Ch-COO-B(2,3-CN)-OC_7H_{15}	15			
C_2H_5-B-C(CN)=CH-B-OC_6H_{13}		20	30	10
C_3H_7-B-C(CN)=CH-B-OC_6H_{13}				10
η(20°C), cP	57	58	69	64
Δε	-4.1	-1.8	-2.1	-1.9
Δ*n*	0.12	0.25	0.14	0.13

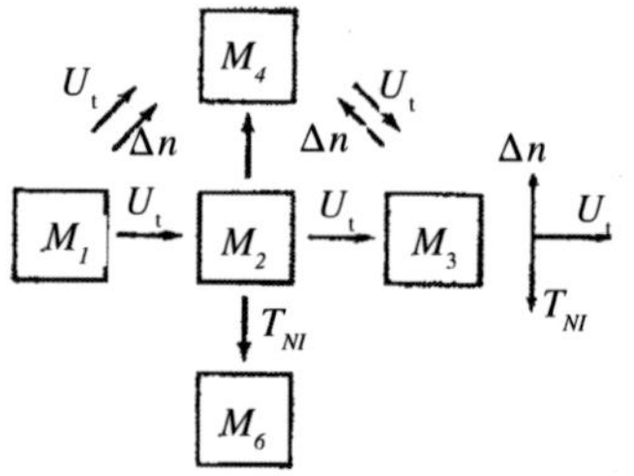

Fig. 6.4.4. Selection of components of binary LC materials $M_1 \div M_6$ for controlling the threshold voltage, birefringence and the clarification temperature of the material.

homogeneity of the thickness of the liquid crystal layer. Taking into account the optical characteristics of these cells, for example, the Gooch–Tarry regime for the twist effect, the optical difference of the path ΔnL for the given thickness of the gap should have a fully determined value (0.86 λ for the first transmission minimum according to Gooch and Tarry, or ≈ 0.5 λ for the supertwist effect, denoted usually as the STN mode). Therefore, additional requirements are imposed on the birefringence of the liquid crystal material developed for a specific application: as the gap between the sheets of the liquid crystal display decreases, the duration of electro-optical

Table 6.9 Composition of the properties of LCM containing substances with *cis*-4-cyanocyclohexyle fragment (-(*e*CN)C_6H_9–) for negative dielectric polarisation

Substance	1	2	3	4
C_3H_7-(*e*CN)C_6H_9-C_6H_{10}-C_3H_7	10	14		
C_3H_7-(*e*CN)C_6H_9-C_6H_4-C_6H_4-C_3H_7			14	
C_3H_7-(*e*CN)C_6H_9-C_6H_4-C_6H_4-C_5H_{11}				9
C_3H_7-C_6H_{10}-$(CH_2)_2$-C_6H_4-(2-F)C_6H_3-C_2H_5		29	29	29
C_3H_7-C_6H_{10}-$(CH_2)_2$-C_6H_4-(2-F)C_6H_3-C_5H_{11}		29	29	29
C_5H_{11}-C_6H_{10}-$(CH_2)_2$-C_6H_4-(2-F)C_6H_3-C_5H_{11}		28	28	28
C_3H_7-C_6H_{10}-C_6H_4-CH_3	20			
C_3H_7-C_6H_{10}-C_6H_4-C_2H_5	17			
C_3H_7-C_6H_{10}-C_6H_4-C_4H_9	7			
C_3H_7-C_6H_{10}-C_6H_4-(2-F)C_6H_3-C_6H_{10}-C_3H_7	3			
C_5H_{11}-C_6H_{10}-C_6H_4-(2-F)C_6H_3-C_6H_{10}-C_5H_{11}	3			
C_3H_7-C_6H_{10}-C_6H_4-(2-F)C_6H_3-C_6H_{10}-C_5H_{11}	3			
C_4H_9-C_6H_4-C≡C-C_6H_4-C_3H_7	5			
C_5H_{11}-C_6H_4-C≡C-C_6H_4-C_3H_7	5			
C_5H_{11}-C_6H_4-C≡C-C_6H_4-OCH_3	5			
CH_3-C_6H_4-C≡C-C_6H_4-OC_2H_5	4			
C_3H_7-C_6H_{10}-C_6H_4-C≡C-C_6H_4-OCH_3	6			
C_3H_7-C_6H_{10}-C_6H_4-C≡C-C_6H_4-OC_2H_5	5			
C_3H_7-C_6H_{10}-C_6H_4-C≡C-C_6H_4-OC_3H_7	7			
T_{SN}, °C	-40			
T_{NI}, °C	80	90	104	104
$\Delta\varepsilon$ (20 °C)	-0.8	-1.0	-0.6	-0.8
Δn (20 °C)	0.1581	0.131	0.150	0.149
η (20 °C), cP	18	29	32	37

Table 6.10 Properties for mixtures for LCD with vertical orientation

Mixture	A	B	C	D	MLC-2037	MLC-2038	MLC-2039
T_{NI}, °C	92	84	70	103	71	80	91
T_m, °C	< –60	< –60	< –60	< –60	< –30	< –30	< –30
Δn (589 nm)	0.071	0.082	0.058	0.10	0.0649	0.1032	0.0821
$\Delta\varepsilon$					–3.1	–5.0	–4.1
γ_1, mPa·s					132	179	163
γ_1/K_{33}, ms/μm^2	18.1	17.1	8.8	18			
U_t, V	1.78	1.70	1.70	1.80			2.1

response also decreases. In the last 15 years the size of the spacers has decreased from 8 to 4.5 and subsequently to 2.7 μm. The next task is to use liquid crystal layers with the thickness of the order of 1 μm. This imposes special requirements on the increase of Δn of the liquid crystal material whilst retaining or improving its viscoelastic characteristics.

In addition, the increase in the extent of application of liquid crystal displays based on active matrices with the thin-film transistors (TFT) requires developers to find new mesogenic components with reduced electrical conductivity.

These two reasons resulted in recent years in increased interest in the liquid crystal substances whose molecules contain 1–2 acetylene groups, and the hydrogen atoms in the benzol rings and/or hydrocarbon end chains are substituted by fluorine atoms.

We examine several examples of the application of liquid crystal materials for the supertwist effect. The authors of [376] developed materials for the STN-mode with the multiplexicity 1/240 with the switching time of 100–120 ms which is 3–4 times shorter in comparison with the conventional liquid crystal materials for this effect. The materials contain new tolane derivatives resulting in the steep curvature of the volt–contrast characteristic (VCC) as a result of the relatively high values of the ratio K_{33}/K_{11}, and also new fluorine-containing components with very low viscosity. These materials have the physical characteristics presented in Table 6.13. It was shown in [377] that the use of substances containing both the acetylene group and fluorine atoms (C_mH_{2m+1}–C_6H_2(2.6–F)–C≡C–C_6H_4–C_6H_4–C_nH_{2n+1}) make it possible to develop a liquid crystal material for a gap of 2.7 μm and use this material for the so-called field-sequential colour displays. The physical and electro-optical characteristics of such a material (Table 6.14, the

Table 6.11 Properties of LCM with low frequency inversion of the Δε, containing substances with 2-chlorophenyl fragment (chlorine atom in the *ortho*-position B(Cl))

substance	1	2	3	4	5	6	7	8	9
C_5H_{11}-Ch-COO-B(Cl)-COO-B-Cl	29								
C_5H_{11}-Ch-COO-B(Cl)-COO-B-OC_4H_9	49							42	
C_3H_7-Ch-COO-B(Cl)-COO-B-OC_5H_{11}						24			
C_4H_9-Ch-COO-B(Cl)-COO-B-OC_5H_{11}		29	26	29	27				
C_5H_{11}-Ch-COO-B(Cl)-COO-B-OC_5H_{11}		31	28	31	29	32			
C_6H_{13}-Nptl-COO-B(Cl)-COO-B-OC_5H_{11}							20	34	20
C_7H_{15}-B-B-COO-B-B-CN							4.5		4.5
C_3H_7-Ch-B-C_4H_9	25								15.5
C_5H_{11}-Ch-B-B-Ch-C_3H_7			10		6.5	6.5			
CH_3-B-COO-B-C_5H_{11}		12	11					7	
C_3H_7-B-COO-B-C_5H_{11}		16	14					10	
C_5H_{11}-B-COO-B-C_5H_{11}		12	11					7	
C_3H_7-Ch-COO-(F)B-C_5H_{11}				10	9.5	9.5	26.5		
C_5H_{11}-Ch-COO-(F)B-C_5H_{11}				16	15	15	26.5		
C_7H_{15}-Ch-COO-(F)B-C_5H_{11}				14	13	13	22.5		
C_3H_7-Ch-COO-(F)B-OC_5H_{11}									30
C_5H_{11}-Ch-COO-(F)B-OC_3H_7									30
T_{NI}, °C	118	75	98	80	93	101	73	139	92
Δε, LF	+1.4	+3.85	+3.6	+2.7	+2.4	+3.0	+1.4	+1.8	+1.5
Δε, HF	-2.7	-1.75	-1.8	-1.9	-2.0	-1.8	-1.0	-2.1	-1.4
f_0, kHz	3.9	12	8.5	9.0	6.0	4.5	4.0	0.64	1.6
η(20 °C), cP_3	89	95	102	90	96		45		59

characteristics of a similar material BDH-TL202 are also given for comparison) make it possible to change the main three colours with the frequency of 180Hz which corresponds to the frequency of change of the full-colour image of 60 Hz. It was shown in [378] that these molecules with two fluorine atoms, directed in the same direction and with simple (without oxygen) bonds greatly reduce of the number and charge of the ions in the working mixture and, correspondingly, reduce of the electrical conductivity and increase important characteristics of the liquid crystal materials, such as the retention ratio.

Table 6.12 Properties of mixtures for binary LC materials [369]

	M1 (TN-701)	M2 (TN-615)	M3 (TN-619)	M4 (TN-621)	M5 (TN-703)	M6 (TN-603)	M7 (TN-623)
T_m, °C	< –10	< –20	< –10	< –20	< –20	< –20	< –20
T_{NI}, °C	62.5	70.5	60.5	76.5	68.5	96.0	101
U_{10}, V	1.10	1.52	2.03	1.58	1.46	1.80	1.69
N_{max} (Θ=0°)	49	51	70	59	51	56	53
η, P	0.58	0.41	0.27	0.45	0.39	0.44	0.66
γ_1, P	2.23	2.03	1.08	2.36	1.70	3.00	3.16
$\Delta\varepsilon$	23.70	12.70	5.80	15.80	12.90	13.20	17.20
K_{11}, 10^{-7} dyne	9.00	10.0	11.1	12.0	9.60	12.1	14.7
K_{33}/K_{11}	1.60	1.39	0.98	1.41	1.52	1.64	1.32
T_{off}, ms	144	83	43	83	78	66	76

Table 6.13 Properties of mixtures for the supertwist effect with increased response speed

Parameter	Mixture A	Mixture B
η, mPa·s	13.4	10.6
K_{33}/K_{11}	1.19	1.20
Δn, (589 nm)	0.215	0.172
$\Delta\varepsilon$	6.8	7.2
T_{NI}, °C	85	75

Another example of a similar compound with a low viscoelastic ratio for application in high-speed electro-optical gates alkyl-alkyl-bis-tolane R1–C_6H_4–C C–C_6H_3(2-Y) –C ≡ C–$C_6H_4R_2$, where Y is hydrogen, fluorine, methyl or ehtyl group, R_1, R_2 is the alkyl group, fluorine or CN-group [379]. The main advantage of these components is the extremely high birefringence ($\Delta n > 0.35$ in the entire visible range, $\Delta n = 0.6$ at 400 nm). The viscoelastic ratio of the component with Y = C_2H_5, R_1= C_5H_{11}, R_2 = C_3H_7 is described by the ratio $\gamma_1/K_{11} = 3.37 \cdot 10^{-6}$ ms/μm² exp (0.429 eV/k_BT)/S.

The properties of substances with different fluorine-containing fragments were presented in Table 5.11. The properties of liquid crystal materials based on these substances were described in [312]. Two mixtures were

Table 6.14 Properties of the material displaced with colour changing with time

Parameter	Mixture A	BDH-TL202
T_{NI}, °C	71	83
Δn (589 nm)	0.201	0.185
$\Delta\varepsilon$	3.1	6.0
γ_1, mPa·s	111	307
U_t, V	2.07	2.00
U_{sat}, V	3.18	3.09
T_{on}, ms	1.8	3.2
T_{off}, ms	4.4	8.0

developed for different supply voltages (5 V and 3.3 V, respectively). These parameters (Table 6.15) are compared with the characteristics of the standard mixtures MLC-6012 and ML C62 to 1-000.

A method for development of liquid crystal materials for the so-called in-plane switching (IPS) was proposed in [380]. An electrical field is applied to two adjacent electrodes placed on the same substrate and, therefore, the field is directed parallel to the substrate plane. Pure torsional deformation is applied in this case, and the dynamics of the deformation is described only by rotational viscosity. The requirements on the low electrical conductivity of the material for active matrix TFT are fulfilled. By selecting the components, the viscosity of the liquid crystal material is reduced by 30–50% in comparison with the liquid crystal material for the twist effect. The most efficient substances from the viewpoint of viscosity and threshold voltage are C_3H_7–C_6H_{10}–C_6H_2–(2,6-F, 1-CN) with U_t = 0.67 V, γ_1 = 100 mPa·s and C_7H_{15}–C_6H_{10}–C_6H_4–F with U_t = 1.02 V, γ_1 = 60 mPa·s. The most efficient low polar diluents are bicyclohexanes with alkenyl substitutes C_5H_{11}–C_6H_{10}–C_6H_{10}–CH = CH_2 and CH_2–CH = C_6H_{10}–C_6H_{10} – CH = CH – CH_2 (γ_1 ~ 50 mPa·s). In addition, in [380] the authors reported on the proportionality of the switching time of the electro-optical response of an IPS cell with the ratio of rotational viscosity to the square of birefringence: T = 12 ms + 4 ms/mPa · s.

A rare example of the liquid crystal material specially developed for non-linear optical effects are the mixtures of the materials of E. Merck Company (Germany) with tolane and azo dyes C_nH_{2n+1}–N(H or CH_3)–C_6H_4–N_6 = N–C_6H_4–NO_2 and C_nH_{2n+1}– NH –C_6H_4– C–C_6H_4–NO_2. As

Table 6.15 Properties of LCM displays with a low supply voltage

Parameter	MLC-6012	Mixture A	MLC-6221-000	Mixture B
T_{NI}, °C	86	87	73	76
$\Delta\varepsilon$	8.2	7.6	9.6	10.6
Δn (589 nm)	0.1016	0.0977	0.0914	0.0928
γ_1, mPa·s	192	127	205	147
U_t, V	1.49	1.52	1.22	1.18
U_{sat}, V	2.34	2.35	1.92	1.84
T_{off}, ms	89	63	133	102

already reported in section 6.2, the addition of dyes increases both viscosity and the viscoelastic ratio, although it also increases birefringence and reduces the threshold voltage of the electro-optical effect. However, the amount and the shape of the substitute may change this relationship. For example, using azo dyes as an example, it was shown that the presence of a volume substitute, directed away from the main chain, at low concentrations (approximately 5%) reduces the viscoelastic ratio of the solution (80 ms/μm^2 for the substance with the NCH_3-group, 25 ms/μm^2 for the substance with an NH-group) in comparison with the viscosity of the nematic matrix (22 ms/μm^2). This corresponds to simulation considerations described in section 6.2 (Fig. 6.2.5).

In [240] and in section 5.1 there is no data on the viscosity properties of some substances for which the literature gives only the data on the switching time. It is not always possible to transfer the time and viscoelastic ratio and, consequently, only the data on the dynamic characteristics of these mixtures are presented below. It was shown in [382] that the addition of 10% cyclohexyl benzotrifluoride C_nH_{2n+1}–C_6H_{10}–C_6H_4–CF_3 reduces the relaxation time of ZLI-1132 in a 10 μm cell from 105 ms to 85 ms at T = 10°C and to 60 ms at T = 22°C. The properties of mixtures containing compounds with the –CH_2O– bridge were described in [383]. The mixtures based on phenylcyclohexanes with alcoxyphenylalkylmethyl esters (CH_3–C_6H_4–CH_2O–C_6H_4–OC_nH_{2n+1}, 25%) were characterised by the cooling time of 65 ms at 20°C and U = 3 V, and those with hydrated analogues (C_6H_{13}–C_6H_{10}– CH_2O–C_6H_4–OC_mH_{2m+1}, 45%) by 86 ms. The shortest switching time (59 ms) was recorded for a mixture with 40% $5H_{11}$–C_6H_{10}– CH_2O–C_6H_4–C_5H_{11}.

The properties of tricyclic derivatives of dioxanes, described by the general formula R–C_6H_4–$C_4H_6O_2$– C_6H_{10}–R', are investigated in [384, 385]. In a cell with a thickness of 14.5 μm (D) the mixture of cyanphenyl dioxanes with 14% C_5H_{11}–C_6H_{10}– $C_4H_6O_2$–C_6H_{10}–CN with T_m = –18÷19°C, T_{NI} = 73÷76°C, gave the threshold voltage of 0.8 V and the switch off time of 170 ms. These compounds are interesting because they have very low values of the melting enthalpy. For example, in the case of C_2H_5–C_6H_4–$C_4H_2O_2$– C_6H_{10}– C_2H_5, ΔH_m = 17.5 kJ/mole. The mixtures with esters of 7-alkyl-perhydrophenanthrene-2-carbonic acids are described in [386]. The initial mixture consisted of cyanphenyl esters of alkylcyclohexanecarbonic acid with the following properties: T_m = 10÷14°C, T_{NI} = 72°C, U_t = 1.4 V, t_{on} = 560 ms, t_{off} = 240 ms. At a thickness of the liquid crystal layer of 20 μm, the addition of 20 mol.% with n = 5, R = –C_6H_4–OC_6H_{13}, reduced T_m to 3÷6°C, increased T_{NI} to 83°C and shortened the switching time to 370 and 184 ms for switching on and switching off, respectively. The same amount of the addition with n = 4 and R = –C_6H_4–CN improved the properties of the mixture to a greater extent: T_m < –10°C, T_{NI} = 101÷103°C, U_t = 2 V, t_{on} = 334 ms, t_{off} = 126 ms. On the other hand, the naphthalene derivative C_4H_9–$C_{10}H_6$–COO–C_6H_{10}–C_4H_9 slightly impaired the properties of the same mixture and also resulted in a small reduction of T_m: T_m < –20° C, T_{NI} = 63.5°C, U_t = 1.8 V, t_{on} = 1.94 s, t_{off} = 0.58 s [387].

In conclusion, it is important to characterise briefly Russian liquid crystal materials developed at NIOPIK from the viewpoint of the effect of their composition on viscosity (see also Fig. 2.4.6).

Various liquid crystal materials, such as ZhK-1277, 1282, 29, 1290, which contain cyanobiphenyl as the main component and alkoxyphenyl esters of alkyl transcyclohexane carbonic acids as a diluent have lower viscosity and higher operation speed in comparison with ZhK-807 and SZhK-1 which consist only of cyano derivative liquid crystals. It should be mentioned that ZhK-1289 has lower viscosity than other liquid crystal materials of the series because this material contains a large amount of low-viscosity pentylcyanobiphenyl (ZhK-609), and other materials contain only relatively viscous alkoxycyanobiphenyls as polar components.

The viscosity of ZhK-1281 is lower than that of ZhK-805 because the former is a mixture of viscous cyclohexane carbonic acids with low-viscosity phenylcyclohexanes and also has a better temperature dependence of the operation speed in comparison with the liquid crystal material from the previous group.

Comparison of viscosity and operation speed of ZhK-1144, which contains only cyano derivatives of phenylpyridines, with the identical characteristics of ZhK-1283, 1285 which contain a diluent (Demus esters), also shows that the liquid crystal material consisting of high- and low-polar components, has better characteristics.

Unfortunately, as regards examples, there is only one case in which the liquid crystal material formed only by polar components would have the comparatively low viscosity and quite satisfactory operation speed, including at subzero temperatures. These properties of ZhK-1133 were achieved as a result of selecting components with the low energy of intermolecular interaction ensuring a weak temperature dependence of viscosity and operation speed. In addition, the presence of bicyclooctanes which contain a volume fragment in the centre, results in less dense packing of the molecules and, correspondingly, lower viscosity.

The proposed method of designing the liquid crystal materials with improved operation speed was applied in the development of ZhK-1350, 1351, 1352, 1390, 1391. Only the cyclohexane derivatives with the lowest viscosity and activation energy, were selected as both polar and non-polar components for these materials. The polar components included substances with short alkyl chains with an odd number of carbon atoms. The non-polar substances included the Demus esters, cyclohexacarbonic acids, alkyl or alkoxyphenyl alkyl cyclohexanes or combination of these compounds. It should be mentioned that the alkoxyphenyl alkyl cyclohexanes have the lowest glass transition temperature T_g of all the liquid crystal materials resulting in high operation speed at subzero temperatures. The value of T_g for FTsG-302 is equal to 159 K [5]. All these liquid crystal materials, formed from cyclohexane derivatives, have low viscosity and high operation speed. It should also be mentioned that the liquid crystal materials which contain only alkoxyphenyl alkyl cyclohexanes as a diluent (ZhK-1351, ZhK-1391) have lower viscosity in comparison with the liquid crystal material which contain the Demus esters (ZhK-1350, ZhK-1390) as a diluent.

References

1. Blinov L.M. and Chigrinov V.G., Electro-Optics of Liquid Crystals, Springer-Verlag, 1994.
2. Chandrasekar C., Liquid crystals, Moscow, Mir, 1980.
3. De Jeu W.H., Physical properties of liquid crystal substances, Moscow, Mir, 1982.
4. Grebenkin M.F. and Ivaschenko A.V., Liquid crystal materials, Moscow, Khimiya, 1983.
5. Ericksen J.L., Mol. Cryst. and Liquid Cryst., 1966, vol. 7, 153.
6. Leslie F.M., Arch. Rat. Mech. Anal. 1968, vol. 28, 265.
7. Parodi O., J. de Phys., 1970, vol. 31, 581.
8. Martin P.C., et al, Phys. Rev. A., 1971, vol. 6, 2401.
9. Forster D., et al, Phys. Rev. Lett. 1971, vol. 26, 1017.
10. Forster D., Phys. Rev. Lett., 1974, vol. 32, 1161.
11. Livshits E.M., et al, Lectures in theoretical physics, vol. 7, Elasticity theory, Moscow, Nauka, 1987, 208-218.
12. Pasechnik S.V., et al, Zhurnal Fiz. Khimii, 1987, vol. 61, 1675.
13. Muschik W., et al, Mol. Cryst. and Liquid Cryst., 1995, vol. 262, 417.
14. Papenfuss Ch. and Nuschik W., Mol. Cryst. and Liquid Cryst., 1995, vol. 262, 473.
15. Ehrentraut H. and Muschik W., Mol. Cryst. and Liquid Cryst., 1995, vol. 262, 561.
16. Gaehwiller Ch., Mol. Cryst. and Liquid Cryst., 1973, vol. 20, 301.
17. Miesowicz M., Nature, 1953, vol. 136, 261.
18. Belyaev V.V. and Nemtsov V.B., Zhurnal Fiz. Khimii, 1992, vol. 66, 2763.
19. Brochard F., et al, Phys. Rev. Lett. 1972, vol. 28, 1681. Pieranski P., et al, J.de Phys. 1973, vol.34, 35.
20. Van Doorn C.Z., J. Phys. Colloq., 1974, vol. 36, C1-261.
21. Chigrinov V.G. and Belyaev V.V., Kristalografiya,1977, vol. 22, 603, Belyaev V.V. Chigrinov V.G., App. Opt., 1993, vol.32, 141.
22. Shmelev O.Ya., et al, Zhurnal Fiz. Khimii, 1995, vol.59, 2036.
23. Kneppe H., et al, Ber Bunsenges. Phys. Chem., 1981, vol. 85, 784; Kneppe H., et al, J. Chem. Phys. 1982, vol.77, 3203.
24. Kneppe H. and Schneider F., Mol. Cryst. and Liquid Cryst., 1981, vol. 65, 23.
25. Belyaev V.V., et al, Kristalografiya, 1985, vol. 30, 1160; Belyaev V.V., et al, Kurzreferäte der VI Konf. Soz. Länder für Flüssigen Kristallen, Halle, DDR, 1985, 82.
26. Bock F.-J., et al, Liq. Cryst., 1986, vol. 1, 239.
27. Van der Meulen J.P. and Zijsltra R.J.J., Physica, 1985, vol. 132B, 153.
28. Malkin A.Ya., Encyclopedia of Physics, vol. 1, Sovietskaya entsiklopediya, Moscow, 1988, 283.
29. Miesowicz M., Nature, 1964, vol. 158, 27.
30. Tsvetkov V.N. and Mikhailov G.M., Acta Physicochim, URSS, 1938, vol.8, 77.
31. Miesowicz M., Mol. Cryst. and Liquid Cryst., 1983, vol. 97, 1.
32. Hennel F., et al, Mol. Cryst. and Liquid Cryst., 1990, vol. 191, 401.

33. Tsvetkov V.A. and Beresnev G.A, Pribory i tekhnika eksperimenta, 1977, vol. 5, 223.
34. Tsvetkov V.A., Mol. Cryst. and Liquid Cryst., 1985, vol. 52, 208.
35. Beens W.W., de Jeu W.H., J. de Phys., 1983, vol. 44, 129.
36. Kuzmin N.G., et al, in: Proc. 5th National conference: Liquid crystals and their application in practice, Ivanovo, 1985, vol. II, part 2, 71.
37. Hess S. and Koo H.M., Non-Equilib. Thermodyn., 1989, vol. 14, 159.
38. Derfel G. and Radomska B., Proc. SPIE, Liquid Crystals: Physics,Technology and Applications, 1997, vol. 3318, 300.
39. Hunniset S.D. and van der Sluijs J.C.A., J.de Physique – Lettres, 1983, vol. 44, 59.
40. White A.E., et al, Mol. Cryst. and Liquid Cryst., 1977, vol. 43, 13.
41. Narasimhan M.N.L. and Eringen A.C., Mol. Cryst. and Liquid Cryst., 1974, vol. 29, 57.
42. Atkin A.J., Arch. Rat. Mech. Anal., 1970, vol.38, 224.
43. Ericksen J.L., Trans. Soc. Rheol., 1969, vol.6, 255.
44. Fischer J. and Frederickson A.G., Mol. Cryst. and Liquid Cryst., 1969, vol. 6, 255.
45. Tseng H.C., et al., Phys. Fluids, 1972, vol. 15, 1213.
46. Kini U.D. and Ranganath G.S., Pramana, 1975, vol.4, 19.
47. Tsikalo A.L. and Shteinberg E.A., in: Proc. 6th National conference: Liquid crystals and their application in practice, Chernilov, 1988, vol.3, 565.
48. Leslie F.M., Mol. Cryst. and Liquid Cryst., 1981, vol. 63, 365.
49. Benicewicz B.C., Mol. Cryst. and Liquid Cryst., 1981, vol. 65, 111.
50. Krynkina S.V. and Malochko V.A., in: Proc. 5th National conference: Liquid crystals and their application in practice, Ivanovo, 1985, vol.1, 199.
51. Hauser A., et al., Crystal Res. and Technol., 1984, vol.19, 261.
52. Umstaetter H., Einfuehrung in die Viskosimetrie. Berlin-Goettingen-Heidelberg. 1952, 91.
53. Meiboom S. and Hewitt R.C., Phys.Rev.Lett., 1973, vol. 30, 261.
54. Derfel G., Mol. Cryst. and Liquid Cryst., 1990, vol. 191, 377.
55. Khazimullin M.V., et al, Mol. Cryst. and Liquid Cryst., 1999, vol. 329, 247.
56. Yablonsky S.V., et al, Proc. SPIE, 1996, vol. 2731, 87.
57. Figueiredo-Neto A.M., et al, Phys. Rev. A., 1991, vol. 44, 825.
58. Vorflusev V.P., et al, Appl.Phys., 1996, vol. 55, 1145.
59. Krekhov A.P. and Chuvyrov A.N., Mol. Cryst. and Liquid Cryst., 1990, vol. 191, 251.
60. Kuzmin H.G., et al, in: Proc. 4th National conference: Liquid crystals and their application in practice, Tsernigov, 1988, vol. 4, 470.
61. Wahl J. and Fischer., Mol. Cryst. and Liquid Cryst., 1973, vol. 22, 359.
62. Waltermann Th., Fischer F., Z. Naturforsch, 1975, Bd. 78, 891.
63. Fischer F., et al, Ber Bunsenges, 1974, Bd.78, 891.
64. Holmstroem S. and Lagerwall S.T., Mol. Cryst. and Liquid Cryst., 1977, vol. 38, 141.
65. Yablonskii C.V., et al, Kristalografiya, 1984, vol. 29, 984.
66. Grupp J., Rev. Sci. Instrum. 1983, vol. 54, 754.
67. Kuss E., Mol. Cryst. and Liquid Cryst., 1978, vol. 47, 71.
68. Diogo A.C., Sol. St. Commun, 1984, vol. 50, 895.
69. Martinoty P. and Canau S., Mol .Cryst. and Liquid Cryst., 1971, vol. 14, 243.
70. Tabidze A.A. and Koshkin H.I., Dep. VINITI 1988, No.2081, vol. 88, 1.
71. Tabidze A.A., et al, Izmerit. Tekhnika, 1981, No.12, 5.
72. Dyachenko B.P., Using the method for measuring viscoelastic properties of liquids at high pressures, Kuibyshev, 1973.

73. Tabidze A.A. and Kazakov R.Kh., Izmerit. Tekhnika, 1983, No.1, 34.
74. Tabidze .A., Koshkin H.I., Zhurnal Fiz. Khimii, 1986, vol. 60, 1501.
75. Langevin D. and Bouchiat M.A., J. Phys., 1972, vol. 33, 101.
76. Langevin D. and Bouchiat M.A., Mol. Cryst. and Liquid Cryst., 1973, vol. 22, 317.
77. McQueen D.H. and Singhal Y.K., J Phys. D: Appl. Phys., 1974, vol.7, 1983.
78. Tsvetkoff V.N., ZHETF, 1939, vol.9, 602.
79. Tsvetkoff V.N., Acta Physicochimica URSS, 1939, vol.10, 555.
80. Prost J. and Gasparoux H., Phys.Lett., 1971, vol 36A, 245.
81. Gasparoux H. and Prost J., J. Phys., 1971, vol.32, 65.
82. Siedler L.T.S., et al, Mol. Cryst. and Liquid Cryst., 1983, vol. 90, 255.
83. Leslie F.M., et al, Chem. Phys. Letters, 1972, vol.13, 368.
84. Kneppe H. and Schneider F., J. Phys. E. Sci Instrum., 1983, vol. 16, 512.
85. Yun C.K., Phys. Lett., 1973, vol. 45 A, 119.
86. Lippman H., Ann Physik, 1958, Bd. I, 157.
87. De Zhen P., Physics of Liquid Crystals, Mir, Moscow, 1977.
88. Brochard F., Mol. Cryst. and Liquid Cryst., 1973, vol. 23, 51.
89. Helfrich W., Phys. Rev. Lett., 1968, vol. 21, 1518.
90. Anisimov M.M. and Lagunov A.S., Zhurnal Fiz. Khimii, 1982, vol. 56, 1316.
91. Bogdanov D.L., et al, Author Cert. URSS, No. 731355, 1980, No.16.
92. Bogdanov D.L., et al, Akust Zh., 1980, vol. 26, 28.
93. Lagunov A.S. and Larionov A.N., Akust. Zh., 1984, vol. 30, 344.
94. Doerrer N., et al, Liq. Cryst., 1986, vol. 1, 573.
95. Gevorkyan E.V., et al, Akust. Zh., 1982, vol. 28, 14.
96. Lagunov A.S. and Larionov A.N., Akust Zh., 1984, vol. 30, 344.
97. Lagunov A.S., et al, in: Proc. 10th National Acoustic Conference, Moscow, 1983, 56.
98. Lagunov A.S., et al, Liquid crystals, Proc., Ivanovo, 1985, 56.
99. Lagunov A.S., Abstracts of 10th Int. Liq. Cryst.Conf., York, 1984, vol. 23.
100. Kozhevnikov E.N., Akust.Zh., 1994, vol. 40, 412.
101. Bogdanov D.L., et al, Zhurnal Fiz. Khimii, 1997, vol. 71, 931.
102. Bogdanov D.L., et al, Abstracts of 17th Int. Liq. Cryst. Conf. Strasbourg, France, 1998.
103. Bogdanov D.L. and Vekovishchev A.M., Proc. All-Russian Scientific and Technical Conference, Voronezh, 1997.
104. Lagunov A.S. and Anikin A.M., Zhurnal Fiz. Khimii, 1986, vol. 60, 417.
105. Bogdanov D.L., Ultrasound and dynamic properties of liquid crystals in magnetic fields of variable orientation at high pressure, Dissertation, Moscow, 1999, 264.
106. Gerber P.R., Appl. Phys., 1981, vol. A 26, 139.
107. Schad Hp., J. Chem. Phys., 1979, vol.71, 127.
108. Peneva A., Frederiks B., Moscow, GIZ, 1926, 16; Frederiks V. and Tsvetkov V., Physik. Z. Sowietunion. 1934, Bd.6, 490.
109. Chigrinov V.G., Kristalografiya, 1982, vol. 27, 404.
110. Stephen M.J. and Straley J.P., Rev. Modern Physics, 1974, vol. 46, 617.
111. Van Dijk J.W., et al, J. Chem. Phys., 1983, vol.79, 3888.
112. Cladis P., Phys. Rev. Lett., 1972, vol. 28, 1628.
113. Leenhouts F. and Dekker A.J., J. Chem. Phys., 1981, vol. 74, 1956.
114. Chigrinov V.G. and Grebenkin M.F., Kristalografiya, 1975, vol. 20, 1240.
115. Schad Hp., J. Appl. Phys., 1983, vol. 54, 4994.
116. Pasechnik S.V., et al, J. de Phys., 1984, vol. 45, 441.
117. Pasechnik S.V., Ultrasound and dynamic processes in phase transitions and structural

transformations in oriented liquid crystals, Dissertation, Moscow, 1998, 271.
118. Leenhouts F., J. Appl. Phys., 1985, vol. 58, 2180.
119. Belyaev V.V., et al, in: Proc. 3rd Scientific and Technical Seminar: Optical Properties and Applications of Liquid Crystals, Leningrad, 1983, 51.
120. Petrov V.F., et al, ibid, Leningrad, 1983, 49.
121. Kuhn W.L. and Finlayson B.A., Mol. Cryst.and Liquid Cryst., 1976, vol. 36, 307.
122. Imai M., et al, Mol. Cryst. and Liquid Cryst., 1995, vol. 262, 267.
123. Hess S. and Pardovitz I., Z. Naturforsch., 1981, Bd.36a, 1224.
124. Hess S. and Z. Naturforsch, 1975, Bd. 30a, 554, Z. Naturforsch, 1975, Bd.30a, 728.
125. Derfel G. and Gajewska B., Proc. SPIE Liquid Crystals: Physics, Technology and Applications, 1997, vol. 3318, 292.
126. Sadovskii V.N. and Usova N.A., Proc. 6th Scientific and Technical Seminar: Optical Properties and Applications of Liquid Crystals, Chernigov, 1988, vol 3, 194, Sadovkii V.N. and Usova N.A., Kristalografia, 1990, vol. 35, 1504.
127. Sadovskii V.N. and Usova N.A., Akust. Zh., 1987, vol. 33, 551.
128. Tabiryan N.V., et al, Mol. Cryst. and Liquid Cryst., 1986, vol. 136, 1.
129. Umeton C., et al, Mol. Cryst. and Liquid Cryst., 1995, vol. 261, 187.
130. Lager L., Sol. St. Comm., 1972, vol. 10, 697.
131. Wang X., Phys. Letters, 1985, vol. 112 A, 402.
132. Geurst J.A., et al, J. Phys., 1975, vol. 36, 653.
133. Allen S.M. and Cahn J.W., Acta Metall., 1979, vol. 27, 1085.
134. Orihara H. and Ishibashi Y., J. Phys. Soc. Japan, 1986, vol. 55, 2151.
135. Ohta T., et al, Phys. Rev. Lett., 1982, vol. 49, 1223.
136. Grigutsch M., et al, Mol. Cryst. and Liquid Cryst., 1995, vol. 261, 283.
137. Lonberg F., et al, Phys. Rev. Lett., 1984, vol. 52, 1903.
138. Orsay Liquid Crystals Group, J. Chem. Phys., 1969, vol. 51, 816.
139. Orsay Liquid Crystals Group, Phys. Rev. Lett., 1969, vol. 22, 1361.
140. Leger-Quercy L., Etude experimentale des fluctuations thermiques d'orientation dans un crystal liquide nematique par diffusion de la lumiere, Thesis, University of Paris, France, 1970.
141. Van Eck D.C. and Perdeck M., Mol. Cryst. and Liquid Cryst., 1978, vol. 49, 39.
142. Van Eck D.C. and Westera W., Mol. Cryst. and Liquid Cryst., 1977, vol. 38, 319.
143. Ivanov S.A., and Vetrov V.Yu., Manuscript deposited at VINITI, No. 4978-80 Dep., Moscow, 1980, 1-12.
144. Ivanov S.A. and Vetrov V. Yu., Kristalografiya, 1982, vol. 27, 1014.
145. Ivanov S.A. and Vetrov V. Yu., Zh. Prikl. Spektroskopii, 1982, vol. 37, 316.
146. Van der Meulen J.P. and Zijlstra R.J.J., J. Phys., 1984, vol. 45, 1627.
147. Cummins H. and Pike E. (eds), Spectroscopy of optical displacement and photon correlation, Moscow, Mir, 1978.
148. Leslie F.M. and Waters C.M., Mol. Cryst. and Liquid Cryst., 1985, vol. 123, 101.
149. Hirakata J., et al, Jap. J. Appl.Phys., 1986, vol. 25, L607.
150. Akiyama R., et al, Jap. J. Appl. Phys., 1986, vol. 25, L769.
151. Miraldi E., et al, Nuovo Cimento, 1980, vol. 60, 165.
152. Hanmer J.M.W. and Coles H.J., Mol. Cryst. and Liquid Cryst., 1995, vol. 262, 235.
153. Belyaev V.V. and Ivanov S.A., Kristalografiya, 1992, vol. 37, 227.
154. Belyaev V.V., in: Physical Properties of Liquid Crystals (eds. Dunmur D., et al), The Institution of Electrical Engineers, London, 2000, 414.
155. Moscicki J.K., in: Physical Properties of Liquid Crystals (eds. Dunmur D., et al), The Institution of Electrical Engineers, London, 2000, 387.

156. Martins A.F., in:m Physical Properties of Liquid Crystals (eds. Dunmur D., et al), The Institution of Electrical Engineers, London, 2000, 405.
157. Imura H. and Okano K., Jap. J. Appl. Phys., 1972, vol. 11, 1440.
158. De Gennes P., Mol. Cryst. and Liquid Cryst., 1969, vol. 7, 325.
159. Helfrich W., J. Chem. Phys., 1972, vol. 56, 3187.
160. Flanders P.J., Mol. Cryst. and Liquid Cryst., 1974, vol. 29, 19.
161. Prost J., et al, J. Physique Lettres, 1976, vol. 37, L341.
162. Maier W. and Saupe A., Z. Naturforsch, 1958, Bd 13a, 564, Z. Naturforsch, 1959, Bd. 14a, 882, Z. Naturforsch, 1960, Bd. 15a, 287.
163. Martins A.F., Portugal. Phys., 1974, vol. 9, 1.
164. Martins A.F. and Diego A.C., Portugal. Phys., 1975, vol. 9, 129.
165. Kumnis C.A. and Kwei T.K., In Diffusion in Polymers, Ed. by Crank J. and Park G.S., Academic Press, New York, 1968.
166. Williams M.L., et al, J. Am. Chem., 1955, vol. 77, 3701.
167. Schad Hp. and Zeller H.R., Phys. Rev., 1982, vol. 26, 2940.
168. Diego A.C. and Martins A.F., Mol. Cryst. and Liq. Cryst., 1981, vol. 66, 133.
169. Eyring H., J. Chem. Phys., 1936, vol. 4, 283. Moscow, IL, 1984, 583, Cohen M.H. and Turnbull D., J Chem. Phys., 1969, vol. 31, 1164.
170. Simöes M., Abstracts of The 18th International Liquid Crystal Conference. Sendai, Japan. July 24-28, 2000. vol. 26D – 97-P, 517.
171. Simöes M., et al, Abstracts of the 6th European Conference on Liquid Crystals. Germany, Halle, March 25-30, 2001, vol. 8, 25.
172. Belyaev V.V., et al, ZHETF, 1984, vol. 86, 955.
173. Pokrovskii B.N., Statistical mechanics of diluted suspensions, Nauka, Moscow, 1978.
174. Pokrovskii V.N., ZHETF, 1976, vol. 71, 1880.
175. Stepanov V.I., Preprint: Statistical theory of thermotropic liquid crystals, Sverdlovsk, Institute of Solid State Mechanics, Uralsk Scientific Centre of the USSR Academy of Sciences, 1982, 39.
176. Doi M., J. Polym. Sci. Phys. Ed., 1981, vol. 19, 229.
177. Kuzuu N. and Doi M., J. Phys. Soc. Japan, 1983, vol. 52, 3486.
178. Stepanov V.I., Statistical and dynamic problems of elasticity and viscoelasticity, Sverdlovsk, Institute of Solid State Mechanics, Uralsk Scientific Centre of the USSR Academy of Sciences, 1983, 39.
179. Stepanov V.I., Abstract of a dissertation, Moscow, VZMI, 1987.
180. Fialkowski M., Phys. Rev. E., 1998, vol. 58, 1955.
181. Tsebers A.O., Magnitnaya gidrodinamika, 1978, No. 3, 3.
182. Marschenyuk A.M., et al, ZHETF, 1973, vol. 65, 834.
183. Tsebers A.O., Physical properties and hydrodynamics of disperse ferromagnetics, Sverdlovsk, 1977, 49.
184. Osipov M.A. and Terentiev E.M., Preprint of the Institute Crystallography, USSR Academy of Sciences, Moscow, 1988, 1-36.
185. Osipov M.A. and Terentiev E.M., Phys. Lett., 1989, vol. 134, 301.
186. Osipov M.A. and Terentiev E.M., Z. Naturforsch., 1989, Bd. 34a 785.
187. Osipov M.A. and Terentiev E.M., Nuovo Cimento, 1990, vol. 12, 1223.
188. Van der Meer B.W. and Vertogen G., In Molecular Physics of Liquid Crystals, Ed. By Luckhurst G.R. and Gray G.W., Academic Press. London, 1979, 149.
189. Tskhai A.A., Flows of anisotropic media, Dissertation, Institute of Thermophysics, USSR Academy of Sciences, Novosibirsk, 1987.
190. Tskhai A.A., Hydrodynamics and acoustics of single- and two-phase flows, Novosibirsk, Institute of Thermophysics, USSR Academy of Sciences, Novosibirsk,

1983, 69.

191. Pokrovskii V.N. and Tskhai A.A., Prikladnaya matematika i mekhanika, 1986, vol. 50, 512.
192. Tskhai A.A. and Pokrovskii V.N., Kolloid. Zh., 1985, vol. 47, 106.
193. Pokrovskii V.N. and Tskhai A. A., in: 5th Congress Mechanics, Varna, 1985, 721.
194. Pokrovskii V.N., in: Current problems of thermophysics, Institute of Thermophysics, USSR Academy of Sciences, Novosibirsk, 1987, 258.
195. Tskhai A.A., VINITI 25.10.88, No. 7610-V88, Inzh.-Fiz. Zh., 1988, vol. 56, 159.
196. Hess S., Z. Naturforsch., 1968, Bd.23a, 1095.
197. Hess S. and Rainwater J.C., J. Chem. Phys., 1984, vol. 80, 1295.
198. Koo H.M. and Hess S., Physica, 1987, vol. 145A, 361.
199. Hess S., Z. Naturforsch., 1976, Bd. 31a. S.
200. Hess S., Z. Naturforsch., ibid, 1984, Bd.39a, 22.
201. Gevorkyan E.V., 8th Liquid Crystal Conference of Socialist Countries. Cracow, 1989, Abstracts, 111.
202. Hess S., Phys. Rev. A, 1980, vol. 22, 2844.
203. Hess S., In : Electro-Optics and Dielectric properties of Macromolecules and Colloids, Ed. By Jennings B.R., Plenum Publishing Corporation, 1979, 321.
204. Forster D., Hydrodynamic fluctuations, disrupted symmetry and correlation functions, Moscow, Atomizdat, 1968,
205. Zubarev D.N., Nonequilibrium statistical hydrodynamics, Moscow, Nauka, 1971.
206. Nemtsov V.B., Theor. i Mat. Fizika,1975, vol. 25, 118.
207. Nemtsov V.B., Physica, 1977, vol. 86A, 513.
208. Nemtsov V.B., Theoretical and Applied Mechanics, Minsk, Vish. Shkola, 1985, vol. 12, 111.
209. Kunts F.M. and Storonkin B.A., Theor. i Mat. Fizika, 1977, vol. 27, 417.
210. Nemtsov V.B. and Chernigov, 1988, Proc.: 6th National Conference: Liquid Crystals and Their Application in Practice, Chernigov, vo.l III, 307, 1988; Nemtsov V.B., Mol. Cryst. Liq. Cryst., 1990, vol. 192, 137.
211. Nemtsov V.B., in: Thermophysics of Condensed Media: Structure and Properties, Minsk, 1990, 38.
212. Tokarchuk, M.V., and Omelyan I.P., Preprint of the Institute of Theoretical Physics, ITF-87-152P, Academy of Sciences of Ukraine, Kiev, 1988, 1-36.
213. Hanley H.G.M., et al, Physica, 1972, vol. 60, 322.
214. Karheck J. and Stell G., J. Chem. Phys., 1981, vol. 75, 1475.
215. Stell G., et al., J. Chem. Phys., 1983, vol. 759, 3166.
216. Zakharov A.V., Phys. Lett., A, 1994, vol. 193, 471.
217. Zakharov A.V., Fiz. Tverdogo Tela, 1998, vol. 40, 1945.
218. Zakharov A.V., et al, Phys. Rev. E., 1999, vol.59, 6802.
219. Zakharov A.V., et al, Phys. Rev. E., 1999, vol. 60, 1142.
220. Gay J.G. and Berne B.J., J. Chem. Phys., 1981, vol.74, 3316.
221. Nemtsov V.B., Dokl. AN BSSR, 1981, vol.30, 1075.
222. Hess S., Phys.Lett., 1984, vol. 105A, 113.
223. Baals D. and Hess S., Z. Naturforsch, 1988, Bd. 43a, 662.
224. Baals D. and Hess S,. Phys. Rev. Lett., 1986, vol. 57, 86.
225. Evans D.J., Mol. Phys., 1979, vol. 37, 1745.
226. Baals D., Thesis, Univ.Erlangen, Nuernberg, 1985,
227. Helfrich W., J. Chem. Phys., 1970, vol. 53, 2267.
228. Sollich H., et al, Mol. Cryst. and Liq. Cryst., 1989, vol. 168, 189.
229. Belyaev V.V. and Grebenkin M.F., Krystalografiya, 1983, vol. 28, 1003.
230. Osipov, M.A., J. Physique Lett., 1984, vol. 45, L823.

231. Belyaev, V.V., Liquid Crystals in Optical Systems for Transformation and Display of Images, Kometa, Moscow, 1996, 324.
232. Leenhouts F., et al., J. Physique, 1979, vol. 40, 989.
233. Kneppe H., et al, Chem. Phys. Lett., 1982, vol. 87, 59.
234. Van Dijk J.W., et al, J. Chem. Phys., 1983, vol. 79, 3888.
235. De Jeu W.H. and Claassen W.A., J. Chem. Phys., 1978, vol. 63, 102.
236. Van der Meulen J.P. and Zijistra R.J.J., Physique, 1984, vol .45, 1347.
237. Leenhouts F., J. Chem. Phys., 1981, vol. 66, 425.
238. Dunmur D.A. and Miller W.H., J. Physique Colloque, 1979, vol. 40, 3.
239. Wu S.-T., Wu C.-S., Phys. Rev. A, 1990, vol. 42, 2219.
240. Belyaev V.V., Uspekhi khimii, 1989, vol. 58, 1601.
241. Bobrow W.J., Lerzyty Naukowe Politechniki Lodzkiey, Fiz. 1981, vol. 6, No. 365, 135.
242. Lavygin I.A. and Skorokhodov I.P., Zhurnal Fiz. Khimii, 1984, vol. 58. 1351.
243. Flapper S.D.P. and Vertogen G., J. Chem. Phys., 1981, vol.75, 3599, Phys. Rev. A 1981, vol. 24, 2089.
244. Belyaev V.V. and Grebenkin M.F., Zhurnal Fiz. Khimii, 1990, vol. 64, 958.
245. Ryzhov V.N. and Lipovskaya T.B., in: Proc. 5th National conference: Liquid crystals and their application in practice, Ivanovo, 1985, vol. 1, part 2, 123.
246. Ryzhov V.N., et al, Zhurnal Fiz. Khimii, 1987, vol. 61, 1549.
247. De Jeu W.H. and van der Veen, Mol. Cryst. and Liq. Cryst.,1977, vol. 40, 1.
248. Pirozov A.I. and Novikov I.V., Zhurnal Fiz. Khimii, 1987, vol. 61, 1754.
249. Leadbetter A.J. and Mehta A.I., Mol. Cryst. and Liq. Cryst., 1981, vol. 72, 17.
250. Mukoh A., et al, Mol. Cryst. and Liq. Cryst., 1983, vol. 94, 155.
251. Belyaev V.V., Zhurnal Fiz. Khimii, 1991, vol. 65, 1679.
252. Askadskii A.A. and Matveev Yu. A., Chemical Structure and Physical Properties of Polymers, Moscow, Khimiya, 1983.
253. Demus D., et al, Mol. Cryst. and Liq. Cryst., 1990, vol. 191, 153.
254. Averyanov E.M., et al, Phys. Tverd. Tela, 1979, vol. 21, 2209.
255. Nguen X.P., Mol. Cryst. and Liq. Cryst., 1990, vol. 191, 61.
256. Bata L., et al, Mol. Cryst. and Liq. Cryst., 1977, vol. 38, 155.
257. Ivanov S.A. and Belyaev V.V., Kristalografia, 1989, vol. 34, 689.
258. Pozhidaev E.P., et al, ZHETF, 1988, vol. 94, 353.
259. Kresse H., et al, Cryst. Res. and Technology, 1981, vol. 16, 1439.
260. Evtyushenko A.M., et al, in: Proc. 5th National conference: Liquid crystals and their application in practice, Ivanovo, 1985, vol. 1, part 2, 197.
261. Agafonov M.A., et al, Kristalografiya, 1986, vol. 31, 528.
262. Agafonov M.A., et al, Opt. Spektroskopiya, 1987, vol. 62, 784.
263. Ryumtsev E.I., et al, DAN USSR, 1984, vol. 270, 1367.
264. Zeller H.R., Phys. Rev. Lett., 1982, vol. 48, 334.
265. Schaefer W., et al, Mol. Cryst. and Liq. Cryst., 1990, vol. 191, 269.
266. Averyanov E.M., et al, Kristalografiya, 1982, vol. 27, 333.
267. Hanson E.G., Shen Y.R., Mol. Cryst. and Liq. Cryst., 1976, vol. 36, 193.
268. Barnik M.I., et al, Kristalografiya, 1978, vol. 23, 805.
269. Belyaev V.V. and Grebenkin M.F., Kristalografiya, 1984, vol. 29, 815.
270. Korzan O.A., et al, Kristalografia, 1985, vol. 30, 750.
271. Belyaev V.V., et al, Mol. Cryst. and Liq. Cryst., 1985, vol. 129, 221.
272. Belyaev V.V., et al, Zhurnal Fiz. Khimii, 1990, vol. 64, 963.
273. Franklin W., Mol. Cryst. and Liq. Cryst., 1977, vol. 40, 91.
274. Sources of Thermodynamic Data on Mesogens, Mol. Cryst. and Liq. Cryst., 1984, vol. 115.

275. Petrov V.F., et al, Kristalografiya, 1988, vol. 33, 1194.
276. Seifert F. and Kuehnel A., Mol. Cryst. and Liq. Cryst., 1990, vol. 192, 161.
277. Belyaev V.V., Izv. RAN, Ser. Fiz.,1996, vol. 60, 12.
278. Belyaev V.V., Zhurnal Fiz. Khimii, 2001, vol. 75, V.3, 545.
279. Ostrovskii B.I., Izv. AN USSR, 1989, vol. 53, 1962.
280. Dozov I. and Kirov N., J. Chem. Phys., 1989, vol. 90, 1099.
281. Horn R. and Faber G., Proc. R. Soc. London, 1979, vol. A368, 199.
282. Szwajczak E. and Szymanski A., Mol. Cryst. and Liq. Cryst., 1986, vol. 139,195.
283. Helfrich W., J. Chem. Phys., 1970, vol. 53, 2267.
284. Chmielwski A.G., Mol. Cryst. and Liq. Cryst., 1986, vol. 132, 339.
285. Saito Y., et al, Proc. of the 8th Int. Display Conf. Japan Display ´89, Kyoto, 1989, 368.
286. Franklin W., Phys. Lett., 1974, vol. 48A, 247.
287. Inukai T., et al, Mol. Cryst. and Liq. Cryst., 1983, vol. 94, 109.
288. Eidenschink R., Mol. Cryst. and Liq. Cryst., 1983, vol. 94, 119.
289. Vuks M.F., Electrical and Optical Properties of Molecules and Condensed Media, Leningrad University, Leningrad, 1984, 334.
290. Belyaev V.V., et al, Zhurnal Fiz. Khimii, 1988, vol. 62, 3087.
291. Gerber P.R. and Schadt M., Z. Naturforsch., A 1982, Bd.37, 179.
292. Haramoto Y. and Kamogawa H., Mol. Cryst. and Liq. Cryst., 1985, vol. 131, 201.
293. Haramoto Y. and Kamogawa H., Mol. Cryst. and Liq. Cryst., 1985, vol. 131, 101.
294. Buchecker R. and Schadt M., Mol. Cryst. and Liq. Cryst., 1987, vol. 149, 359.
295. Margerum J.D., et al, Mol. Cryst. and Liq. Cryst., 1985, vol. 122, 97.
296. Takatsu H. and Takeuchi K., Mol. Cryst. and Liq. Cryst., 1986, vol. 138, 231.
297. Schadt M., et al., Mol. Cryst. and Liq. Cryst., 1986, vol. 139, 1.
298. Schadt M., et al, IEEE Transactions on Electron Devices, 1986, vol. ED-33, 1187.
299. Petrov V. F. and Grebenkin M.F., Zhurnal Fiz. Khimii, 1991, vol. 65, 1356.
300. Literature of Dai Nippon Japan, DIC Recent Liquid Crystal Materials, Dai Nippon, Japan, 1988.
301. Eidenschink R., Mol. Cryst. and Liq. Cryst., 1983, vol. 94, 119.
302. FRG Patent 3 420 110 (C.A. 1986, vol. 103, 162 746 m).
303. Europe Patenz 90 548 (C.A. 1984, vol. 100, 59 667 j).
304. Literature of E. Merck, Germany, (Liquid Crystal Materials, E.Merck, Germany, 1988).
305. Eidenschink R., Chemie in Unserer Zeit, 1984, No. 6, 168.
306. Chan L.K.M., et al, Mol. Cryst. and Liq. Cryst., 1987, vol. 147, 113.
307. Petrov V.F., et al, Zhurnal Fiz. Khimii, 1991, vol. 65, 1359.
308. Takatsu H., et al, Mol. Cryst. and Liq. Cryst., 1984, vol. 112, 165.
309. FRG Patent 3 317 921 (C.A. 1984, vol. 100, 128 628 q).
310. Europe Patent 87 102 (C.A. 1984, vol. 100, 15 416 x).
311. Goto Y., et al, Proc. of the 6th International Display Research Conference Tokyo, Japan, 1986, 324.
312. Tarumi K., et al, Proc. of the 15th International Display Research Conference, Asia Display ´95, Hamamatsu, Japan, October 16-18, 1995, 559.
313. Grebyonkin M.F., et al, Mol. Crys. and Liq. Cryst., 1985, vol. 129, 245.
314. Kneppe H. and Schneider F., Mol. Cryst. and Liq. Cryst., 1983, vol. 97, 219.
315. Constant J. and Raynes E.P., Proc 8th Int.Liquid Crystal Conf. Kyoto, Japan, 1980.
316. Sarkisyan A.Ts., et al, Zh. Phys. Khimii, 1981, vol. 55, 1796.
317. Grebyonkin M.F., et al, Mol. Cryst. and Liq. Cryst., 1983, vol. 103, 1.
318. Grebyonkin M.F., et al, Kristalografia, 1975, vol. 20, 984.

319. Caccamo C. and Malescio G., J. Chem. Phys., 1989, vol. 90, 1091.
320. Korzhenevskii V.M., Trudy GOI, 1987, 121.
321. Kodabakas N.B.and Pirogov A.I., Dep.v ONIITEKH im. G. Cherkassi, No.303-KhP-89, 1989, 11.
322. Novikov I.V. and Pirogov A.I., Dep. V ONIITEKH im. G.Cherkassi, No. 58-KhP-89, 1989, 16.
323. Pirogov A.I., et al, Zhurnal Fiz. Khimii, 1985, vol. 59, 2813.
324. Pirogov A.I. and Novikov I.V., Khimiya Khim. Tekhnologiya, 1985, vol. 28, No.6, 33.
325. Pirogov A.I. and Novikov I.V., Khimiya Khim. Tekhnologiya, 1987, vol. 30, No.3, 63.
326. Pirogov A.I., Novikov I.V., Khimiya I Khim. Tekhnologiya, 1987, vol. 30, No.10, 63.
327. Kodabakas N.B., Dissertation, Institute of Chemical Technology, Solutions, Ivanovo, 1989.
328. Novikov I.V., Dissertation, Institute of Chemistry of Non-Aqueous Solutions, Ivanovo, 1989.
329. Afanasev V.N. and Mershchikova V.Yu., Zhurnal Fiz. Khimii, 1987, vol. 61, 232.
330. Demus D., et al., US Patent 4 767 565.
331. Scheuble B.S., et al, SID Digest, 1984, vol. 14, 176.
332. Hara M., et al, J. Appl. Phys., 1984, vol. 23, 1420.
333. Elashvilli Z.M., et al, J. Physique, Colloque, C3, 1979, vol. 40, C3-22.
334. Belyaev V.V., et al, Liquid Crystal Reserch and Applications, Ed. by L.Bata, Pergamon Press, Oxford and Budapest, Akademiai Kiado, 1980, 673.
335. Chigrinov V. G., Dissertation, Institute of Crystallography, Moscow, 1988.
336. Chigrinov V.G. et al., ZHETF, 1979, vol. 77, 2081.
337. Marignan J., et al, J. Physique, 1976, vol. 37, 365.
338. Belyaev V.V., Dissertation, Moscow, NIOPiK, 1980.
339. Belyaev V.V., Opt. Zh., 2000, vol. 68, 57.
340. Toriyama K. and Nakagomi T., BRD Offenlegungsschrift 31 22 529, UK Patent 2 077 286, C.A. 1982, vol. 96, 226684 v.
341. Grebenkin M.F., Liquid Crystal Materials for the Twist Effect, Dissertation, Moscow, NIOPiK, 1991.
342. Eidenschink R. and Pohl L., BRD Offenlegungsschrift 23 46 838.
343. Hasegawa S., et al., BRD Offenlegungsschrift 32 25 432, UK Patent 2 101 626, C.A. 1983, vol. 83, 189047 h.
344. Hasegawa S., et al., BRD Offenlegungsschrift 32 25 457, C.A. 1983, vol. 98, 135318d.
345. Literature of E. Merck Co., Germany, 1987.
346. Literature of Hoffman-La- Roche, Switzerland, 1985.
347. Literature of Hoffman-La- Roche, Switzerland, 1987.
348. Literature of British Drug House (BDH), England, 1984.
349. Liquid Crystal Materials, Literature of NIOPiK, 1985.
350. Balkvill P., et al, Mol. Cryst. and Liq.. Cryst., 1985, vol. 123, 1.
351. Literature of E. Merck Co, Germany, April 2000.
352. Geelhaus J., et al., Recent LC Material Developments and Future Technology Trends for Advanced LCDs, MICRO, tec. 2000, September 25-27, 2000, EXPO 2000, Hannover, Germany.
353. Seiberle H., et al, Proc. of The 19th International Display Research Conference Euro Display ´99, 1999- Germany, Berlin. Sept. 6-9, 121.
354. Eidenschink R., et al, BRD Offenlegungsschrift 24 32 024.

355. BRD Offenlegungsschrift 33 15 618, C.A. 1984, vol. 100, 59676m.
356. Gerber P.R. and Schadt M., Z. Naturforsch., 1982, Bd. 37a, 179.
357. Scheuble B.S. and Baur G., Proc. of the 2nd Int. Display Conf. Japan Display ´83, 1983, 224.
358. Scheuble B.S., Weber G. BRD Offenlegungsschrift 34 04 117, C.A. 1986, vol. 104, 59500g.
359. Literature of Chrisso, Japan.
360. Gooch G.H. and Tarry H.A., J. Phys., 1972, vol. 5, 25.
361. Weber G., et al, BRD Offenlegungsschrift 37 05 071, C.A. 1989, vol. 110, 31534w.
362. Boller A., et al, UK Patent 2 063 288, C.A. 1982, vol. 96, 13521c.
363. Roemer A., et al, BRD Offenlegungsschrift 24 92 976.
364. Kitamaru T., et al, Japan Patent 57/034 176, 1982, vol. 97, 136687q.
365. Scheuble B.S., et al, BRD Offenlegungsschrift 36 31 415, C.A. 1988, vol. 109, 14947v.
366. Scheuble B.S. and Geelhaar T., Proc. of the 19th International Display Research Conference EuroDisplay ´99, 1999, Germany, Berlin, Sept. 6-9, 15.
367. Hong S.Z., et al, Proc. of The 6th International Display Workshops, December 1-3, 1999, Japan, Sendai, 73.
368. Barnik M.I. and Yudin S.G., Pribori i Tekhnika Eksperimenta, 1981, vol. 14, 27.
369. Belyaev V.V. and Chigrinov V.G., Kristalografiya, 1978, vol. 23, 811.
370. Belyaev V.V., et al, ZHTF, 1980, vol. 6, 845.
371. Fujimura K., Patent 53 025 Japan, C.A. 1983, vol.97, 123765j.
372. Smith R.A., McDonnel D, European Patent 50 023, C.A. 1982, vol. 97, 153980x.
373. Blinov L.M., et al, Mol. Cryst. and Liq., Cryst., 1981, vol. 123, 310.
374. Habbard R.L. and Liang J.C.H., BRD Offenlegungsschrift 33 37 316, C.A. 1984, vol. 101, 101316s.
375. Schadt M. and Gerber P.R., Proc. of the SID, 1982, vol. 23, No.1, 29.
376. Kagawa H., et al, Proc. 15th International Display Research Conf. Asia Display ´95, Hamamatsu, Japan, October 16-18, 1995, 177.
377. Nakajima S., et al, Society for Information Displays, International Symposium, Long Beach, California, USA, May 16-18, 2000. Digest of Technical Papers, vol. XXXI, 242.
378. Nakazano Y., et al, Conference Record of the 1997 International Display Research Conf., Toronto, Canada, Sept. 15-19, 1997, 65.
379. Wu S.-T., et al, Society for Information Displays International Symposium, San Jose, California, USA, May 16-18, 1999, Digest of Technical Papers. Vol. XXX, 706.
380. Klement D. and Tarumi K., Society for Information Displays International Symposium, Annaheim, California, USA, May 17-22, 1998, Digest of Technical Papers, vol. XXIX, 393.
381. Wu S.-T., et al, Proc. 15th International Display Research Conf. Asia Display ´95, Hamamatsu, Japan, Oct. 16-18, 1995, 567.
382. Liang J.C. and Kumar S., Mol. Cryst. and Liq. Cryst., 1987, vol. 142, 77.
383. Carr N., Gray G.W., Mol. Cryst. and Liq. Cryst., 1985, vol. 124, 27.
384. Demus D., et al, BRD Offenlegungsschrift 32 00 414, C.A. 1983, vol. 99, 194969k.
385. Sethofer N.L., USA Patent 4 325 830, C.A., 1982, vol. 97, 47238b, BRD Offenlegungsschrift 31 50 763.
386. Demus D., et al, Patent 210 920 DDR, C.A. 1985, vol. 102, 103738a.
387. Demus D., et al, BRD Offenlegungsschrift 33 05 013, C.A. 1984, vol. 100, 34290t, Patent 203561 DDR.

Index